KU-616-453

1
BBC Books, an imprint of Ebury Publishing
20 Vauxhall Bridge Road, London SW1V 2SA
BBC Books is part of the Penguin Random House group of companies
whose addresses can be found at global.penguinrandomhouse.com

First published by BBC Books in 2008
This edition published in 2020
www.penguin.co.uk
A CIP catalogue record for this book is available from the British Library
ISBN 9781785947001
Printed and bound in Italy by Lego S.p.A

Alan Titchmarsh

Grow Your Own Fruit and Veg

BOOKS

CONTENTS

INTRODUCTION 6

THE GROUND RULES 8

THE DIRECTORY 62
INTRODUCTION 64
VEGETABLES 66
FRUIT 186
HERBS 262

WHERE TO SEE GREAT FRUIT AND VEG GARDENS 306
SUPPLIERS AND SPECIALISTS 307
INDEX 308
ACKNOWLEDGEMENTS/PICTURE CREDITS 312

WHY GROW YOUR OWN?

'Growing your own' has never been more popular, more possible, more fashionable or more fun than it is today. All sorts of people are jumping on the bandwagon, and there are lots of good reasons why.

Some are embracing growing their own for the health benefits. Fresh fruit and vegetables provide fibre, antioxidants and also vitamins and minerals which, it's thought, do you more good when they come from fresh ingredients than from a jar of pills. We are all encouraged to eat at least five portions of fruit and veg every day, and a real health-buff would try to incorporate all the various colours – red, green, purple and orange or yellow – to be sure of topping-up on the full range of natural plant phytochemicals that are nature's disease-busters.

If you grow your crops organically – which you really should, because they'll taste better – you'll be avoiding having pesticides in your food, too. Don't overlook the benefits of the good all-round physical work out that fruit-and-veg gardening gives either; it's great for toning and tightening nature's baggy bits, and the fruits of your labours are low calorie. Who needs the gym when you've got a productive kitchen garden?

Some people grow their own to save money, and certainly if you go in for cultivating crops which are expensive to buy in the shops, such as asparagus, artichokes and soft fruit, you can make some very worthwhile savings. But think of the *convenience* too; for anyone who works late or lives miles from the shops it's handy to have a well-stocked, living larder in the garden. Managed carefully, you can have something tasty and fresh available all year round, without needing to fall back on the freezer or phone for a pizza.

For some people, growing their own food is a quality-of-life thing; it's an activity that the whole family can do together on summer evenings or at weekends. Allotment gardening has really taken off in recent years, and in many cases it's young families and professional people who've swelled the ranks of folk who are found pottering away on their plots. These same people to whom quality of life is important are also often those who are concerned about the environment and want to do their bit for the cause. They appreciate that growing your own is good for the planet and global warming; since it cuts down food miles, reduces the use of fossil fuels and lets you recycle rubbish that might otherwise end up in landfill sites.

Then, of course, there are the food fanciers. Anyone who's serious about good food very often falls into growing their own so that they can enjoy fresh produce while it's *really* fresh – not fresh from a plastic bag full of gas, or pre-prepared so that half the flavour has leaked out or dried up. And it's a great way of sourcing 'special' ingredients that you can't easily find in the shops – such as fresh kaffir lime leaves or exotic species of Thai basil. For foodies the big buzz is currently for 'fresh local produce', and it doesn't come any more fresh or local than from your own plot with the dirt still on.

Although this isn't a cookery book, I couldn't miss the opportunity to pass on a few of my favourite ways of using garden crops. When you grow your own fruit and veg it's almost impossible not to be interested in the way that you cook and eat them and, however well you plan, you'll usually end up with a surplus of certain crops at plentiful times of year. To my mind it's far better to have a few recipes that really make the most of this glut instead of just shoving it all in the freezer, which I reckon should be a last resort.

So take up the spade and put aside your worries; you're unlikely to regret it.

'Growing your own' has never been more popular, more possible, more fashionable or more fun than it is today.

The Ground Rules

Now, I don't mean to tub-thump, but when your time and space are valuable – which they are for everyone these days – there's no point in taking a 'leave it to nature' approach to kitchen gardening. It's not like digging a hole and bunging a shrub in. No, you need to be much more hands-on with fruit and veg – it demands regular, on-going commitment and attention to detail. But that doesn't mean you have to slave away all hours on your plot; it pays to get right some 'key points', then you only need spend time and effort where it counts most in order to produce real results.

So that's what I'm going to emphasize in this book. Where people usually go wrong is that they take on far more than they can manage then can't keep up with the work. The result is that their crops are ruined and their time has been wasted. If you were to ask me for my top kitchen-gardening tip, I'd say that you'd do far better to grow half the amount, but grow it twice as well.

SITE AND SITUATION

To produce a good, healthy crop, fruit, vegetables and herbs need to grow fast and steadily, without any setbacks. Good growing conditions are vital for a kitchen garden, so choose a reasonably sunny but sheltered situation with well-drained ground that's 'in good heart'. (This is gardener-speak for healthy, fertile soil that's chock full of well-rotted organic matter and which is brimming with worms, beneficial bacteria and natural nutrients.) Chill winds, heavy shade or soil that becomes waterlogged periodically or dries right out, all spell disaster for your harvest.

Some fruit, veg or herbs also have their own personal peculiarities and requirements which need to be observed in order to keep them happy and healthy. If you want to grow 'exotics' such as peaches, nectarines, almonds, posh pears and Mediterranean herbs, a particularly warm, sunny, sheltered and well-drained spot is essential. These sun-worshippers do best against a south-facing wall or on a sun-trap patio where a lot of other crops would 'fry'. In the same vein, root veg such as carrots aren't worth growing on ground that's stony or that has recently been heavily manured, because such conditions will cause their roots fork, making them impossible to peel and also inedible.

So, as a general rule, give edible crops the best part of your garden or else improve the conditions where you would like them to grow. You can do this simply by planting a natural windbreak, such as blackberries grown on a wire-netting fence, by improving the soil with lots of organic matter or, better still, by making a few intensive veg or salad beds with seriously well-enriched soil. (See page 30 for how to improve soil.)

DESIGN

Traditionally, fruit and veg were always grown down the end of the garden, out of sight, along with the shed, the compost heap, the bonfire and other messy working areas. Nowadays it's far more fashionable to grow your own in an attractive way, particularly in a small garden where a plot can be seen from the house. To my way of thinking this is a big step forward – a spot of designer flair is no bad thing, even with edible crops.

Planning a fruit and veg patch isn't so very far removed from designing an ornamental border; the big difference is that fruit and veg are traditionally grown in rows. This is done for a practical rather an aesthetic reason: it is simply because it is more convenient to look after them that way as it's a lot easier to run a hoe between plants that are growing in straight lines than those scattered all over the place.

Treat this task like any other garden design job: be practical and ask yourself a few basic questions about what you and your family really want from the plot, what you'll eat lots of, or just what you might fancy occasionally. Think about what's involved with the various ways of growing veg, how much time you want to spend doing it, and what will happen when you take your summer holidays – it's no good planting masses of fruit that ripens in August if you're always away then.

Most importantly, assess the site; see how much space you can spare and if the soil or situation will suit some crops more than others and – I know I keep banging on about it – don't forget that *all* fruit and vegetables need good growing conditions.

As for looks, there are many ways in which you can include edible crops in a garden. You might want a conventional plot with fruit and veg laid out in rows, or, if you have a fair-sized space and a garden that is already rather compartmentalized, you might create a separate kitchen garden-within-a-garden. On the other hand, if space is short and the area is always 'on show' you could create a decorative potager – a 'flower bed' filled with edible plants instead of bedding (see box, right) – or you could incorporate edible crops into your family garden; with a good-looking and productive salad bed conveniently close to the house, a few 'special' crops in pots on the patio and trained fruit trees round the walls.

Whichever approach you choose, and just as with any design changes you want to make to your garden, it pays to start with a plan. Begin by making a sketch of the space available to you to use as a vegetable bed, drawn roughly to scale, and start pencilling in roughly what you want to grow, how much space it occupies and where it could go.

If possible, try to do your planning in autumn or winter so you have plenty of time to undertake any construction, soil preparation and fruit planting before the 'busy season' starts.

Potagers

A potager is basically a decorative vegetable bed. The idea started in the gardens of grand chateaux in France, and the best example still to be seen today is probably the one at Villandry.

But despite its grand origins, the idea does adapt very well to a small scale, and you can be as practical or as inventive with it as you like. The shape doesn't have to be the standard rectangle; some enthusiasts make circular beds divided into segments like the spokes of a wheel, or you could go modernist and octagonal, or even multi-storey.

A potager can be planted along the lines of traditional geometric herb beds so that crops grow in blocks rather than long rows, (though the plants in the blocks are still planted in straight lines, for easy hoeing). If you go for the traditional style of planting at ground level only, you can add height to the display by growing runner and climbing French beans up symmetrically-placed obelisks of bamboo, timber or metalwork.

The big difference between a potager and a flower bed is that while ornamental plants will still flower if they have a bit of competition from weeds or close neighbours, edible crops must be grown without a check or any setback, otherwise you won't have much of a harvest.

You can also make a potager in an informal flower bed shape with veg and salad plants dotted about in drifts just like flowers – but be warned, if you take this route you'll have your work cut out keeping on top of the weeding, and you might not like the look of the gaps that are left behind when you harvest.

GROWING VEG IN BEDS

We're not talking 'interior sprung' here; the 'bed' system is for intensive vegetable growing and will allow you to cram a lot of plants into a small space.

The idea is to make rectangular beds of very deep, rich and fertile soil and plant crops at closer spacings than usual. This approach works because the roots can grow down deeply instead of spreading out widely just below the surface, as they do on more compacted earth. As a result the thickly planted beds need very little weeding, since the closely spaced crops quickly cover the ground and their shade prevents lurking weed seeds germinating.

These intensive beds are really best for herbs, salads and other veg, but they also suit the more compact forms of fruit, such as strawberries, cordons and step-over trees. But growing in beds is a bit different from cultivating crops in large-scale plots. Once you've made your bed, the trick is to keep off it so that the soil stays light, fluffy and uncompacted, enabling plant roots to work their way through it quickly and easily and settle themselves in.

From a maintenance point of view, you'll need to feed and water such beds more frequently than other veg patches because there are a lot of plants packed in a small area, draining it of moisture. But when it comes to protecting plants from pests, it's far easier and more efficient to cover intensively-grown crops with pest-proof netting or to surround beds with copper strips to deter slug and snails. Which means that beds are well suited to organic growing.

When the ground is vacant in winter, fork beds over to re-fluff up the soil. This is also the time to work in more organic matter (see page 33). However, in the places you'll be planting brassicas (which need firm, compacted soil) don't fork the soil, simply spread organic matter on the surface instead.

Making a raised bed

Raised beds are good for gardens with wet, heavy or clay soils that are 'claggy' in winter, as they improve drainage and warm up faster in spring, but they are also ideal for anyone with thin, shallow soil and for those gardeners who find bending difficult. They look good in a small garden, especially when you want a more decorative 'designer' or 'potager' effect.

The best time to build beds is in winter, since it's a job that takes a little time to do, and that way they'll be ready for use at the start of the season, in spring.

First mark out the area. A good size for intensive beds is 3 x 0.8m (10 x 2.6ft), since you can reach the middle easily without stepping on the soil. If you need a bigger area, it's better to make several beds with gravel or paved paths between them. (A very long or wide bed just tempts you to step on it, which compresses the ground and ruins much of the benefit of the bed system.)

Fork the ground over as deeply as possible and remove roots, weeds or rubble. Ideally you should work in some well-rotted organic matter too, but don't bring any underlying clay or chalk subsoil to the surface.

Next, construct the raised edges for the beds. Use wooden planks 30cm (1ft) wide (secondhand scaffolding planks from a salvage yard are ideal). Treat them on both sides and on the edges with a plant-friendly wood preservative and let it dry. (For a more cheerful look, apply the sort of coloured wood preserving paint you'd use on garden fences.) Knock 5 x 5cm (2 x 2in) wooden posts in at each corner and also at 90cm (3ft) intervals along the inside of the framework. Nail or screw your planks to these posts (top illustration). For a decorative potager effect, use corner posts 15cm (6in) taller than your edging planks and top them with wooden knobs or finials. A longer-lasting bed can be made with railway sleepers stacked on top of each other – making the sides three sleepers high is most convenient. Drill through the sleepers and secure them to each other with long screws.

Fill the beds up to the top with a 50:50 mix of good topsoil and well-rotted compost or manure (bottom illustration). Finish by spreading gravel or laying paving slabs between the beds to make good all-weather paths.

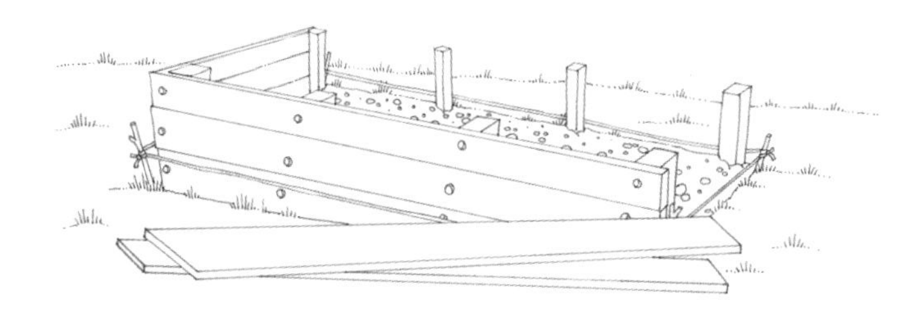

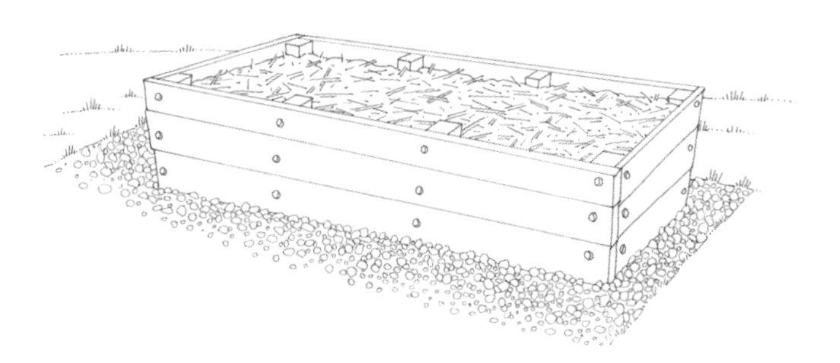

Making a deep bed

This system was popularized by my old mate Geoff Hamilton, and was widely aired on *Gardeners' World* during his time as presenter of the programme. With this system it's essential to have ground that is naturally well drained, otherwise any winter crops will just get bogged down and rot.

Dig down to two spades' depth, loosening the soil and working in large quantities of well-rotted organic matter as you go. As with raised beds, it pays to make deep beds narrow enough that you can work from both sides without needing to step on the soil. Three metres wide and 0.8 metres long (10 x 2.6ft), is about right – if you make longer beds, put in the occasional plank as a 'bridge' so you can cross without stepping on the soil.

If you need to fork the beds over between crops you can do it from the paths alongside, or put down a temporary plank to avoid squashing down the soil in the bed with your feet.

GROWING IN CONTAINERS

A good many vegetables, most herbs and a fair few fruits are well worth growing in containers. This might be a suitable method for you if your garden just doesn't offer good enough growing conditions, or if you only have a tiny space, or – the big attraction for some people – if you want to avoid doing any digging. But don't take on too many 'edible containers', as they do need a lot of regular watering during the warm summer months and crops will suffer badly if they dry out; just a few days' forgetfulness can mean you lose the lot. Large containers, such as tubs and growing bags, are the best for fruit or veg since they don't dry out so quickly, but some crops are even suitable for hanging baskets and window boxes.

As a general rule, small crops are the most suitable for containers; salad leaves, mini cos lettuce, herbs, strawberries and edible flowers are fine even for window boxes, but some of the larger kinds, including runner beans, chillies, outdoor tomatoes and even courgettes, do very well in large containers and they can look very attractive on a patio.

Tubs are also ideal for growing fruit which need acid soil (use ericaceous – lime-free – compost), such as cranberries and blueberries, and for patio peaches and nectarines, which stay small and make ideal flowering/fruiting plants that can spend their whole lives in tubs on a sunny patio.

Figs are another excellent container fruit; by keeping the roots restricted the trees stay smaller and they will also fruit better. You can even grow apple and pear trees in pots – either trained as upright cordons or as normal trees on dwarfing rootstocks (see page 188). Don't bother trying this with cane fruit such as raspberries, blackberries and their hybrids, though, as they don't do well in containers.

Since container-gardening conditions are very different from those found in a kitchen garden (where crops grow freely in the ground), it does require a slightly different growing regime in order to succeed. The way you manage your container crops also depends on whether you are growing them as temporary summer visitors to be dispensed with at the end of the season, or as longer-term residents.

SUMMER VEG IN CONTAINERS

Fill containers with peat-free, multi-purpose compost, then sow or plant crops into them at the same time as you'd normally sow or plant in the garden, if you are keeping them outdoors. Keep crops regularly watered and once they've been planted for four weeks, begin liquid feeding them weekly. (Use a general-purpose feed for leafy crops such as spinach and lettuce, and a high-potash tomato feed for tomatoes and other fruiting crops, such as courgettes.)

When you've grown fast-maturing crops such as lettuce, salad leaves or spinach, it's quite all right to grow something else in the same compost afterwards – just clear out the old crop debris and pull out the roots first. Then you can sow or plant something else, but make sure it's a different kind of crop from a different family: for example, don't follow lettuce with more lettuce, but planting dwarf beans or spring onions next would be fine. This way you can re-use the same compost several times over during the growing season.

Crops such as tomatoes or chillies will occupy the same container all season, so start them off in fresh compost that's not been used for something else first, and at the end of the season tip the containers out onto the compost heap. Although it's tempting to try and save money, don't reuse the same compost for more than one season, since the nutrients will have run out, the texture will have broken down, and there's always the risk of disease passing from the old crop to the new one through the compost.

FRUIT IN CONTAINERS

Since fruit stays in the same pots for several years on end, you need to use a specific compost to grow it well – I like to mix John Innes No.3 potting compost with an equal amount of peat-free, multi-purpose compost and 10 per cent of sharp grit. This mixture opens up the texture of the John Innes (which can become very compacted), holds on to nutrients better than multi-purpose compost would on its own, and increases the weight of the soil, which means pots don't blow over so easily in windy weather.

Bonica

It's best to pot up fruit into containers in spring, but if you buy pot-grown plants in summer you can move them into containers when they are in leaf or even carrying fruit, provided you take care not to break up the root-ball and you keep them very well watered afterwards. When planting fruit in pots, use the largest container you can; a 38–45cm (15–18in) tub or a half-barrel is best, since the greater capacity means plants are less likely to dry out in hot weather and are slower to become pot-bound. Support tall plants by tying them to a cane or stake.

Fruit in containers needs a *lot* of watering from April to September or October; this is particularly crucial while the plants are cropping – if they dry out badly even once during this time they will shed their fruits and you won't get another chance to 'get it right' until the following year. Potted fruit also needs weekly liquid feeding from late April to mid-August. Use liquid tomato feed for most, but use an ericaceous feed for lime-hating kinds such as cranberries or blueberries.

Every spring you'll need to top-dress potted fruit by scraping away the top five centimetres (an inch or two) of compost and replacing it with fresh stuff. You can make up the same mixture as before (see page 16), only this time you can add a little slow-release fertiliser to replace missing nutrients. (But you'll still need to liquid feed in summer as well, since fruit is notoriously hungry.) After 3–4 years the plants will need repotting; do this in spring, about mid-March, just before they start growing again. Either repot them into a slightly larger container, or remove a little of the outside of the rootball and pot them back into the original container with some fresh compost added.

In winter, move plants close to the house or a wall for shelter; you may need to tie them up to trellis to prevent them blowing over and getting damaged in strong winds. Stand the pots up on pot feet or bricks to ensure good drainage. You may still need to water potted fruits occasionally if the plants are sheltered by nearby walls and rain can't reach them. Check them every week or so.

If a long spell of freezing weather threatens the containers will need insulating in order to stop the roots freezing solid. Strangely enough, if plants do get caught by frost, it's not the temperature that kills them but drought, since all the water in the compost turns to ice and cannot be absorbed. To prevent this, wrap the pots individually in sacking material or bubble wrap to provide insulation, or push all your containers close together in a corner and tip bark chippings into the gaps between the pots. Alternatively, move the containers into an outbuilding such as a shed or car port; even if it's fairly dark they'll be okay for a week or two while they are dormant. Be sure to bring them back outside as soon as it's safe to do so, though, or by spring they'll be ex-fruit.

Container-gardening takes a different growing regime to succeed … The way you manage your container crops depends on whether you are growing them as temporary summer visitors or longer-term residents.

TRADITIONAL FRUIT AND VEG PLOTS

These are designed to house a bit of everything – some fruit, veg and herbs – and even though crops are grown in conventional rows, a well-tended plot that's kept meticulously weeded and with militarily straight rows can still look attractive.

On Victorian and Edwardian country estates the walled kitchen garden was designed like a St George's flag: very formal, with four identical squares of ground for veg intersected by paths that crossed in the middle. The paths were outlined with showy herbaceous flower beds so that the veg were viewed through airy sprays of blooms, and the surrounding walls housed trained fruit trees. The general effect was pretty, but practical. A stunning walled kitchen garden was a great status symbol which, together with its glasshouses and outbuildings, was often shown off to house-guests during a stroll round the grounds, yet it provided enough food for the whole estate, with truck loads being sent up to London to stock the town house, too.

You can give a small 'garden-within-a-garden' a traditional touch by designing it on similar lines, but instead of herbaceous borders I'd suggest edging the paths with 'cutting beds' stocked with annual flowers grown especially to be cut and displayed indoors.

If you don't happen to have a walled garden (and few of us do), then you'll need to alter the layout. But when designing any formal vegetable patch, it pays to keep annual veg separate from fruit and perennial veg, since they need different management. Fruit trees stay put for a long time, so think carefully about where you plant them. You could have a row along the back of your plot, but don't plant tall fruit trees where they are going to grow up and cast the whole patch into shade, otherwise your veg won't grow well. Some people grow soft fruit and strawberries in a fruit cage, as this makes them easier to protect from birds without the bother of constantly rolling and unrolling netting every time you want to do some weeding or pick the ripe fruit.

If you plan on growing perennial veg, such as globe artichokes and asparagus, or perennial herbs, it's worth keeping them in permanent beds of their own so they don't interfere with the smooth running of the annual vegetable bed, which needs digging every year.

Fitting fruit into a family garden

Fruit trees are increasingly being grown in the decorative part of the garden as dual-purpose flowering/fruiting specimens, and they are well worth finding space for.

All have superb spring blossom that rivals many purely flowering species, and some also have autumn colour and good shapes that make them valuable year-round attractions. What's more, modern dwarfing rootstocks mean they often stay smaller than many purely decorative trees.

Trained fruit trees come in various forms that make them easy to slot into small spaces in decorative gardens; fans and espaliers are stunning on south- or west-facing walls or for growing on trellis, cordons make productive fruiting hedges or they can be trained over arches or used as 'accents' at the back of a border, while 'step-over' trees are short, single-tier espaliers that make novel 'rails' to edge a path or flower border (see page 190 for training information).

Some fruits, such as thornless blackberries, can be trained over arches and others, such as strawberries, are ideal for growing in containers.

Redcurrants and gooseberries are often grown as 'U'-shaped double cordons, trained against a wall or on trellis, and both can also be trained as small standard trees for growing in tubs or in flower borders.

ALLOTMENTS

Once you've cut your teeth on a small fruit and vegetable patch you may feel like spreading your wings – in which case, why not think about taking on an allotment? Here you'll have the room to grow crops that take up lots of space, such as pumpkins, maincrop potatoes, sweetcorn, or fruit trees and bushes that don't fit in your garden. But one word of advice – don't give up on your small herb and salad patch at home where it's handy. You'll find it's just too much bother to pop over to the allotment to pick a sprig of mint or water a row of baby spinach leaves, but while you're working there at weekends it's quite practical to dig enough spuds for the week, or bring back a bucketful of leeks or a carrier bag full of cooking apples.

If you're toying with the thought of taking on an allotment, don't think about it for too long. Since 'growing your own' became fashionable, there's been a big demand for allotments and some local authorities have long waiting lists. Allotments change hands in the autumn, so phone the council ahead of time to find out where your nearest site is and if any vacancies are coming up. If a whole allotment is too big for you (and they *are* pretty hefty), you can always share one with neighbours or a group of friends, and in some popular areas where allotment space is scarce, such as London, it's quite normal to find half- or quarter-sized plots on offer.

But allotment growing is very different from gardening at home. For a start, allotment sites are far less secure than domestic back gardens, so don't leave anything valuable there – even locked sheds aren't totally burglar-proof. Access to water is another problem; there may be a few standpipes, but it's rare to have a tap for every plot, so you may be reliant on lugging watering cans or taking turns with a hose.

The water question is one reason why allotment crops are traditionally grown at much wider spacings than those in the garden, as this gives each plant a larger area for its roots to 'forage' in. The downside to this is that crops take far longer to cover the ground, so you'll need to do more weeding, but by leaving wide paths between rows of veg it's a lot easier to walk through the crop to hoe efficiently without doing any damage. Some veg enthusiasts use a wheeled hoe, which you just push between the rows at walking pace; I won't say it's entirely effortless (and it takes a bit of getting used to), but when you have a large area to hoe it certainly makes light work of it.

Weeds can be a big issue if you take over an allotment site that's been neglected by a previous tenant, so be prepared to spend your first winter clearing up, digging, removing roots and working in lots of manure. And since there's plenty of room, it pays to follow a proper crop rotation plan. (See 'Planning', page 24).

If you need further persuading to sign up for an allotment, you'll find that they are very sociable places and allotment-gardeners are a friendly lot who are usually very helpful with advice, encouragement and will even happily give young plants and seeds to beginners to start them off.

If a whole allotment is too big for you … you can always share one with neighbours or a group of friends.

PLANNING

The kitchen gardening season really starts in mid-winter, when the catalogues plop onto your doormat and you start working out what you fancy growing next year. Now, what with long winter nights and a shortage of serious gardening to do, it's easy to get carried away and order masses of seeds or plants that you don't have room to grow. We've all done it, then spent the summer trying to shoe-horn everything in while watching overcrowded crops go to waste. So it pays to have a planting plan early on.

Draw a plan of your plot, and mark in what you intend to plant in spring, roughly how long the crops will occupy the ground, and what you'll then follow them with. (You'll find information about sowing/planting and harvesting times in the fruit and veg directories in this book, and information will also be given on the backs of seed packets and in catalogues – don't ignore it!)

MAKE THE MOST OF YOUR PLOT

Your aim is to make the best use of your space to produce a good range of crops you can eat all year round without ending up with a glut. If you're clever about it, you can get two or more crops from much of your plot during a single season.

You'll have a natural talent for kitchen-garden planning if you're good at Sudoku, playing three-dimensional chess or scoring highly in the sort of computer games that involve going on a quest through several levels while fending off aliens. Even so, you'll find it helps to 'park' the various crops into groups that have similar needs and time scales, so they slot in together to fill the space available without any vacant or double-booked rows.

Some crops occupy the ground for the best part of a year – such as Brussels sprouts, sprouting broccoli and winter caulis – while some kinds are very fast growing, taking only 10–15 weeks from planting to eating, such as baby spinach leaves, lettuce and salad leaves. If you sow a row of these between bigger but slower-growing crops, the salads will have come out by the time their bigger neighbours need more space. This is a practice known as 'intercropping'.

Some crops are sown or planted early in spring but have been harvested by midsummer, so then you can clear away the rubbish and re-use the space for something else. Early potatoes are ready for digging up in June, which leaves time to plant frost-tender veg such as courgettes, sweetcorn or French beans and also oriental vegetables, which bolt if they are sown too early. Crops that are cleared in mid- or late summer, such as broad beans, can be followed by spring cabbage or fast-maturing veg or salad crops, which can be covered with fleece in autumn to give late crops.

It may sound complicated, but it gets easier the more you do it and you can fine-tune your basic masterplan year after year. Honest!

The big mistake beginners tend to make is to sow far too much of crops that will all come to maturity in the space of two or three weeks, such as lettuce, and create a glut you can't possibly use before it's run to seed and ruined (and salads can't even be put in the freezer). Sowing little and often (say 90cm/3ft of row each week) is a good way of using space productively and keeping your larder sensibly stocked at all times.

Don't forget to plan for your summer holidays; it's no good putting in lots of veg that'll suffer when you're not there, or growing the sort of fruit that ripens and spoils while you're away. But if you are planning to go away early in the season, there's a lot of late veg you can sow when you come home which will crop in autumn under fleece, or you can choose varieties of fruit which ripen before or after your holiday time. (Again, you'll find out all you need to know about harvest times in the fruit directory, and in specialist nursery catalogues.)

Aim to make the best use of space to produce a range of crops without ending up with a glut.

LETTUCE 'ROMANA ROSSA'

CROP ROTATION

Another thing to build into the planning equation is crop rotation. I know that by now you are probably thinking 'I'm a gourmet, get me out of here!', but bear with me. Once all the 'rules' of rotating crops through your beds have sunk in it becomes almost instinctive, and it really isn't nearly as complicated as it sounds.

The idea behind this technique of vegetable growing is to avoid planting the same crop in the same bit of ground more than one year in four, which is a great natural way of preventing root diseases. It also makes the maximum use of resources such as manure, which is only needed for certain crops and not others (don't use it when growing root crops, for instance).

The traditional way to rotate crops was to divide the plot in four; which meant growing potatoes on one quarter, root veg in another, peas and beans in one, and brassicas (the cabbage family) in another, with salads and other quick crops planted in the gaps between rows of bigger, slower crops. (The old intercropping wheeze, remember?)

Now this doesn't really take into account the demand for 'modern' crops such as sweetcorn and courgettes (tuck them in where you can, or grow them on the potato plot if you aren't growing spuds), but if you stick to the general principle you won't go far wrong.

Even if you grow in small beds where it's not practical to follow a strict rotation, just try to make sure you don't grow the same crop on the same patch of ground any more often than you can possibly avoid. And that's where making a proper plan helps – especially if you keep your old ones, then you can refer back to them year after year whenever you're working out what to grow, and where.

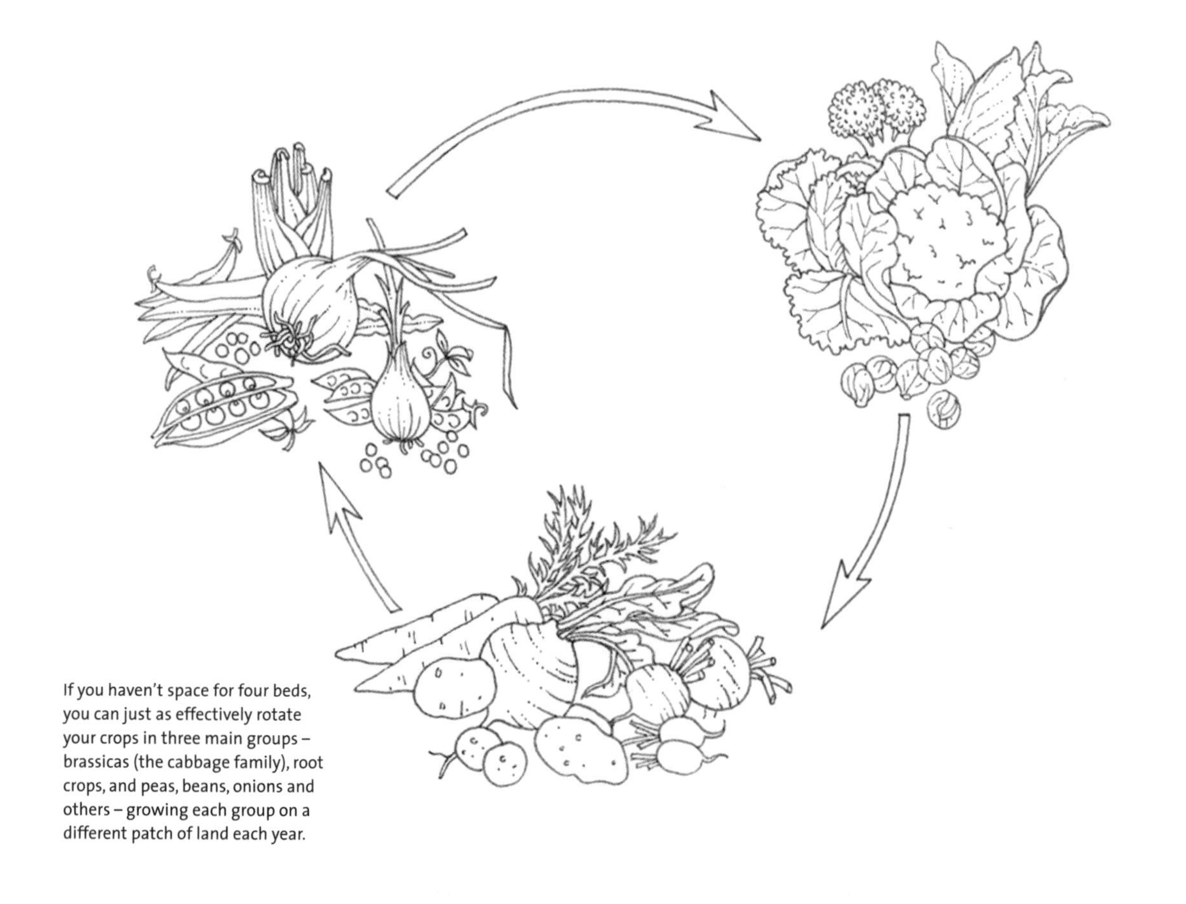

If you haven't space for four beds, you can just as effectively rotate your crops in three main groups – brassicas (the cabbage family), root crops, and peas, beans, onions and others – growing each group on a different patch of land each year.

CHOOSING VARIETIES

The types of veg you finally decide to grow must be a personal choice, based as much as anything on what your family likes to eat. But when it comes to choosing particular varieties, there are several options.

Tried-and-tested varieties: These are the old favourites you've grown before. It's worth making these your mainstay as you know you can rely on them to do well and that the family will like them.

New varieties: It's fun trying something new, but be aware that the seed will be more expensive – especially if they are F1 hybrids, which are the really reliable, get-what-you-pay-for seeds – and you don't get so many per packet.

Unusual veg: These are often unusual for the good reason that they are not to everyone's taste – think of them as things to experiment with rather than to devote much space to.

Heritage veg: There's a big trend towards reviving old favourites, particularly with 'cult' crops such as tomatoes, potatoes and beans. Some are very good – the ones that have survived have usually done so because of their outstanding flavour – though they'll often be less productive than modern varieties, or lack modern benefits such as disease resistance or dwarf habits.

Baby veg: These are special varieties that produce properly proportioned veg from very early on so you can pick them when small while they are very tender, (most regular varieties grow long and slim before starting to fill out). Many seed catalogues have a page devoted to baby veg, or they'll mark varieties as such on the appropriate page or on the packet.

Early cropping: Some varieties of veg, such as carrots and peas, are especially suited to sowing under cover to produce early crops. These are faster growing than other varieties, so they are ideal for small gardens where you need a quick turnover to make the best use of space, and they are also good for sowing near the end of the summer to crop in autumn under fleece or cloches. Again, you'll find suitable varieties marked as such in catalogues and on the packet.

Disease resistance: A good many of the newer varieties have been specially bred to have natural resistance to pests or diseases that are common to that particular crop – which means you can avoid spraying, so these are ideal for organic growers and anyone too busy to be bothered with pest and disease control.

Dwarf habit: Dwarf varieties of crops such as peas and French beans will save you time, since you don't have to put up supports or tie them up and they'll usually perform better in windy weather. Dwarf varieties are often quicker to start cropping than tall varieties as they don't have to grow a large superstructure first, but, conversely, they often stop cropping sooner, so you need to sow more often, and the total yields are smaller than from tall varieties.

Frequently asked questions

Q What's most worthwhile for growing in a small flower bed or in containers close to the house?
A Quick-growing crops that are always useful and best eaten absolutely fresh – such as lettuce, salad leaves, baby spinach, dwarf French beans, radish, spring onions and herbs.

Q What's decorative and useful to grow in tubs on a warm sunny patio?
A Tomatoes, runner beans, climbing French beans (including the sort with pretty pods used for drying), chillies, courgettes (particularly the kinds with golden fruit), Mediterranean herbs (which like sun and heat) and patio peaches and nectarines (which are naturally dwarf so they can stay in large pots or tubs for life).

Q What's likely to do best on an allotment where there's plenty of room but limited access to water, and which won't spoil if we only visit once every week or two outside the main harvest times?
A Potatoes, sweetcorn, pumpkins and squashes, onions, garlic, brassicas – including sprouting broccoli and Brussels sprouts – leeks, fruit trees and larger canes such as blackberries, loganberries and tayberries.

PREPARING SOIL FOR A NEW FRUIT AND VEG PATCH

When you want to create a new fruit and veg garden on a patch of ground that's currently covered by lawn or rough grass, or that's been left uncultivated so that it's full of weeds, you should allow plenty of time for clearing up first.

The mistake a lot of people make is to hire a rotavator and just push it around to chop up weeds or turf, then bury the lot and start planting straight away. But this is one shortcut that can be disastrous – it simply creates problems for years to come. Strange as it sounds, rotavating actually *propagates* perennial weeds; it does so by chopping their roots into hundreds of 'root cuttings' which all turn into new plants, and it does nothing to discourage destructive soil pests which are usually prevalent in old grassland. So don't do it; it pays to take your time and do the job properly. It seems like hard work at the time – and it is – but it'll save you hours of grief in the years to come, I promise.

If the area you have earmarked for your veg patch is covered by grass, your first job is to strip off the turf. Use a spade and skim it off the ground in slices 4cm (1½in) thick so you remove most of the roots. Stack the pieces upside down in an out-of-the-way corner of your garden for 18 months and they will turn into good fibrous loam that you can recycle back onto the garden. If instead of grass you have rough or overgrown ground to contend with, dig out brambles, tree seedlings, perennial weeds and any annual weeds with seedheads, but you can leave smaller, non-seeding annual weeds, as they can safely be dug back in.

Next, dig the ground over roughly – a fork is often best at this stage – taking out large stones, roots and any other rubbish you might find. Then dig it over again, this time using a spade, working in as much well-rotted organic matter as you can lay your hands on. (A barrow-load per square metre/yard isn't too much when you're breaking in new ground for the first time – a bucketful per metre/yard is enough for ground that's previously been well cultivated but allowed to run wild for a time.)

Leave the ground rough for several months at this stage so that birds and other natural predators can remove soil pests and slugs or snails. If you can spare the time, it's a good idea to fork over the area occasionally to expose more pests so that predators can continue to reduce the population. This also helps to reduce your weed population, as it brings dormant weed seeds to the surface where sunlight 'triggers' them to germinate. Hoe off these new crops of weed seedlings or flame-gun them before they have time to set seed or become established. The more of this you can do before you start to cultivate a new plot, the easier it will be to manage the patch later.

Plan your clean-up campaign so that the ground is ready to start using at planting time – for fruit trees and bushes the best planting time is over the winter (any time from late October to mid-March), while the veg-growing season starts in spring, around March/April.

Ground that's been thoroughly cleared in this way then only needs a little attention immediately before sowing or planting. At that time, fork over the ground, removing any last-minute weeds, sprinkle an organic general-purpose fertiliser (such as pelleted poultry manure or blood, fish and bone) evenly all over the area and rake it in, removing any stones and roots as you go so the ground ends up clean, level and ready to go.

Plan your clean-up campaign so that the ground is ready to start using at planting time.

ROUTINE SOIL PREPARATION

Although you'll only need to do a *big* soil preparation job when you first start a new kitchen garden, the ground that's being used for annual vegetables still benefits from a little attention every time it falls vacant.

Each autumn, as soon as summer crops have been cleared, spread well-rotted organic matter on those areas that won't be used for growing root vegetables (such as parsnips or carrots) or potatoes next year. Continue this digging and soil improvement over the winter and early spring as winter crops are cleared, so that by spring the whole plot is ready to go.

Even in summer it's worth doing a little light soil improvement every time you clear rows of salads or other short-term crops, to prepare it for the next crop. Clear away the old roots and leaves, plus any weeds, taking care not to disturb adjacent rows of plants, then sprinkle some organic general-purpose fertiliser over the bare soil and work it lightly into the ground with a rake before sowing or planting your next crop. This replaces lost nutrients and makes sure growing conditions are back up to scratch so the whole space stays as productive as possible all season.

The weedkiller alternative

You might consider using weedkiller instead of clearing new ground by hand if you have a serious perennial weed problem, you want quick results, you are not organic, or if you don't mind using chemicals to clear up first but then plan to turn organic later.

If this is the way you want to go, take care to choose the right product; you need a total weedkiller (which kills everything it touches – grass, weeds and everything else) but make sure it is not a residual type (such as path weedkiller or sodium chlorate) as this will linger in the ground. Choose a product based on glyphosate (such as Round-up or Tumbleweed).

When applying weedkiller, spray it onto weeds or use a watering can fitted with a fine rose or 'trickle bar'. These chemicals should be applied while weeds are growing vigorously in late spring or early summer. (Wait for a still day to do this, since spray drift can affect nearby plants if the product splashes or blows onto them). Follow the instructions on the packet and don't exceed the maker's recommended dose.

Don't expect instant results; this product is absorbed through the leaves and slowly kills the roots as well, so it'll be several weeks before you start to see the tops dying off. When this starts to happen you can clear those away by hand or let them dry out on the spot and burn them off with a flame-gun. Wait a month or six weeks to see if re-growth occurs from the remaining roots – this may happen with weeds that are very well established. If it does, repeat the treatment as soon as the weeds are a few centimetres high, so that there's enough foliage to absorb the product, but not so much that the weeds have time to regain their strength. You might need to re-treat strong and long-established colonies of problem weeds, such as ground elder, three or four times until weeds stop re-shooting.

Allow six or eight weeks after apparent death to be absolutely certain of eradication before cultivating the area.

ORGANIC MATTER AND FERTILISER

People often confuse the roles of organic matter and fertiliser. The two are completely different, and for a thriving kitchen garden you need a bit of both.

Organic matter is the soil's 'roughage'. It's spongy stuff which, when mixed with soil, holds onto moisture so the ground doesn't dry out too fast. But at the same time it creates lots of tiny air spaces that allow excess rain to drain away faster, *and* it makes the soil 'softer' and 'looser' so that roots are able to spread through it more easily. That's why plants grow so much better in ground that contains plenty of organic matter. (A lot of books talk about plants needing soil that's moisture retentive but well drained, which people always think is a contradiction in terms – well, it's not, and it's organic matter that makes it that way). But for organic gardeners there's an extra bonus from using organic matter; it houses flourishing colonies of beneficial soil bacteria which break the material down to humus and release valuable trace elements – often severely lacking in ground that's been cultivated using only fertilisers.

There are several sources of organic matter. You can buy in manure or spent mushroom compost, or make your own garden compost from 'green' garden waste, but whichever you choose it must be well rotted before it's used – it should look almost like fluffy, fibrous soil with 'bits' in it.

However wonderful it undoubtedly is, organic matter isn't enough to improve soil on its own. When you are growing plants intensively (which is, after all, what veg and fruit growing is all about) the breakdown of organic matter doesn't generate enough of the *main* nutrients – nitrogen, potash and phosphates – to keep hard-working, heavy-cropping plants properly supplied, so you need to use fertiliser as well. The shelves in garden centres are stocked with dozens of different types of fertiliser, and organic gardening supply catalogues list a lot more, but a few are all you'll need – depending on what you grow and whether you're 100 per cent organic or just 'careful'.

Before you go much further, you will have to decide whether you want to be an organic gardener or not. True organic gardeners do not use chemical sprays or fertilisers derived from inorganic sources. They are the 'vegans' of the gardening world and, in spite of being a meat-eater, I am one of them. Other gardeners avoid sprays but do not mind using inorganic fertilisers. You must make your own choice. But remember that organic fertilisers will utilize soil bacteria (and thus keep soil alive) when they are being 'broken down' for absorption by plants' roots; inorganic ones can be absorbed in solution with no bacterial involvement and, as a result, soil can become more lifeless.

I don't want to preach; you must choose your own style of gardening, but I will say that I feel happier with the organic approach – especially in the kitchen garden where I will be eating the plants I grow.

'Solid' fertilisers – granules, powders and pellets: Use solid fertilisers when preparing soil before planting, but only for plants growing in the ground, not those in containers.

General purpose fertiliser: A product such as blood fish and bone (organic) or Growmore (inorganic) provides roughly equal quantities of all the three main nutrients: nitrogen (N), phosphates (P) and potassium (K). As a rule of thumb, nitrogen promotes leafy growth, phosphates are good for root development, and potassium (or potash) encourages flower and fruit production. Use a general fertiliser when preparing the ground for sowing or planting vegetables, and again after you've cleared a row of crops in summer ready for replanting the same patch with something else. Use it, too, at the start of the growing season (mid- to late April) where permanent crops such as fruit, nuts or perennial veg are growing. Sprinkle it evenly all over the soil underneath the plants and lightly work it into the surface. Follow the instructions on the packet, but as a rough guide use about a handful per square metre.

Bonemeal: This is a fertiliser rich in phosphates, but as most of them are locked up by the soil on application it is not of as much value as many people believe. For best results, stick to using it when it is accompanied by blood and fish.

Sulphate of potash: This is a high-potash feed that's especially appreciated by heavy-cropping, fruiting plants. It can be used in small quantities (usually about 15–25g (½–1oz) per square metre/yard), along with general fertiliser, when feeding fruit trees in spring. Notoriously greedy plants, such as vines, need it most. Use it too when preparing the ground where tomatoes are to be grown, since they have a high need for potash. Although it's not strictly an organic feed, it's quite difficult to find genuine organic alternatives. However, if you are a 100 per cent organic gardener, you can get organic potash from organic gardening suppliers.

Chicken manure pellets: These pellets are a useful high-nitrogen feed to apply before sowing or planting green veg or to 'boost' brassicas about halfway through their growing season. Use them with, or instead of, general-purpose feed for crops like these. Completely organic gardeners would want to assure themselves that the manure came from a welfare-friendly source and not from factory-farmed hens.

Hoof and horn: This slow-acting, high-nitrogen, organic fertiliser is used when preparing ground where leafy veg crops are to be grown. It is best applied very early in spring so it has time to start breaking down by the time crops need it. Organic suppliers offer an animal-product-free, high-nitrogen alternative to hoof and horn (which is made from soya) which is acceptable to vegetarians. Use either as suggested for chicken manure pellets.

Epsom salts: Chiefly used to feed tomatoes, Epsom salts are a source of magnesium. You need only use this product sparingly, little and often, during the growing season to treat obvious deficiencies (such as leaves yellowing between veins that remain green). It is also good when sprinkled over ground that is being prepared for planting tomatoes, in greenhouse soil borders or in beds outdoors. If you use it this way, you need to scatter 15g (½oz) per square metre/yard.

Liquid and soluble feeds: Use these types of products to feed plants growing in containers and for anything that needs a quick boost, since liquid feeds are instantly absorbed.

Liquid seaweed extract: This is a good natural source of trace elements, useful for spraying or watering onto plants as a quick pick-me-up after a setback of some kind, such as a spell of bad weather during the summer, or heavy rains that wash nutrients through the soil. It's also a valuable tonic for high-yielding veg or fruit plants.

Tomato feed: Both liquid or soluble tomato feeds are a good source of potash and magnesium for tomatoes, in pots and in the ground, and all fruiting plants in containers. Completely organic growers will find various organic alternatives in the catalogues of specialist suppliers.

General-purpose feed: This is a liquid or soluble feed which is higher in nitrogen than potash, and is recommended for use on leafy crops. Organic suppliers will stock organic alternatives.

Ericaceous feed: This is a specialized liquid or soluble feed which is intended primarily for lime-hating plants, such as rhododendrons, when in containers but it is also the best product for ericaceous fruit such as cranberries and blueberries, which need acid conditions. Often needs sourcing from a specialist supplier.

The secret of making good garden compost

The right container

For small quantities use a compost bin – it looks tidy and it makes small amounts of green garden rubbish rot down quickly and efficiently since the solid walls trap the heat inside. For large quantities, you're better off using a compost heap as results will be better and the heap will stay tidier. Make a big container (minimum 1.2 x 1.2m/4 x 4ft) with wooden slatted sides for storing compostable materials. If you don't want to build a container, you can just leave a pile of garden rubbish out in the open, but you'll need to 'turn it' so that the sides are moved to the middle of the heap, allowing all of the material to rot down evenly. Besides making more work, this also doubles the time before the finished compost is ready to use. Whichever method you use, ideally you need two bins or heaps – one that's being filled while the other is rotting down.

The right ingredients

There's a knack to making good compost that rots down properly instead of merely 'mummifying'. For best results use a mixture of materials; things you can successfully compost include annual weeds, complete with roots (leave a little soil on them), soft fallen leaves (not from evergreens or from plants that have been affected by disease such as potato blight or peach leaf curl), crop debris, including trimmings and outer leaves when veg or salads have been cut, and plant remains when the ground is cleared ready for replanting. You can also add grass clippings, kitchen peelings and plain, unprinted paper (such as white paper kitchen roll). Don't add woody materials such as prunings unless you put them through a shredder first – they take too long to rot down. Most folk also rot tree leaves separately as they take longer to decompose than soft vegetable waste.

The right method

Add new material in 15cm (6in) deep layers, firm it down well, and dampen it if it is dry. Spread a few shovelfuls of soil or manure over the top before adding the next 15cm (6in) layer (soil and manure both provide for free beneficial bacteria that act as very efficient compost starters). Between making new deposits, put the lids on bins or cover heaps with old carpet, a tarpaulin or something similar to conserve heat and stop the compost drying out.

The right time

When the container is full to the top, cap it with 15cm (6in) of soil or manure, cover it securely and then leave it alone. Don't add new material, even when the old stuff sinks down. Compost should be ready to use three months later in summer and after six months in winter (because when it's colder decomposition slows down).

Use homemade garden compost to dig into soil to enrich it ready for planting, or spread a layer about 5cm (2in) deep under fruit trees and bushes as a mulch in spring. This suppresses weeds and helps the soil to retain moisture. You'll feel incredibly virtuous if you have a working heap, and you'll savemoney, too.

SOWING SEEDS

DIRECT SOWING

Some vegetable seeds are sown straight into the ground where they are to grow and crop; this technique is used particularly for root crops, such as beetroots and carrots, since they don't transplant well, and on the herb side this also applies to the likes of parsley and chervil.

Since these seeds tend to be small you'll need to prepare the ground well, leaving a finely raked, stone-free surface. (But don't sow root crops onto a patch of ground that's been freshly manured or they are likely to 'fork'.) Make a straight shallow groove or drill with the corner of a hoe drawn against a taut line, or else lightly press a bamboo cane into the surface of the soil to make a shallow indentation. Sprinkle the seeds very thinly along the drill, then cover the seeds to roughly their own depth – don't bury them too deeply. You can use plain garden soil to cover seeds if it's good, fine and stone-free, but if you garden on heavy clay or soil which stays wet for a while after rain, you'll have better germination if you use sharp sand or horticultural vermiculite because it gives seedlings better drainage and is easy for the tiny plants to push up through.

When your seedlings appear, keep the soil between the rows well weeded so they aren't swamped. As soon as they are big enough to handle easily, thin out the seedlings, pulling up the ones you do not want. Do this in several stages – to allow for a few natural deaths among the ones you leave behind – until they are left at the correct final spacing.

RAISING PLANTS UNDER COVER

Some crops are normally sown under cover. This is the best way to raise frost-tender vegetables such as sweetcorn, French and runner beans, courgettes and pumpkins. These plants need an early start, in around mid-April, and since the seeds can't be sown outside until after the last frost (usually around mid-May) you extend their growing season by a month or more by being able to plant them out then as young plants rather than seeds. But the same technique is sometimes

used for producing early crops of other veg, including lettuces, when growing conditions are cold or wet outside, or when pests such as pigeons or mice are a problem (they'll often pinch early-sown pea or bean seeds).

To sow seeds under cover you need to fill seed trays or small individual pots with seed compost. Sow small seeds such as lettuce thinly and prick out the seedlings singly into small pots when they are big enough to handle. Sow large seeds, such as peas or sweetcorn, at the rate of one or two per pot and 'weed out' the weakest seedling later. Plant out your young plants when the pot is full of roots or, in the case of frost-tender plants such as tomatoes and courgettes, after the last frost has safely passed. Harden off the seedlings carefully first (see box below right). If necessary, re-pot the plants into bigger pots so that they can be kept under cover for longer without suffering if the weather isn't quite up to scratch.

Plant pot-grown plants carefully, without breaking up their ball of roots. Most kinds should be replanted so that the top of the rootball is slightly buried, but some plants (tomatoes and brassicas) do best if they are planted a little deeper as this gives their stems more stability and, in the case of tomatoes, allows the stems to take root as well, which makes the plants more sturdy.

RAISING PLANTS OUTSIDE IN A SEEDBED TO TRANSPLANT

Frost-hardy crops are best sown in a seedbed outside. Leeks, lettuces and brassica plants are most commonly grown this way.

Choose a convenient, semi-sunny, sheltered corner where plants can be given special care – a cold frame is ideal but not essential. Prepare the ground well by removing weeds and working in lots of very well-rotted organic matter, then, just before sowing, fork it over, sprinkle on a dressing of general fertiliser and rake thoroughly to remove stones and roots to leave a smooth, level, cake-crumb-like surface.

Make short, shallow grooves (known as drills) with the corner of a hoe or a bamboo cane, and sow seeds thinly. When they come up, thin out the seedlings in several stages until they are 5–7.5cm (2–3in) apart. When they are big enough to handle, with 2–3 'true' leaves as well as their first pair of baby 'seed leaves', dig up the seedlings and transplant them with as much root as possible into well-prepared soil where you want them to grow, placing them at their final spacings.

Replant the seedlings only very slightly deeper than they were originally growing in their seedbeds, and firm them in lightly. New transplants, especially of plants such as lettuce, often flop for a while when first moved, but they should soon perk up. Plant brassicas in very firm ground so they heart up properly, and plant them slightly deeper so that the first set of leaves are only just above the surface of the soil. Planted like this, they stand up straight away – otherwise the plants will never be strong. You'll know if you've planted brassica transplants correctly: if you hold onto a leaf and pull it gently it should tear and the plant should not come out of the ground.

Hardening off

Because frost-tender plants are raised in warm, still, humid conditions on windowsills indoors or in a heated propagator, they have very soft, delicate foliage that doesn't take kindly to being put straight outside into the rough-and-tumble of outdoor conditions. Plants that are treated this way usually turn up their toes straight away, even when they aren't exposed to a frost.

The idea is to accustom them slowly to outdoor conditions, starting two or three weeks before you want to plant them out – a technique known as 'hardening off'.

If you have a cold frame, stand the young plants in that and simply close the lid at night and open it every day. On cold, windy or rainy days, leave the lid closed all the time for protection. If you don't have a cold frame, stand plants out in the garden on fine days and bring them in at night – an unheated greenhouse or conservatory is fine if you have one, otherwise it will have to be a shed, utility room or even a windowsill indoors (but choose a cool room, or they'll never get acclimatized).

After 2–3 weeks, once the weather forecast or local knowledge suggests that all the frosts are over in your area, plant them out. But, as belt and braces, keep some fleece handy so you can give them a bit of protection if there's a spell of bad weather or frost is forecast.

PLANTING FRUIT

The very best time to plant fruit is while the plant is leafless and dormant – between late October and mid-March. (This is when mail-order nurseries send out bare-rooted plants, and it's the only time that they *can* be planted.) But it's vital that soil conditions are suitable – don't try to plant when the ground is very soft and muddy or when it is frozen solid. In those conditions it's far better to 'heel' the plants in (see box opposite).

Pot-grown plants are also best planted during the dormant season. They *can* be planted in summer, even when they are in full leaf, flower or even fruit, but because there is more stress on the plant at that time it's essential to keep summer-planted fruit really well watered for the rest of that season. Treat pot-grown fruit in the same way as your bedding plants, just for that first summer, and you won't go far wrong.

If you are creating a new fruit garden, prepare the ground thoroughly over the whole area – the way I've described it earlier on page 29 – and plant it all up at the same time. But if you are just adding a plant or two to an existing fruit garden or other well-cultivated area, instead of preparing a large area of soil you only need to dig individual planting holes. Make a hole that is much bigger than the rootball and work plenty of well-rotted compost into the bottom of it. Mix more compost in with the soil you'll use to fill in around the roots.

To plant bare-root fruit, spread the roots out well over the bottom of the hole and, as you backfill, check that the neck of the plant is level with the surface of the surrounding soil – you might see a soil ring on the bark which shows the original planting depth, or place a stick across the hole and position the plant so the mark is at the level of the stick. Plant a pot-grown plant without breaking up the rootball – unless it is badly potbound, in which case tease a few coils of roots out. Stand the plant in position and use the soil/organic matter mixture to fill the gaps round the edge of the planting hole. Firm the soil down gently and water well.

Don't forget supports. Fruit trees will need staking; trees on dwarfing rootstocks need strong stakes for life,

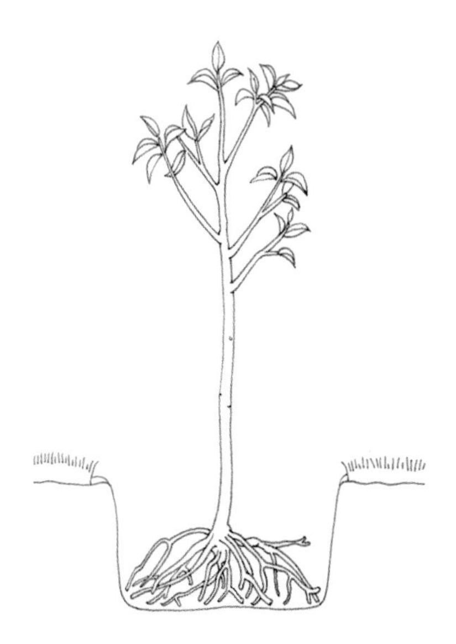

1 Dig a generous-sized hole and put the removed soil to one side. Work compost or manure into the bottom of the hole, then spread out the tree's roots over the top.

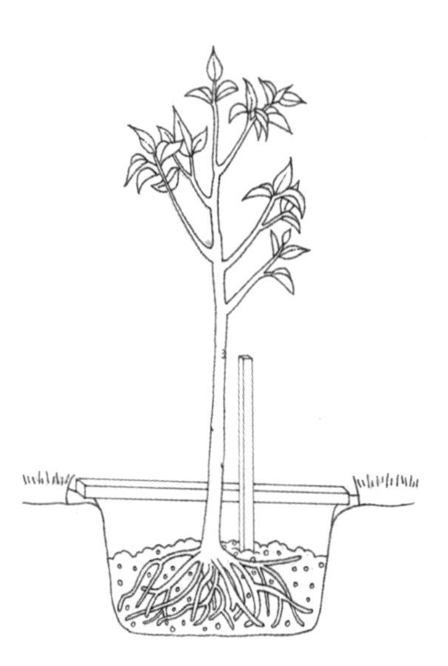

2 Insert a stake so that its top sits (30cm/12in) below the lowest branch. Return the removed soil, mixed with manure or compost and firm the tree around the roots. Place a stick across the hole to check the planting depth.

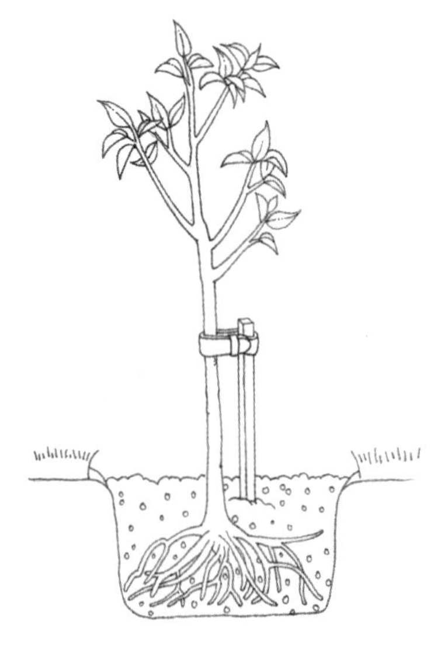

3 Secure the tree to the stake with a proprietary tree tie at the very top. If the trunk is more than 1.5m (5ft) tall, place another tie 30cm (1ft) above soil level.

because their roots aren't that strong. Trained trees, such as espaliers, fans and cordons, need supporting on trellis or post-and-wire fences with canes fixed in place for each individual branch to be tied in to. Cordons can be trained along stout canes or poles or tied up to trellis, depending on whether they are being grown against a wall or freestanding out in the open.

Most cane fruit also needs training against a fence of some sort – a row of posts with two or three strong wires stretched horizontally between them is ideal if you don't have a handy wooden fence you can use. Bush fruits, such as currants and gooseberries, that aren't trained as cordons don't need any support.

It's also worth sinking a pipe vertically into the ground alongside each new fruit tree or bush as you plant it. This is useful for watering in summer when the plants are carrying fruit, since you can use it to 'funnel' water straight to the roots without any wastage. (This is ideal if you live in an area prone to hosepipe bans, or if you have a water meter.) Use a 60cm (2ft) length of plastic pipe that is slightly wider than a garden hose, and 'plant' it at one edge of the planting hole with the top 5cm (2in) sticking out of the ground. When you need to water, simply stick the end of your hose down the tube. (If you are using recycled water during a hosepipe ban, push a bit of hose over the nozzle of your watering can to make it easier to hit the spot and to save bending.)

Heeling-in

If bare-root fruit trees, bushes or canes are delivered when the soil is unfit for planting, unpack them and heel them in.

Dig a large hole in a sheltered spot where the ground is well drained and workable. Pile the soil you've removed to one side, so the heap slopes down into the hole, then put the plant bundles with their roots in the hole and their stems lying up the slope and cover the roots with soil. You can also do this in the soil border of a greenhouse or walk-in polytunnel – or you can even plant them temporarily in a large pot of garden soil or potting compost. You can keep plants like this for several weeks, but plant them properly as soon as you can.

FEEDING

Edible crops need a lot more feeding than any other plants in the garden because they are constantly taking a lot out of the ground. It's not just the fruit or vegetables you eat that deplete reserves of nutrients in the soil; think of all the old plants you pull out and the stems you prune off. Crops are like children; to grow up strong and healthy they need the right nourishment, and plenty of it. But in kitchen gardens the 'diet' varies from one type of crop to another.

VEGETABLES

At the start of each growing season, use general-purpose fertiliser (blood, fish and bone, Growmore or a general-purpose organic equivalent) when you're getting the ground ready to sow or plant. You can use this on outdoor ground, including seedbeds, and also on border soil in the greenhouse or polytunnel. Follow the instructions on the back of the packet, but a handful per square metre/yard is a handy rule of thumb.

During the summer, the same feed can usefully be reapplied at about half the usual rate to replenish nutrients when you've cleared crops and you're getting the ground ready to plant something else. You can also use a general-purpose feed (again at about half the usual rate) during the growing season to boost 'greedy' crops that occupy the same ground for a long time, such as maincrop potatoes, artichokes or brassicas. If you have a selection of fertilisers in stock in your shed, though, the very best mid-season feed for leafy crops, such as brassicas, is chicken manure pellets, or a similar quickly available, high-nitrogen feed.

For tomatoes growing in containers – and indeed any fruit, herb or veg crops in pots – always use liquid or soluble feeds, since solid fertilisers could easily scorch the roots. These plants will need feeding once a week all summer. Use liquid tomato feed for 'fruiting' plants such as courgettes, tomatoes and peppers, and so on. Use general-purpose liquid feed (which contains more nitrogen and less potash) for small leafy plants such as basil, salad leaves and spinach – it helps to keep them a good deep green and delays them bolting prematurely and 'running to seed'.

HERBS

When you're growing herbs in the open ground, prepare the soil in the same way as you would for vegetables. Annual herbs won't need any extra feed during the summer when grown this way, but perennial herbs fall into two camps – Mediterranean, or not.

Mediterranean herbs, such as rosemary, thyme and sage, are thought by some to taste better if they aren't fed at all, although personally I give mine a general feed every spring – in late April or early May – but nothing else afterwards. Leafy herbs, such as chives and tarragon, *do* need feeding though, and mint is particularly greedy. So besides their start-of-the season general feed, I'd give these plants a drop of liquid feed every few weeks or top them up with a little more general fertiliser halfway through the summer, and water it in well.

All herbs grown in containers need regular feeding. Annual herbs planted in pots of multi-purpose compost will need weekly liquid feeding after the plants have been growing well for 4–6 weeks, since this type of compost does not contain a lot of nutrients. If you are growing perennial herbs in containers, use a loam-based John Innes compost mixed with an equal amount of peat-free, multi-purpose compost and a little extra grit added to open it up. This mixture contains enough feed for the first year, but in subsequent years they'll need regular liquid feeding every two or three weeks from April to September.

FRUIT

Each spring, at the start of the growing season, sprinkle general-purpose feed on the soil round fruit canes and bushes and underneath fruit and nut trees. Heavy-cropping or heavily pruned fruit also appreciate 15–25g (½–1oz) of sulphate of potash per square metre/yard, which should be applied at the same time as the feed.

Container-grown fruit needs weekly or fortnightly liquid feeding from April to late August or early September. Use ericaceous feed for lime-hating fruit such as cranberries and blueberries, but a liquid tomato feed is ideal for everything else. Some types have specific needs, so check the growing instructions.

WATERING

Watering is crucial for the healthy development of fruit and veg in all but the wettest summers; it's often the limiting factor in what's achievable at home, but on allotments being able to water regularly can be a real problem, which is why it pays to choose the crops you grow with care.

When water is in short supply, prioritise your plants' needs so that the most deserving cases get it first.

TOP PRIORITIES – VEGETABLES, SALADS AND STRAWBERRY PLANTS

All these are very shallow rooted and you can't afford to let them dry out at all during their growing season; some kinds bolt (run to seed) or turn tough and woody if they suffer any water stress. Lettuce and celery are particularly touchy about getting enough water, and if they don't get what they need they will bolt and quickly run to seed, which ruins them for eating, while strawberries will sulk by producing small crops of tiny, hard fruits.

In a hosepipe ban, your best solution is to use rainwater which has been sourced from the roof and stored in water butts. Set up a water butt (or two) in early spring and attach it to a downpipe on the house, greenhouse or shed and it will quickly fill up after a good bit of rain. Don't use recycled bathwater (which can contain bacteria) or washing up water (which may be dirty or greasy), since you won't be able to prevent splashing it over the part of the plant that you'll eat.

All these crops mentioned here need watering every day or two in dry spells to keep the soil moist near the surface, which is where most of their roots are.

HIGH PRIORITIES – SOFT FRUIT AND FRUIT CANES

Gooseberries, currants, raspberries and the like can survive slight water shortages quite well at most times of year, but if they go short while they are carrying fruit they'll shed their immature crop, so they are the next priority for watering. Soft fruit and canes are more suitable candidates for being given recycled water, since you water the roots without splashing the edible parts. Give them a good soaking once a week; in a long dry spell sink a pipe down 45cm (18in) alongside each plant so you can direct water straight to the roots.

MEDIUM/LOW PRIORITIES – FRUIT TREES

These are last in the pecking order when it comes to watering. Mature standard trees can usually tolerate dryish summers quite well without help, but trained trees and those on dwarfing rootstocks *do* need watering in dry spells. These forms have small shallow root systems and when they are carrying crops they'll drop the lot prematurely if they dry out. Trained trees growing against walls or fences are especially likely to go short of water in a dryish summer, as the structure deflects rain and foundations or builders' rubble at the base soak up a lot of available moisture.

Since fruit trees are the most deep-rooted of kitchen garden crops, give them a very thorough soaking only in prolonged dry spells (ideally use a pipe to direct water straight down to the roots – see page 39) and keep the ground mulched around them. But any fruit that's newly planted or growing in containers needs regular watering to keep it going.

When water is in short supply, prioritise your plants needs so the most deserving get it first.

WEEDING

Weeds don't just make a fruit and veg patch look neglected; they actually *harm* edible plants and reduce – or ruin – your crops, and they do so in several ways.

Weeds overtake and smother small seedlings so that they die from lack of light, and if the weeds are left to grow large enough they'll do the same with strawberry plants and newly planted fruit bushes. Weeds also harbour pests, particularly slugs and snails, which are then perfectly placed to scoff crops growing at ground level; strawberries and soft young veg and salad plants are especially at risk. Some weeds suffer from the same diseases as veg crops – cruciferous weeds belong to the same family of plants as cabbages so they can carry club root, which will affect all brassica crops.

But even if they do none of these things, weeds compete with edible crops for water and feed. This may not sound that serious, but believe me it can mean the difference between good crops and iffy crops – or no crops at all. Even fruit trees can be hampered by weeds left to thrive beneath them. So all this is why you need to take weed control seriously.

There are several ways of keeping weeds at bay.

The veg and salad patch needs hoeing every week or two while crop plants are small. For best results, hoe on a dry, sunny day while weed seedlings are tiny, then hoeing takes hardly any time to do – what's more, the seedlings shrivel up in the sun so you don't even have to bother clearing them up. If the weeds are a bit bigger you can still hoe them out, but it takes longer and you need to clear them up afterwards, since they can often root back into the soil, especially in showery weather.

If you leave it far too late, there's no alternative but to dig weeds out individually with a trowel. This takes an awful lot longer and there's always the risk of disturbing veg roots because they are so near the surface, or to find you've accidentally pulled out the odd veg plant as well. So regular hoeing early on really does save you time and effort in the long run.

The good news is that you'll only need to hoe most veg crops at the start of their working life. Once they grow big enough to cover the soil between rows they will largely smother annual weeds, so you only need to whip out the odd one that struggles through. (Intensive beds need the least weeding as crops are planted so close together, and allotment plots need most weeding because crops are spaced a lot further apart.) But plants such as onions never create enough shade to do the job, so you will need to regularly hoe around them from sowing or planting right through until you harvest them. That's why it pays to space them a good hoe's width apart when first planting them.

Fruit trees, bushes and canes are easiest to keep weed-free if you mulch round them each spring. Mulching prevents weed seeds in the soil from germinating by keeping them in the dark, but it won't give you 100 per cent anti-weed cover because weed seeds will blow in from elsewhere over the summer and germinate in the fresh layer of well-rotted organic matter. But if your mulching material is soft and loose, it's much easier to weed thickly mulched soil than hard, dry, and unimproved ground.

I'm often asked 'why not just grow fruit in grass and mow round them, then you don't have to do any weeding?' Well, that's fine in the case of large, well-established fruit trees – the sort you can sit under in summer or sling your hammock from – but not so easy if you're planting trees on semi-dwarfing rootstocks – those you grow in the lawn or plant as a small orchard. In this situation it's a good idea to leave a circle of bare soil round the trunk for the first 3–4 years, while they find their feet, after which you can allow grass, spring bulbs or low-growing perennials to colonize.

But when you grow fruit bushes, cane fruit and fruit trees on dwarfing rootstocks (including trained forms) they have small, weak root systems that can't cope with any competition for water or nutrients, so with these plants it's essential to keep the ground bare underneath them *throughout their life.*

Mulching

It's really worthwhile mulching any crops that occupy the same ground for years on end, such as fruit trees, canes and bushes, as well as perennial vegetables such as globe artichokes. A layer of mulch saves lots of work during the growing season, besides making for far better growing conditions, which in turn means better crops.

The technique is exactly the same as the one you use in mixed beds and borders in the ornamental parts of the garden. Each spring, remove any weeds and spread around 5cm (2in) of mulching material over all the exposed soil. For this you can use well-rotted manure, garden compost, spent mushroom compost or bark chippings (which last longer before they need to be topped up). If you like to get ahead, you could spread your mulch any time from February onwards, but try to get it done before mid-March when fruit trees and bushes start coming into leaf or bloom, so there's less risk of damaging new growth as you work.

The important thing is that the soil needs to be moist when you spread the mulch, since besides inhibiting weeds, one of the main reasons for mulching is that it helps to keep the soil moist in summer by minimizing evaporation.

PROTECTING CROPS

It's no good leaving edible crops to fend for themselves; they need protecting from birds and bad weather, and whether you are a 'green-ish' gardener or fully organic, you'll need to protect them from pests and diseases without using chemicals. The trick is to protect crops *before* they need it, instead of hoping to salvage the situation when it's too late. This makes life easier in the long run, and it also means you don't lose half your crop while you're 'getting round to it'.

BIRDS

Birds are a well-known problem around soft fruit and cherries, but they'll go for all sorts of tree fruit if it's allowed to stay on the tree until it's overripe. It's not just fruit, veg can be affected too – pigeons can be very destructive of brassica crops in winter and early spring, and they'll also pinch early sown pea and bean seeds, given half a chance.

Protect at-risk plants *before* birds start taking an interest. Covering them with bird netting is the answer, but avoid using thin, cheap, plastic types since they are very easy for birds to get their feet entangled in, which makes them easy prey for predators. Birds can often injure themselves trying to struggle free, too, and they get so caught up they are difficult to disentangle. The only solution is to cut the net, and they'll be quite distressed by the time you let them go. Thick, heavy-duty netting is far more bird-friendly.

Ideally, don't just drape netting over the tops of plants as birds can sit on top to weigh the netting down, which allows them to reach through with their beaks to get at the crop, and they can also sneak underneath unsecured edges. It's far more effective to construct a framework and hang the netting over that.

Temporary structures can be made from garden canes and special balls with holes in them that act as connectors for quick self-assembly 'cages', but a permanent fruit cage with netting sides, roof and door is the very best way to protect soft fruit bushes, cane fruit and strawberries. Proper cages stay put all year round, and you can just walk in, work and come out again, without fiddling around with netting. As a bonus, the netting 'walls' also give the plants some protection from wind, without keeping pollinating insects out or depriving them of rainfall.

Where trained fruit trees are grown against a wall you should go for the next best thing to a cage: fix a wooden batten along the top of the wall and use it to suspend netting from. This can easily be rolled up when you want to pick the fruit, and replaced afterwards.

BAD WEATHER

Fruit is particularly susceptible to bad weather because so many varieties flower in early spring, and late frosts and early blossom don't mix. The trees or bushes are perfectly hardy, but if their flowers are hit by a frost they'll turn black and die off, which means pollinating insects give the blooms a miss and any flowers that have already been pollinated will fail to develop into fruits.

The simple solution is to drape the plants with horticultural fleece, or even old net curtains, overnight when a frost is predicted. Horticultural fleece is a white, woven, polypropylene fabric which is sold off the roll in garden centres or in packs from mail-order organic supply firms; it can be re-used several times, until it tears. Being porous, water can permeate through so you don't need to remove fleece for watering; it gives a few degrees of frost protection and also shelters crops from wind, besides softening the force of driving rain. But it won't allow insects to get through, so unless the weather stays below freezing, uncover the plants by day so that pollinating insects can reach them – even self-fertile varieties need to have the pollen moved from one flower to another in order for them to set fruit.

If you grow *very* early flowering fruits, such as almonds, nectarines and peaches, they'll often flower before there are enough pollinating insects around to do the job, so you'll need to pollinate the flowers by hand. (You'll also need to do this if you grow fruit under cover where insects can't easily reach them.) To hand-pollinate, take a clean, dry, soft, artist's watercolour brush and dab it briefly into all the flowers that are fully open, so that you work your way right round each one. Do this every day during the flowering season. The best time is

around lunchtime, when conditions are warmest and brightest and pollen is being shed naturally.

Veg crops can also benefit from protection from the weather, and these days it's quite common to extend the growing season in spring and autumn by covering early and late crops with fleece. All in all, horticultural fleece is a very handy kitchen garden accessory. It's almost as good as growing in a cold frame, greenhouse or polytunnel, but it costs a heck of a lot less. It's also instant, so it's worth keeping some in stock for emergency use.

PESTS

These days, as an alternative to using sprays, a lot of gardeners use a sheet of fine, insect-proof mesh to cover crops and protect them from insect pests. (It looks like white net woven from very fine plastic strands. Don't be tempted to use fleece for this job, since in summer it traps too much heat and crops can literally 'cook'.) The main use for this mesh is to protect crops from hungry insects, but as a secondary benefit the same material will also keep birds and cats off crops, and it protects plants from hail, wind or over-strong sun.

Rain passes easily through the material, so there's no need to take it off if you need to do any watering, though it does make a bit of extra work as you need to roll it back to do any weeding, then replace it again afterwards.

To use it, lay the mesh over the top of the crop then bury the edges in a shallow trench dug all round the sides, or weigh them down evenly with stones. To be most effective at screening out pests, it's best to cover crops as soon as they are sown or planted to prevent insects such as greenfly or carrot fly finding a way in. It's no good waiting until the pests are already at work on the crop, as then the mesh just traps them inside. As the crop grows, the fabric billows up like a white cushion over the top, so it's best used for covering low-ish crops like lettuce, spinach, salad leaves and carrots, and left over them until they are harvested. If you use it over taller crops, bear in mind that they will eventually outgrow it, but at least you're giving them a head start.

If you use the mesh for plants that require pollinating – such as broad beans – then you'll need to uncover them at flowering time and deal with any pests that strike by other means.

Insect-proof mesh is available from mail-order organic goods suppliers, and even though it is more expensive than fleece, it lasts an awful lot longer (I still have some I've been using for ten years) so it represents excellent value.

Growing in a greenhouse or polytunnel

If you want to grow crops such as tomatoes, cucumbers, peppers, chillies, aubergines or melons that really need warmer, stiller and more humid conditions than you find outside in most UK summers, then growing them in a greenhouse or in a walk-in polytunnel is the most reliable way of ensuring a good crop. What's more, since the climate is better under cover, you'll be able to keep harvesting over a much longer season than is possible outside.

Most people lay out a greenhouse or tunnel with a soil border on each side with a path through the middle for easy access, but if you want some staging (basically benches set at waist-height which make watering and caring for seedtrays and pots easier), then run it down one side and put a soil border along the other – the area will still be quite productive.

Every autumn, when the last summer crops have been pulled out and the border becomes vacant, prepare the soil for the next season by working in lots of well-rotted organic matter. Then, before sowing or planting each crop, apply a dressing of general-purpose fertiliser, and use liquid feed regularly during the summer. Heavy feeding is essential here as you are making the same patch of soil work very hard – it's like an intensive veg bed, but under cover.

A greenhouse or tunnel might seem like a luxury, but once you've taken the plunge you'll find it's very useful for providing a wide range of out-of-season produce for the rest of the year, as well as for growing all the popular summer crops. It really is a worthwhile investment, and a whole heap of fun, too.

Protect crops before they need it, instead of hoping to salvage the situation when it's too late.

COMMON KITCHEN GARDEN PESTS AND DISEASES

Pests and diseases are seen by a lot of gardeners as the biggest obstacle to growing healthy and edible fruit and veg, but in practice weeds and weather are responsible for far more crop failures. True, there are some pests and diseases that you can safely expect to put in regular appearances, but many are mostly a problem only to a limited range of crops. (These are covered in the pest and disease section under individual entries in the directory part of this book.)

Despite there being a number of pests and diseases that plague some fruit and veg, there are many effective 'green' remedies for them that you can use without needing to resort to chemicals. (In fact, of the few chemicals left on the market these days, only a handful are suitable for use on edible crops, so read the small print if you feel inclined to try.) As a general rule, if you want to avoid using chemicals you need to keep crops healthy by good cultivation, and spray or feed plants with seaweed extract to boost their natural defences. You can also pick off pests or the occasional diseased leaf or flower by hand before a problem spreads. Try to avoid being a control freak; most crops can shrug off mild attacks if they are growing lustily.

But the big advance is in new varieties that have been bred to have natural resistance to certain pests and/or diseases, so choose these if you want an easy life – they are especially suited to organic growing for the very reason that you don't need to spray. These new varieties are appearing all the time and catalogues make much of this big new benefit, which means you'll have no trouble spotting them.

Biological control

Becoming more and more popular in these days of organic gardening, biological control uses natural predators to attack pests such as whitefly (1) (predated by the chalcid wasp (2) *Encarsia formosa*) and red spider mites (3) (controlled by another mite, *Phytosieulus persimilis* (4)) when they appear under glass. Vine weevils (5) and slugs (6) can also be kept in check by 'natural' means, which are available from specialist suppliers (see list on page 307). It is important that conditions are suitable for their application, but the predators all come with full instructions for use. They are generally most effective when temperatures are rising in spring.

PESTS

Slugs and snails (1) are probably the most serious single pest of vegetables; they'll clear whole rows of young plants or newly emerged seedlings overnight, making them disappear as if by magic, but they'll also reduce leaves of older plants to skeletons and small slugs or snails will slither down inside the hearts of veg such as lettuces or cabbages.

Instead of using slug pellets, deter the beasties by using a selection of the various organic remedies available – don't rely on one method alone. For best results, start using 'alternative' remedies in and around the veg garden a few weeks before you start sowing or planting. Besides deterring future intruders, they will eliminate residents who would otherwise pounce on your new seedlings and scoff the lot while they are at their most vulnerable.

Copper tape is best fixed round the outside of raised veg beds and containers; it works by giving molluscs a slight electric shock when they touch it. The sort with serrated 'teeth' is especially effective as it makes it even less likely that they'll climb over the top. Various remedies based on absorbent granules are also available; these dehydrate slugs and snails so they can't glide along on their own layer of slime but instead dry up in the sun. You may also find deterrents based on bitter-tasting yucca extract or garlic. Sharp grit, holly leaves and other prickly leaves have a limited effect, but rather better are old-fashioned beer traps made from yoghurt pots or jam jars sunk into the ground – stale beer is okay, but low-alcohol lager seems to be their real favourite.

There is also a biological control available which uses nematodes (microscopic beneficial eelworms). Although they can be expensive to buy, the method is quite cost effective on small, intensively cultivated veg patches. You'll need to order them by post in spring; mix the powder with water and apply with a watering can to the affected area – (it won't work well if you try to cut costs by spreading it too far). The tiny nematodes seek out slugs and kill them by spreading a sluggy disease (which won't affect us). It's not so effective against snails, though, as they can climb up into plants where the nematodes can't reach.

Don't underestimate the value of regular hand picking where snails are concerned, either, and remember that if you simply lob them over the fence, they do have a homing instinct.

Greenfly and blackfly (2 and 3) (otherwise known as aphids) affect fruit and veg mainly in spring, although they can be a nuisance on veg all summer, too. On fruit they most frequently attack the young shoots of apples and plum trees, often resulting in curled, deformed and distorted growth. Although this looks alarming, the culprits have usually gone by the time you spot the trouble because by then they will have been controlled naturally by blue tits and similar predators; so my advice would be to ignore them. The new growth soon returns to normal; even quite badly affected fruit trees aren't harmed and there's no adverse effect on crop yields.

On vegetable crops aphids are mostly found near the tips of soft young shoots and on the undersides of young leaves. A small infestation won't do much harm, though a larger one can weaken plants. The greatest harm they do is to spread virus disease between plants.

1

2

3

Natural predators of aphids include hoverflies, ladybirds, lacewings, spiders (including money spiders) and several species of parasitic wasp, so minor infestations can usually be left to take care of themselves, but if you feel you want to *do something*, aphids are quite easily wiped off by hand with a damp tissue or cotton wool.

Plants grown under fine, insect-proof mesh are unlikely to be affected unless they were already carrying a few aphids when they were covered by the mesh, but they are far more likely to be a problem on crops such as peppers which are grown under glass. If all else fails, spray with an organic greenfly remedy, but check the small print to make sure that it's suitable for edible plants.

Soil pests (4) such as wireworms and cutworms strike underground, as the name suggests. Wireworms bore holes into root veg, in particular potatoes, which often encourages rotting, especially in a damp season. Cutworms will shear the roots off newly planted veg seedlings such as lettuce; if your crop suddenly wilts, give a plant a gentle tug and if it lifts out of the ground minus the roots you should suspect cutworm. (You may even find the culprit nearby – a fat, greyish-buff caterpillar curled into a 'c' shape.)

Soil pests are mostly a problem in ground that's previously been down to grass, so when making vegetable beds in a patch of lawn, strip off the turf in autumn and dig the ground, turning it over several times through the winter to expose pests to birds and wildlife. If you keep hens, give them a treat and let them loose on the area – they'll make a good job of it.

Some organic gardeners like to collect up black beetles and 'corral' them inside low soil walls round

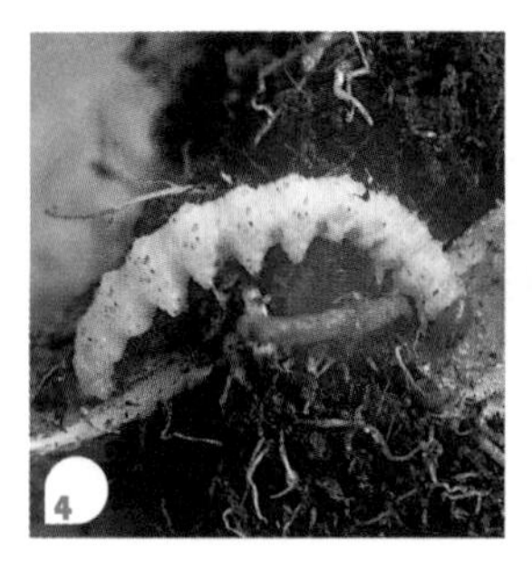

brassica crops, as they feed on cabbage root fly larvae. They are also good natural predators of lots of other ground-level pests, as are centipedes, so don't rush to evict things you'd probably think of as 'creepy-crawlies'. Generally speaking, if an insect moves slowly it eats plants; if it moves fast it eats other insects.

Caterpillars (5) can be found as several different species on soft fruit bushes, and 'cabbage whites' (caterpillars of the large white and small white butterflies) are notorious for stripping brassica plants. It's not essential to identify the precise species, unless you are particularly interested, as it makes no difference to the 'cure'. Natural predators such as blackbirds and robins will take small caterpillars, and even wasps will do so early in the season before they discover plums and picnics, but it's also worth checking over at-risk plants regularly and removing any caterpillars or suspect clusters of eggs on the undersides of leaves by hand.

If caterpillars are a regular problem it might be a good idea to use insect-proof netting to protect plants from egg-laying butterflies, but in the case of fruit, don't forget to remove it to allow pollinating insects access at flowering time.

There are some pests and diseases that will put in regular appearances, but many are mostly a problem only to a limited range of crops.

DISEASES

Potato (1) and tomato (2) blight are common in damp summers – if it rains in late June and July, anticipate an attack and start spraying every fortnight with Bordeaux mixture or risk losing your crop. (The active ingredient is largely copper, so it isn't totally off limits to organic growers.) Affected plants develop brown spots, especially round the edges of the leaves, and these quickly spread to larger 'rotting' areas. If not treated swiftly, the stems turn brown and dry out, tomatoes develop brown rotting patches and potatoes develop scabby cankers that lead to maroony-brown patches inside the tubers, which soon rot. If you experience an outbreak, don't put foliage, fruit, tubers or peelings from affected plants onto the compost heap. When choosing your crops for the following year, look out for the many new blight-resistant varieties of potatoes and tomatoes coming onto the market.

Botrytis (3) (grey mould) is a common problem under glass and in polytunnels, especially early in the season when conditions are cold, dull and damp. It usually looks like its name – grey mould – and can affect leaves (often when they have been damaged by physical injury or other problems first), flowers (especially ones that are going over) or fruit (botrytis is responsible for the translucent spots known as 'ghost spot' that sometimes appear on green or red tomatoes). The disease also appears as grey mould on strawberries in a wet summer when the fruit has been splashed.

Poor air circulation is often responsible for this problem, so try to improve ventilation as a preventive measure. General-purpose fungicides usually tackle botrytis and other common fungal diseases, but modern products come and go, so see what's available in the garden centre. If you do resort to buying a chemical, first check the instructions carefully; you can only use it on edible crops if that is specifically stated in the instructions – and avoid using it on any crops listed as unsuitable.

Virus (4) isn't one disease but lots of them, which produces a variety of symptoms; usually a variation on yellow blotches or marbled patterns on foliage. Viruses can be quite a problem with cucumbers and courgettes, which develop yellow mottles or blotches on the leaves and also the fruit, in some cases. Growth slows down in a plant infected with a virus and eventually it will stop cropping.

There's no cure for virus – affected plants are best pulled out and destroyed before the disease can be spread to other plants (greenfly are the commonest source of infection as they puncture plants to suck sap then move to a new plant which is consequently 'injected' with the disease organism).

STORING SURPLUS CROPS

To my mind, the whole reason for growing your own fresh fruit and veg is to enjoy them the way nature intended – fresh. And fresh means *really fresh* – ideally straight out of the garden, reaching your plate within an hour or so of picking. (Though potatoes, onions, apples and pears keep perfectly well in a shed or garage for some time.)

Any means of preserving most crops has to be second best. People tend to keep stored crops far too long, forgetting that they deteriorate slowly in store; dried herbs lose their colour and flavour, and the texture of frozen veg slowly breaks down – in any case half the flavour runs out with the water when they thaw. Crops such as tomatoes and strawberries change their character completely when they've been frozen, and with a very few exceptions you can't use crops that have been frozen in the same way as you'd use them fresh.

Freezing does have its uses, though – frozen fruit can be used for making pies and all sorts of cooked desserts, and it's ideal for making jam and similar preserves because it's handy to pick the crop as it ripens and 'save it up' to do your jam making later, when you have time. (For jam making, grow your fruit without fungicides.)

But all too often people make the mistake of freezing crops that are less than perfect – such as calabrese heads full of tiny green caterpillars, or runner beans that have started to turn stringy – imagining they'll somehow improve in the freezer. They won't. You still won't fancy eating them in two years time, and you're just wasting electricity keeping them frozen.

Herbs can be more successfully preserved, since they can be stored in various ways to use later in cookery. Popular kinds, such as parsley, can be frozen as whole bunches kept in plastic bags in the freezer. When you want a little garnish, a handful is simply crumbled off the bunch, which is far easier than trying to separate some and chop it. An even better method, though it needs more advance preparation, is to chop fresh herbs into ice cube trays with a few teaspoonfuls of water and freeze them – then you only need to drop one or two whole frozen cubes into whatever you are cooking.

Oils can be easily flavoured with herbs, too, by putting a few sprigs of herb into a bottle of olive oil and leaving it for the flavour to infuse, or you can air-dry herbs by hanging bunches up in a cool, airy place out of sunlight until they are brittle, then crumble them into airtight jars to store.

A real enthusiast can invest in a dehydrator, which not only dries herbs beautifully but can also be used to sun-dry tomatoes or veg such as carrots, celeriac and peppers, which are perfectly good to use in stews and casseroles. You can also use the gadget to dry slices of apples and apricot halves. A dehydrator is not cheap, so it's probably only worthwhile buying one if you have a lot of crops to tackle.

If you only have a few tomatoes at the end of the summer you can often 'sun-dry' them in the oven, turned to the very lowest setting and left on for a couple of hours with the door just ajar. But ask yourself if the cost of the electricity is worth it and if there isn't some other way of using a surplus.

For the crops that don't keep you'll need to find ways to use them up quickly. A sociable thing to so would be to invest in a few good seasonal cookery books that offer brilliant recipes that would be prohibitively expensive to make if you had to buy the main ingredients from the shops (Jane Grigson's *Vegetable Book* and her *Fruit Book* are particularly good for this job). Then all you need do is spend a little time in the kitchen and invite your friends round for dinner. The 'foodies' among them will be most impressed at the home-grown fare!

One of the best, and healthiest, ways of using surplus crops in bulk is to invest in a juicer and convert tomatoes, carrots, apples and other fruit into delicious drinks. These need to be drunk straight away, though, as the juice quickly oxidizes and changes colour or turns cloudy so that they then look less appetizing and lose certain vitamins. For anyone with lots of apple trees, a fruit press of the sort sold for home winemakers is a good way of producing your own apple juice, which is one thing that can be frozen and brought out when needed over the next few months.

But, as ever, the big trick with 'growing your own' is to avoid enormous gluts in the first place by being careful about how much you sow or plant at any one time – and that's back to planning again!

WORKING CALENDAR

MARCH

MUST-DO JOBS

► Clear ground around fruit trees, canes and bushes and mulch generously.
► Complete winter digging of vegetable beds.
► Prepare ground for sowing and planting: fork over, remove any weeds, apply general fertiliser, remove stones and rubbish and rake smooth.
► Sow tomatoes, chillies, peppers and aubergines indoors in warmth; sow early varieties of veg such as peas indoors, and sow early varieties of veg under cover or outside under fleece.
► Complete any fruit pruning and planting before the middle of the month.
► Protect early blossom from cold and wind with horticultural fleece.
► Chit seed potatoes.

VEGETABLES/SALADS

Sow indoors in warmth:
Greenhouse tomatoes, sweet peppers, mangetout peas (to grow under cover for early crops), lettuces, edible flowers, celeriac, celery, chillies, cauliflowers (autumn-heading varieties), Brussels sprouts (early varieties), French beans (dwarf), broad beans, aubergines.

Sow under cover:
Turnips, outdoor tomatoes, spinach (summer varieties), sorrel, globe artichokes, salad leaves (mixed), rocket, radish, mangetout peas, lamb's lettuce (corn salad), calabrese, broccoli (summer-sprouting varieties).

Sow outside:
Watercress, land cress, spinach (summer varieties), rocket, radishes (edible-podded), radishes, purslane, shelling peas (early varieties), parsnips, lettuces, carrots (early varieties), spring onions, baby leeks, leeks, calabrese, cabbages (summer- and early-autumn-hearting varieties), red cabbages, Brussels sprouts, beetroots (early varieties).

Plant outside:
Onion sets, garlic, cauliflowers (summer-heading varieties), shallots, globe artichokes, onions, early potato varieties (last week in March).

Plant under cover:
Early potato varieties.

Harvest under cover:
Mangetout peas, lamb's lettuce (corn salad).

Harvest outside:
Leeks (late varieties), kale, endive, dandelions (for salad leaves), cauliflowers (winter-heading varieties), sprouting broccoli.

HERBS

Sow in warmth indoors:
Summer and winter savory, parsley, lemon grass, chervil.

Plant outside:
Horseradish.

Harvest:
Thyme, bay, lemon grass, rosemary.

FRUIT

Sow indoors in warmth:
Passionfruits, Cape gooseberries (Physalis).

Plant outside:
Strawberries.

Prune:
Complete pruning of winter-pruned fruit (see winter, page 61).

Harvest:
Forced rhubarb.

APRIL

MUST-DO JOBS

► Sow frost-tender veg, such as runner and French beans, courgettes and pumpkins, on warm windowsills indoors or in a heated propagator.
► Start sowing fast-growing salad crops 'little and often' through the season to keep you supplied in small quantities.
► Towards the end of the month, start hardening off frost-tender crops ,such as tomatoes, which will be planted outside after mid-May.
► Plant potatoes.
► In mid- or late April, feed fruit trees, canes and bushes.
► Top dress or re-pot pot-grown fruit.
► Order biological control for slugs and apply late this month or early in May.
► Hoe and weed regularly.

VEGETABLES/SALADS

Sow indoors in warmth:
Sweetcorn, squashes, pumpkins, Florence fennel, outdoor cucumbers, greenhouse cucumbers, courgettes, marrows, summer squashes, runner beans, beans (for drying such as flageolet, haricot), French beans (climbing), French beans (dwarf), asparagus peas, tomatoes, peppers, aubergines.

Sow under cover:
Globe artichokes, pak choi.

Sow outside:
Watercress, land cress, turnips, Swiss chard, New Zealand spinach (late April), perpetual spinach, spinach (summer varieties), sorrel, salad leaves (mixed), rocket, radishes (edible-podded), radishes, purslane, mangetout peas, snap peas, shelling peas (early varieties), shelling peas (maincrop varieties), spring onions, lettuces, baby leeks, kohl rabi, edible flowers, dandelions (for salad leaves), carrot (maincrop varieties), leeks, kale, cauliflowers (winter-heading varieties), calabrese, cabbages (winter-hearting varieties), Savoys, cabbages (summer- and early-autumn-hearting varieties), red cabbage, Brussels sprouts, broccoli (summer-sprouting varieties), sprouting broccoli, beetroots, broad beans.

Plant under cover (late April):
Greenhouse tomatoes, French beans (dwarf).

Plant outside:
Onion sets, lettuce, cauliflowers (summer-heading varieties), cauliflowers (autumn-heading varieties), globe artichokes, asparagus, early potato varieties (first week of April), second-early potato varieties (first or second week of April), maincrop potato varieties (second and third week of April), edible flowers, calabrese, broccoli (summer-sprouting varieties), broad beans.

Harvest outside:
Spinach (baby leaves/summer varieties, late April), asparagus

(late April), rocket, kale, dandelions (for salad leaves), cauliflowers (winter-heading varieties), cabbages (spring-hearting varieties).
Harvest under cover:
Spinach (baby leaves/summer varieties), salad leaves, rocket, mangetout peas (autumn-sown), lamb's lettuce (corn salad).

HERBS
Sow indoors in warmth:
Marjoram, oregano, parsley, sweet cicely, summer and winter savory, salad burnet, lemon grass.
Sow outside:
Fennel, dill (for seeds), dill (for leaves), chervil, caraway, borage.
Plant outside:
Thyme, sage, rosemary, mint, lovage, lavender, chives, chamomile.
Harvest outside:
Thyme, sage, rosemary, lemon grass, chervil, bay.

FRUIT
Sow indoors in warmth:
Melons.
Plant outside:
Strawberries, lingonberries, cranberries, blueberries.
Prune:
Lemons, kumquats, oranges, limes and other citrus fruit, if needed.
Harvest:
Forced rhubarb.

MAY

MUST-DO JOBS

- ► Continue hardening off frost-tender plants.
- ► Once the last risk of frost is past, plant outside tomatoes, runner beans and other frost-tender crops, but delay planting the most tender kinds such as aubergines, peppers and chillies until late May or early June.
- ► Keep a close watch on newly planted fruit and any grown in pots and start watering as needed. Begin liquid feeding fruit and perennial herbs in containers.
- ► Keep an eye on the pest situation and step in only if need be – leave nature to deal with greenfly and other insects as far as possible.
- ► Continue sowing small quantities of salads.
- ► Hoe and weed regularly.
- ► Transplant veg seedlings.

VEGETABLES/SALADS

Sow in warmth indoors:
Florence fennel.
Sow outside:
Watercress, land cress, turnips, Swiss chard, New Zealand spinach, sweetcorn, swedes (late May), perpetual spinach, spinach (summer varieties), salad leaves (mixed), rocket, radishes (edible-podded), radishes, purslane, mangetout peas, snap peas, shelling peas (maincrop varieties), Chinese mustard greens, mizuma, spring onions, lettuces, endive, baby leeks, kohl rabi, edible flowers, dandelions (for salad leaves), carrots (maincrop varieties), kale, cauliflowers (winter-heading varieties), calabrese, cabbages (winter-hearting varieties), Savoys, broccoli (summer sprouting varieties), sprouting broccoli, beetroots, runner beans, beans (for drying such as flageolet, haricot), French beans (climbing), French beans (dwarf), asparagus pea.
Plant under cover:
Greenhouse tomatoes, sweet peppers, Florence fennel, greenhouse cucumbers, chillies, aubergines.
Plant outside:
All month: lettuces, edible flowers, cauliflowers (autumn-heading varieties), Brussels sprouts (early varieties), kale, cauliflowers (winter-heading varieties), calabrese, cabbages (summer- and early-autumn-hearting varieties), red cabbages, broccoli (summer-sprouting varieties).
After the last frost (around mid-May): outdoor tomatoes, squashes, pumpkins, outdoor cucumbers, courgettes, marrows, summer squashes, celeriac, celery, runner beans, beans (for drying such as flageolet, haricot), French beans (climbing), French beans (dwarf), asparagus pea.
Harvest outside:
Watercress, land cress, spinach (summer varieties), asparagus, salad leaves (mixed), rocket, radishes (edible-podded), radishes, purslane, early potato varieties (late May), onions (over-wintering varieties), winter-hardy spring onion varieties, dandelions (for salad leaves), cauliflowers (summer-heading varieties), cabbages (spring-hearting varieties), beetroots (early varieties).
Harvest under cover:
Turnips, radishes, early new potatoes, mangetout peas, French beans (dwarf).

HERBS

Sow indoors in warmth:
Lemon grass, basil.
Sow outdoors:
Salad burnet, parsley, fennel, dill (for seeds), dill (for leaves), coriander (for seed), coriander (for leaves), chervil, borage.
Plant outside:
Thyme, tarragon, sweet cicely, sage, rosemary, mint, marjoram, oregano, lovage, lemon verbena, lavender, chives, chamomile, bay.
Harvest:
Thyme, sweet cicely, summer and winter savory, salad burnet, sage, rosemary, parsley, mint, marjoram, oregano, lemon grass, fennel, dill (for leaves), chives, chervil, bay.

FRUIT

Plant outside:
Blueberries.
Plant under cover:
Melons, Cape gooseberries (Physalis), passionfruits.
Prune:
'Finger-prune' fan-trained peaches, nectarines and almond trees from May onwards over the summer. Prune apricots if needed.
Harvest:
Rhubarb, culinary gooseberries, lemon grass.

JUNE

MUST-DO JOBS

- ► Start watering crops of all kinds regularly in dry spells.
- ► Liquid feed container and greenhouse crops regularly.
- ► Hoe and weed regularly.
- ► Transplant veg seedlings.
- ► As the sun grows stronger, apply greenhouse shading to the outside of the glass to prevent crops scorching and damp down on hot days to keep the temperature from exceeding 29°C (85°F) – the point at which plants 'shut down' temporarily. There's no

need to do this to polytunnels, since the condensation on the inside does both jobs naturally, but open the ends on warm days for ventilation.
► Plant out the most cold-sensitive crops, such as chillies, aubergines and peppers.
► Frost-tender plants such as citrus and lemon grass can be moved from the conservatory out onto the patio for the summer.
► Net strawberries.

VEGETABLES/SALADS

Sow outside:

Watercress, land cress, turnips, Swiss chard, sweetcorn, New Zealand spinach, swedes (early June), perpetual spinach, salad leaves (mixed), rocket, radishes (edible-podded), radishes, purslane, mangetout peas, snap peas, shelling peas (maincrop varieties), shelling peas (early varieties) end of June, Chinese mustard greens, mizuma, pak choi, spring onions, lettuces, baby leeks, kohl rabi, endive, dandelions (for salad leaves), chicory, Chinese cabbages, carrots (maincrop varieties), kale, beetroots, runner beans, beans (for drying such as flageolet, haricot), French beans (climbing), French beans (dwarf), asparagus peas.

Plant under cover:

Sweet peppers, chillies, aubergines.

Plant outside:

Outdoor tomatoes, squashes, pumpkins, sweet peppers, Florence fennel, outdoor cucumbers, courgettes, marrows, summer squashes, celeriac, celery, chillies, kale, leeks, cauliflowers (winter-heading varieties), calabrese, cabbages (summer- and early-autumn-hearting varieties), red cabbages, Brussels sprouts, broccoli (summer-sprouting varieties), sprouting broccoli, runner beans, beans (for drying such as flageolet, haricot), aubergines.

Harvest outside:

Watercress, land cress, turnips, New Zealand spinach, spinach (summer varieties), sorrel, asparagus (until mid-June), salad leaves (mixed), rocket, radishes (edible-podded), radishes, purslane, early potato varieties, second-early potato varieties (mid-June onwards), mangetout peas, snap peas, shelling peas (early varieties), Chinese mustard greens, mizuma, onions (overwintering varieties), winter-hardy spring onion varieties, spring onions, lettuces, baby leeks, garlic (autumn-planted), edible flowers, dandelions (for salad leaves), courgettes, marrows, summer squashes, carrots (early varieties), cauliflowers (summer-heading varieties), calabrese, broccoli (summer-sprouting varieties), French beans (dwarf), broad beans, asparagus peas.

Harvest under cover:

Turnips, early new potatoes (early June), green sweet peppers, mangetout peas, pak choi, Florence fennel, French beans (dwarf).

HERBS

Sow indoors in warmth:

Basil, fennel.

Sow outside:

Dill (for leaves), coriander (for leaves), chervil, borage.

Plant outside:

Tarragon, sweet cicely, salad burnet, sage, rosemary, parsley, mint, marjoram, oregano, lovage, lemon verbena, lavender, chives, chamomile, bay, basil.

Harvest:

Tarragon, sweet cicely, summer and winter savory, salad burnet, sage, rosemary, thyme, parsley, mint, marjoram, oregano, lovage, lemon grass, fennel, dill (for leaves), chamomile, chervil, borage, bay, basil.

FRUIT

Plant:

Passionfruits, melons (outside and under cover).

Prune:

'Finger-prune' fan-trained peaches, nectarines and almond trees from May onwards throughout summer.

Harvest:

Strawberries, rhubarb, raspberries (late June), redcurrants, whitecurrants, culinary gooseberries.

JULY

MUST-DO JOBS

► Continue picking soft fruit; cover the next batch of varieties to start ripening with bird netting.
► Increase watering and liquid feeding of crops growing in containers and under cover as temperatures rise, and water outdoor crops in dry weather. Pay particular attention to fruit plants carrying crops.
► Summer-prune trained forms of apple and pear trees late this month or very early in August.
► Summer-prune trained soft fruit.
► Weeding will tail off as crops cover the ground in the veg patch, but be ready to clear early crops that have come to an end and replace them with other crops.

VEGETABLES/SALADS

Sow outside:

Watercress, land cress, turnips, Swiss chard, perpetual spinach, salad leaves (mixed), rocket, winter radishes (including Chinese/Japanese/Mooli), radishes, purslane, shelling peas (early varieties) to mid-July, Chinese mustard greens, mizuma, pak choi, spring onions, lettuces, endive, chicory, Chinese cabbages, carrots (early varieties), French beans (dwarf) (early July).

Plant outside:

Late-planting new potatoes (specially prepared seed potatoes), Florence fennel, kale, cabbages (winter-hearting varieties), Savoys, sprouting broccoli.

Harvest outside:

Watercress, land cress, turnips, Swiss chard, New Zealand spinach, outdoor tomatoes (late July), sweetcorn (late July), perpetual spinach, sorrel, shallots, globe artichokes, salad leaves (mixed), rocket, radishes (edible-podded), radishes, purslane, early potato varieties, second-early potato varieties, mangetout peas, snap peas, shelling peas (early varieties), shelling peas (maincrop varieties), Chinese mustard greens, mizuma, onions (overwintering varieties), spring onions, lettuces, baby leeks, kohl rabi, garlic, edible flowers, dandelions (for salad leaves), outdoor cucumber, courgettes, marrows, summer squashes, chillies, carrots (early varieties), cauliflowers (summer-heading varieties), calabrese, broccoli (summer-sprouting varieties), beetroots, runner beans, French beans (climbing), French beans (dwarf), broad beans, aubergines, asparagus peas.

Harvest under cover:

Greenhouse tomatoes, green sweet peppers, pak choi, Florence fennel, greenhouse cucumbers, chillies, aubergines.

HERBS

Sow indoors in warmth:
Basil.

Sow outside:
Fennel, coriander (for leaves), borage.

Plant outside:
Salad burnet, sage, rosemary, lavender, chives, bay, basil.

Harvest:
Tarragon, sweet cicely, summer and winter savory, salad burnet, sage, rosemary, thyme, parsley, mint, marjoram, oregano, lovage, lemon verbena, lemon grass, lavender, fennel, dill (for seeds), dill (for leaves), parsley, annual marjoram, coriander (for leaves), chives, chervil, chamomile, borage, bay, basil.

FRUIT

Prune:
Summer-prune cordon-trained redcurrants, whitecurrants and gooseberries, and trained apple trees late July/early August. 'Finger-prune' fan-trained peaches, nectarines and almond trees from May onwards throughout the summer.

Harvest:
Tayberries, tummelberries, summer strawberries, perpetual-fruiting strawberries, rhubarb, redcurrants, whitecurrants, raspberries, plums, peaches, loganberries, lingonberries, citrus fruit (lemons, kumquats, oranges, limes), dessert gooseberries, greenhouse figs, cherries, boysenberries, blueberries, blackcurrants, early blackberry varieties, apricots, early apple varieties.

AUGUST

MUST-DO JOBS

► Pay particular attention to watering. If you are going away on holiday, arrange for a friend or neighbour to drop in regularly and take over this job for you. As an incentive, tell them to help themselves to any crops that are ready while you're away – it's worth it to make sure beans and courgettes are picked so you don't come home to a few uneatable whoppers and plants that have stopped producing tender new veg.

► Start picking early apple varieties; use straight from the tree.

► At the end of the month, remove yellowing lower leaves from outdoor tomato plants and nip out the growing tips at the top of each plant to encourage green fruit to swell and ripen before the end of the season.

VEGETABLES/SALADS

Sow under cover:
Turnips (late August), winter purslane (Claytonia), pak choi, carrots (early varieties).

Sow outside:
Spinach (spring/autumn varieties), salad leaves (mixed), rocket, radishes, winter purslane (Claytonia), Chinese mustard greens, mizuma, pak choi, onions (overwintering varieties), winter-hardy spring onion varieties, Chinese cabbages, carrots (early varieties).

Plant outside:
Late-planting new potatoes (specially prepared seed potatoes) early August.

Plant under cover:
French beans (dwarf).

Harvest outside:
Watercress, land cress, turnips, outdoor tomatoes, Swiss chard, sweetcorn, New Zealand spinach, perpetual spinach, sorrel, shallots, globe artichokes, salad leaves (mixed), rocket, radishes (edible-podded), winter radishes (including Chinese/Japanese/Mooli), radishes, purslane, second-early potato varieties, mangetout peas, snap peas, shelling peas (maincrop varieties), Chinese mustard greens, mizuma, pak choi, onions, spring onions, lettuces, baby leeks, kohl rabi, garlic, Florence fennel, edible flowers, dandelions (for salad leaves), outdoor cucumbers, courgettes, marrows, summer squashes, self-blanching celery, Chinese cabbages, chillies, green peppers, onion sets, lettuces, cauliflowesr (summer-heading varieties), cauliflowers (autumn-heading varieties), calabrese, cabbages (spring-hearting varieties), cabbages (summer- and autumn-hearting varieties), red cabbages, broccoli (summer-sprouting varieties), beetroots, runner beans, French beans (climbing), French beans (dwarf), aubergines, asparagus peas.

Harvest under cover:
Greenhouse tomatoes, red and green peppers, greenhouse cucumbers, chillies, aubergines.

HERBS

Sow indoors:
In pots to harvest from the windowsill in autumn: parsley, annual marjoram, dill (for leaves), coriander (for leaves).

Sow outside:
Sweet cicely, summer and winter savory.

Harvest:
Tarragon, sweet cicely, salad burnet, sage, rosemary, thyme, parsley, mint, marjoram, oregano, lovage, lemon verbena, lemon grass, fennel, dill (for seeds), dill (for leaves), coriander (for seed), chamomile, caraway, borage, bay, basil, coriander (for leaves), chives, chervil.

FRUIT

Plant outside:
Strawberries.

Prune:
Summer-fruiting raspberries; 'finger-prune' fan-trained peaches, nectarines and almond trees from May onwards throughout the summer.

Harvest:
Tayberries, tummelberries, perpetual-fruiting strawberries, redcurrants, whitecurrants, autumn-fruiting raspberries, plums, greengages, peaches, nectarines, passionfruits, mulberries, melons, loganberries, citrus fruit (lemons, kumquats, oranges, limes), figs (greenhouse and outdoor), Cape gooseberries (Physalis), morello cherries, cherries, boysenberries, blueberries, blackcurrants, blackberries, apricots, early varieties of apples.

SEPTEMBER

MUST-DO JOBS

► Start on the first of the autumn veg, such as cauliflowers and early varieties of leeks and sprouts. Start using maincrop potatoes and carrots but leave the rest in the ground for now.

► Pick early varieties of apples and pears.

► Stop feeding container-grown fruit and herbs and reduce the watering slightly.

► Remove lower leaves and the growing tips of indoor tomatoes to encourage any green fruit to swell and ripen before the end of the season.

Towards the end of the month, be ready to cover late outdoor crops with fleece at night if it turns cold, and move tender plants such as citrus and lemon grass back into the conservatory.

VEGETABLES/SALADS

Sow in warmth indoors:

Crops to grow on windowsills; rocket.

Sow under cover:

Turnips (early September), salad leaves (mixed), rocket, mangetout peas, Chinese mustard greens, mizuma.

Sow outside:

Spinach (spring/autumn varieties), rocket, winter-hardy spring onion varieties, lamb's lettuce (corn salad).

Plant outside:

Onion sets (overwintering varieties), cabbages (spring-hearting varieties).

Harvest outside:

Watercress, land cress, turnips, outdoor tomatoes, Swiss chard, sweetcorn, swedes, New Zealand spinach, perpetual spinach, spinach (spring/autumn varieties), sorrel, globe artichokes, salad leaves (mixed), rocket, radishes (edible-podded), winter radishes (including Chinese/Japanese/Mooli), radishes, purslane, second-early potato varieties, maincrop potato varieties, mangetout peas, snap peas, shelling peas (maincrop varieties), shelling peas (early varieties) in late September; Chinese mustard greens, mizuma, pak choi, onions, spring onions, lettuces, baby leeks, leeks (early varieties), kohl rabi, Florence fennel, endive, edible flowers, dandelions (for salad leaves), outdoor cucumbers, courgettes, marrows, summer squashes, celeriac, self-blanching celery, Chinese cabbages, chillies, peppers, carrots (maincrop varieties), onion sets, lettuces, cauliflowers (summer-heading varieties), cauliflowers (autumn-heading varieties), calabrese, cabbages (summer- and autumn-hearting varieties), red cabbages, Brussels sprouts (early varieties), broccoli (summer-sprouting varieties), beetroot, runner beans, beans (for drying such as flageolet, haricot), French beans (dwarf), aubergines.

Harvest under cover:

Greenhouse tomatoes, red and green sweet peppers, greenhouse cucumbers, chillies, French beans (dwarf), aubergines.

HERBS

Sow indoors:

In pots to harvest from the windowsill in autumn/winter: parsley, annual marjoram, dill (for leaves), coriander (for leaves).

Sow outside:

Sweet cicely.

Harvest:

Tarragon, summer and winter savory, salad burnet, sage, rosemary, parsley, mint, marjoram, oregano, lovage, lemon verbena, lemon grass, horseradish, fennel, dill (for seeds), coriander (for seeds), chives, chervil, caraway, borage, bay, basil.

Harvest indoors:

In pots sown in warmth on windowsills indoors and kept inside: parsley, annual marjoram, dill (for leaves), coriander (for leaves).

FRUIT

Plant:

Strawberries.

Prune:

Tayberries, tummelberries, boysenberries, loganberries, blackberries (late September/early October).

Harvest

Perpetual-fruiting strawberries, autumn-fruiting raspberries, plums, greengages, early varieties of pears, peaches, passionfruits, cobnuts, filberts, mulberries, melons, loganberries, lingonberries, citrus fruit (lemons, kumquats, oranges, limes), greenhouse grapes, figs (greenhouse and outdoors), damsons, cranberries, Cape gooseberries (Physalis), morello cherries, blueberries, blackberries, apricots, apples.

OCTOBER

MUST-DO JOBS

► Protect late crops of salads, courgettes, baby carrots and French beans with fleece on cold nights and leave them covered all the time if the weather turns bad later in the month.

► Start closing the greenhouse down early in the afternoon to trap heat and keep the last of the summer crops going for as long as possible. Pick the last tomatoes, peppers, aubergines then pull out the old plants, clean out the greenhouse, wash the shading paint off the glass and work well-rotted organic matter into the soil border before sowing winter crops.

► If you live in an exposed area, knock a stake in alongside Brussels sprouts and sprouting broccoli plants to keep them upright in windy weather.

► Pick apples and pears mid-month before windy weather sets in, and store them.

► Towards the end of the month, move tubs of outdoor fruit trees close to the house for protection from the worst of the wind, secure tall plants to trellis if possible, and raise the pots up on bricks or 'pot feet' to improve drainage in wet weather.

VEGETABLES/SALADS

Sow under cover:

Mangetout peas, lamb's lettuce (corn salad).

Sow outside:

Winter-hardy spring onion varieties, lamb's lettuce (corn salad).

Plant outside:

Onion sets (overwintering varieties), cabbages (spring-hearting varieties).

Harvest outside:

Watercress, land cress, outdoor tomatoes, Swiss chard, swedes, squashes, pumpkins, New Zealand spinach, perpetual spinach, spinach (spring/autumn varieties), globe artichokes, rocket, winter radishes (including Chinese/Japanese/Mooli), radishes, winter purslane (Claytonia), late-planting new potatoes (specially prepared seed potatoes), maincrop potato varieties, shelling peas (early varieties), parsnips, Chinese mustard greens, mizuma, pak choi, lettuce, leeks (early varieties), endive, courgettes, marrows, summer squashes, hearting chicory/radicchio, celeriac, self-blanching celery, Chinese cabbages, carrots (maincrop varieties), carrots (early varieties), onion sets, lettuces, cauliflowers (summer-heading varieties), cauliflowers (autumn-heading varieties), cabbages (summer- and autumn-hearting varieties), red cabbages, Brussels sprouts (early varieties), broccoli (summer-sprouting varieties), beetroots, runner beans, beans (for drying such as flageolet, haricot).

Harvest under cover:

Turnips, greenhouse tomatoes, salad leaves (mixed), winter purslane (Claytonia), pak choi, greenhouse cucumbers, chillies, French beans (dwarf), aubergines.

HERBS

Sow indoors:

In pots to harvest from the windowsill in autumn/winter: parsley, annual marjoram, dill (for leaves), coriander (for leaves).

Harvest outside:

Tarragon, summer and winter savory, sage, rosemary, thyme, parsley, lemon grass, horseradish, borage, bay.

Harvest indoors:

Pots sown in warmth on windowsills indoors and kept inside: parsley, annual marjoram, dill (for leaves), coriander (for leaves).

FRUIT

Plant:

Strawberries.

Prune:

Blackberries.

Harvest:

Perpetual-fruiting strawberries, autumn-fruiting raspberries, melons, quinces (late October), medlars (late October), pears, passionfruits, cobnuts, filberts, sweet chestnuts, almonds, walnuts, citrus fruit (lemons, kumquats, oranges, limes), kiwi fruits, greenhouse grapes, greenhouse figs, damsons, cranberries, Cape gooseberries (Physalis), apples.

WINTER – NOVEMBER TO EARLY/MID-MARCH

MUST-DO JOBS

► Clear and prepare the ground ready to make a new fruit or veg patch.

► Clear the crops that are over and do winter digging in the veg bed; work carefully round areas of winter veg.

► Plant new fruit trees and bushes; bare-root plants can only be planted while they are dormant when they aren't in leaf, but this is also the very best time for planting pot-grown fruit. Heel plants in temporarily if the ground isn't in a fit state to plant.

► Sow windowsill crops, including pots of herbs and salad leaves. Grow sprouting seeds, start a mushroom kit.

► Force chicory and rhubarb.

► Prune standard apples and pear trees, blackcurrant bushes and bush gooseberries and redcurrants. Prune grape vines and figs only in the very middle of winter when they are fully dormant, so they don't 'bleed'.

► In January, chit the earliest seed potatoes, for planting under cover for early crops.

► In early spring (Feb), sow peppers, aubergines, chillies and tomatoes in warmth indoors at room temperature to give you the earliest plants for growing under glass.

VEGETABLES/SALADS

Sow or start in warmth indoors:

All year round: mushroom kits, sprouting seeds such as mung beans, alfalfa.

November–March: lamb's lettuce (corn salad), watercress, land cress.

January–February: cauliflowers (summer-heading varieties), chit seed potatoes of early varieties to grow under cover.

February: greenhouse tomatoes, chit seed potatoes of early varieties to grow outdoors; February–March chit second earlies.

Plant outdoors:

February: shallots and Jerusalem artichokes.

November or February–March: garlic.

Harvest under cover:

November: turnips and rocket.

November–December: Chinese mustard greens.

November–March: mizuma; in December and March harvest mixed salad leaves sown in November and February.

Throughout the winter: winter purslane (Claytonia), lamb's lettuce in pots on windowsills, sprouting seeds, mushroom kits.

Harvest outside:

November: maincrop potato varieties, carrots (maincrop varieties), winter radishes (including Chinese/Japanese/Mooli), onions, lettuces, cauliflowers(summer-heading varieties), cauliflowers (autumn-heading varieties), lift roots of forcing chicory for use December to March.

October–December: hearting chicory, parsnips, late planting new potatoes (specially prepared seed potatoes).

November–December: spinach (spring/autumn varieties) (cover with fleece), leeks (early varieties).

December–March: leeks(late varieties).

November–February: cabbages (winter-hearting varieties) and Savoys.

November–March: Brussels sprouts.

Mid-January–end March: sprouting broccoli.

February–April: cauliflowers (winter-heading varieties).

Throughout the winter (November–March): kale; swedes, celeriac, Jerusalem artichokes, winter purslane (Claytonia), Swiss chard and perpetual spinach (both can keep cropping through the winter until May, given a mild winter), endive.

HERBS

Sow in warmth indoors:

In pots to harvest from windowsills: parsley, annual marjoram, dill (for leaves), coriander (for leaves), chervil, basil.

Plant outside:

Horseradish (February).

Harvest outside:

Thyme, rosemary, horseradish, bay.

Harvest indoors:

Pot-grown plants kept under cover: salad burnet, mint, chives.

In pots sown in warmth on windowsills indoors: parsley, annual marjoram, dill (for leaves), coriander (for leaves), chervil, basil.

FRUIT

Plant outside:

Plant most fruit throughout this period, avoiding times when the ground is not in workable condition due to excess wet, or when frozen.

Prune:

Throughout winter from November to early/mid-March: prune quinces, medlars, free-standing apple and pear trees, blackcurrants.

November–end of February: kiwi fruits, free-standing bush redcurrants, whitecurrants, gooseberries.

December: outdoor and greenhouse grape vines, figs.

January: outdoor grape vines, figs.

February: autumn-fruiting raspberries, passionfruits, cobnuts and filberts (late February).

Harvest:

November: walnuts, greenhouse grapes.

February–March: forced rhubarb.

The Directory

The gardening reference books I find most useful are those set out in a straightforward and sensible way and in which the information I am looking for can be located quickly. The Directory aims to help you do just that.

This section of the book is divided into three parts – vegetables, fruit and herbs – and in each of these sections the crops are listed alphabetically. Within each entry the information should be similarly easy to find. There is first an at-a-glance calendar of sowing, planting and harvesting times, so that you can find out just what you should be doing and when, followed by detailed information on how to grow each individual crop and recommendations of what I consider to be the most reliable and tasty varieties. There is also a run-down on the possible problems you might encounter with each crop: from pests and diseases to weather conditions, and even mistakes on the part of the grower.

I am an organic gardener, so you will not find comprehensive lists of chemicals you can spray at the first sight of any beast or bug that raises its head.

Instead, I have emphasised more environmentally responsible means of gaining the upper hand. There are three very good reasons for this: first, good cultivation and the choice of disease-resistant varieties can prevent most pests from attacking in epidemic proportions. Second, there are now very few pesticides available to the amateur gardener. (But surely the joy of growing your own food is in being able to eat fresh and healthy produce that is untainted by chemicals? If you want your food of even shape and colour, covered in wax and preserved by chemicals, go to the supermarket!) Third, and most importantly, the vast majority of wildlife in your garden adds to its value and is essential for the wellbeing of your crops. Pesticides upset the balance of nature, and while you might successfully control one particular pest with the application of a chemical, you will at the same time be removing the food source of another creature.

It is perfectly possible to harvest tasty and succulent fruit, vegetables and herbs and also have a garden that is true to nature. The Directory will show you how.

Vegetables

Artichoke, globe

Cynara scolymus

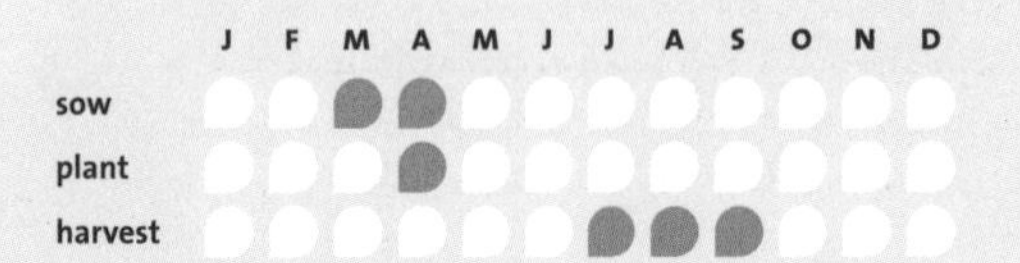

This is one veg in the gourmet league that always strikes me as a great excuse for putting aside the low-calorie spread and overdosing on melted butter. If you feel guilty about it, tell yourself that you use up far more calories wrestling with the artichoke than you gain by eating it.

Globe artichokes are striking architectural, thistle-like plants with jagged silvery leaves and which grow up to 1.5m (5ft) high. They are good planted at the back of a perennial border if you don't have anywhere else to grow them. And in case you were wondering, the bit you eat is the flower bud – well, parts of it.

HOW TO GROW

Degree of difficulty: Dead easy and not much work – treat it just like any herbaceous perennial plant.
Sow: March/April under glass or in a seedbed outdoors and plant out a year later. Most people buy young plants, though, rather than grow from seed.
Plant: April.
Spacing: 90cm (3ft) apart with 90cm–1.2m (3–4ft) between rows. Unless you are a real artichoke enthusiast, or you have an allotment where there's room to grow a row, one or two plants are usually enough for a few good meals over the summer. Each plant should yield six to ten buds per season. If you need to save up enough same-sized heads for the whole family, artichokes will keep for a week in the salad drawer of the fridge.
Routine care: Water in after planting and during dry spells in summer when plants are carrying a crop. Every April, feed by sprinkling organic fertiliser round each plant and mulch generously. No supports are needed. In autumn, cut down the old stems in the same way as for herbaceous perennial flowers in your borders.

SECRETS OF SUCCESS

Artichokes like a warm, sunny spot with some shelter, and light, well-drained soil is essential.

Allow plants a year to get established before you start pruning, but don't allow flowers to develop during the first summer – snip off baby buds as soon as they are seen. Plants are most productive in their first four to five years, so in year three remove some of the small offsets that form round the base of mature plants, then pot them up to start a new row next year. These will establish themselves while the old plants are in their final year.

VARIETIES

1 **'Green Globe'** – common and widely available, also sold as seed, but not the most outstanding for flavour.
2 **'Gros Camus de Bretagne'** – French variety with very large heads and superb flavour; the gourmet's choice, but it can be hard to find plants.
3 **'Gros Vert de Lâon'** – French variety with good flavour, available fairly easily as plants.
4 **'Violetto Di Chioggia'** – available as seed or plants with pretty mauve-purple buds; great for flower gardens, they have a fair flavour but crops are not that heavy.

Cutting garden

If you don't pick artichoke buds at the right stage for eating they continue to develop into flowers. These open out into attractive giant thistle flowers with purple spikes in a 'crewcut' at the top of the prickly green base. They look good in the garden, but if you cut them just as the colour starts to show they can be dried and used in winter flower arrangements.

HARVEST

Throughout the summer (from July to the end of September), cut off half-grown flower buds when they are roughly the size of your fist, using secateurs to snip through the stem about 2.5cm (1in) below the base of the bud. Take care, they are prickly.

PROBLEMS

Blackfly often attack flower buds; a light attack is no problem as the outer scales are removed when preparing artichoke heads for cooking, but a bad infestation is most off-putting on the plate. So if you see blackfly while crops are growing, wipe them off developing buds with a damp cloth or a soft hand-brush dipped in water.

Culinary tip: cooking globe artichokes

Remove the very outer layers of scales and cut off the stalk, then drop the 'globe' into boiling, lightly salted water for 40 minutes, until soft enough for the scales to pull away easily.

To eat hot, serve one head per plate and peel off individual scales with your fingers, dipping them in melted butter or hollandaise sauce. Then scrape off the flesh at the base of each scale with your teeth. It's a lot of bother for very little sustenance, but when you reach the middle you can remove the fluffy 'choke' to reveal the base of the bud (the 'heart'), which is much better value. Eat it with more melted butter. (This is not a vegetable you can eat with dignity; you'll end up with butter down your chin and bits of fluffy choke all over the place but, oh, it's worth it!)

To eat cold, with salads or as hors d'oeuvres, remove the outer scales and central fluff and remove the heart in one piece. Serve with vinaigrette dressing.

Artichoke, Jerusalem

Helianthus tuberosus

	J	F	M	A	M	J	J	A	S	O	N	D
plant		●	●									
harvest	●	●	●							●	●	●

The poor old Jerusalem artichoke's greatest claim to fame is that it is often recommended as a screen for the compost heap. It is not the most essential of vegetables, I grant you, nor one of the prettiest (growing to 3m /10ft with tall, sunflower-like stems), but the plants are rugged, weatherproof and not at all fussy about soil. Yes, they are good for making a windbreak at one end of the allotment or veg patch – just make sure you plant them where they aren't going to cast their shadow over anything else – but they also make a great winter soup.

With Jerusalem artichokes the bits you eat are the plump underground tubers, which are ready from October onwards. These should be scraped, then boiled or roasted rather like potatoes, and as well as making a decent soup, they can also be used in casseroles. Save some of the best tubers to replant next year.

HOW TO GROW

Degree of difficulty: Very easy and less bother than potatoes.
Plant: Egg-sized tubers 15cm (6in) deep in February or March.
Spacing: Plant 40cm (15in) apart, with 90cm (3ft) between rows. Plant a double or triple row for a more efficient windbreak, and so the plants can help support each other.
Routine care: Use a general fertiliser when preparing the soil, but no extra watering or feeding is needed, making this a good allotment crop. You'll have a bigger crop if you draw earth up round the plants when they've reached about 45cm (18in) high. In a windy area, or if you have a very neat and tidy disposition, bang a stake in at each end of a row and run long strings or plastic-coated wire either side of the plants to give them some support.

SECRETS OF SUCCESS

Experts recommend removing the flower buds throughout the growing season: the technique is said to produce best results and give a bigger crop, but in practice most people don't bother, as the plants are tall and the 'sunflowers' make the plants look far more attractive. You still get a lot of artichokes per plant, anyway. Expect to reap on average 2kg (4lbs) from each tuber you originally planted.

VARIETIES

'Fuseau' – has less knobbly tubers than usual, which makes them much easier to prepare for cooking. As there's no great difference in flavour between varieties, this is the best reason for choosing one over another!

HARVEST

When the plants start dying off in autumn, cut them back to 30cm (12in) high. Dig up individual plants from October right through the winter until March any time you want some tubers. Leave the rest in the ground, as they store better there than anywhere else. Any tubers still in the ground by March will start growing again, so dig and use them before that.

Asparagus

Asparagus officinalis

	J	F	M	A	M	J	J	A	S	O	N	D
plant				●								
harvest				●	●	●						

If you're an asparagus fan, this is a vegetable that really pays to be homegrown – you can save yourself a fortune on shop prices. Even though the English asparagus season is quite short, you can be cutting a couple of bunches a week from late April to the middle of June. After that you have to stop cutting to let the plants grow and complete their life cycle. Although it's not generally recommended, I find in practice that it's perfectly okay to cut a little of the 'fern' from well-established plants to use as foliage for flower arranging, if you do so very sparingly.

The main drawback with growing asparagus is the space it needs – it's a perennial crop which occupies the ground full-time, and since it takes several years to settle in, you can't pick much for the first three years. But, oh, the anticipation!

HOW TO GROW

Degree of difficulty: Not difficult, but it does need proper care.

Plant: April, in well-drained ground or in slightly raised beds which have had lots of organic matter dug in during the winter. Work in some general-purpose fertiliser just before planting. Dig a generous planting hole for each crown, spread the roots out well then cover with 5cm (2in) of soil. As shoots start to appear above the ground, gradually earth up the plants until they are growing along ridges up to 15cm (6in) high.

Spacing: 30–45cm (12–18in) apart, with 90cm (3ft) between rows.

Routine care: Water new plants in well. Early each April, sprinkle general fertiliser over asparagus beds; water if the weather is very dry during the cropping season. Hand-weed beds regularly; don't let weeds grow large or use a fork for removing them, since asparagus plants are shallow rooted and dislike disturbance.

Do not cut any emerging spears for the first two summers after planting, and then only take a light crop for the first few weeks of the cutting season in year three. From then on you can harvest as much as you like until mid-June (or the very end of that month if you have old-established plants and don't mind pushing it a bit). From mid- to late-June onwards it's then essential to stop cutting to allow the 'fern' to grow up.

Cut yellowed or browning fern down to 5cm (2in) above ground in autumn, then weed over the bed and mulch the soil generously with well-rotted organic matter.

VARIETIES

1 **'Connover's Colossal'** – this old favourite is readily available, but a batch will contain both male and female plants and females are less prolific – they also shed seed via berries that ripen towards autumn, and self-sown seedlings are nuisance 'weeds' in an asparagus bed.

2 **'Jersey Giant'** – similar to 'Jersey Knight', but earlier; cropping starts a couple of weeks before most varieties and continues until mid-June as usual, giving a slightly longer 'season'.

3 **'Jersey Knight'** – a strong-growing, heavy-cropping, all-male F1 hybrid that is readily available from the veg pages of mail-order seed firms. It has thick spears and superb flavour.

HARVEST

Start cutting asparagus spears from the time they first start to appear: towards the end of April, and they will crop until mid-June. Cut them as soon as they reach about 15cm (6in) tall and look like the ones you see in the shops. Use a strong knife to cut spears off 5cm (2in) below the soil surface and then fill the hole this leaves with soil, to prevent pests getting in.

PROBLEMS

If self-sown asparagus seedlings are left alone they grow quickly and are soon indistinguishable from the plants that are meant to be growing in a bed and, being 'mongrels', these seedlings do not produce good yields and their flavour is often much less impressive. They also overcrowd beds, which reduces the yield from your named variety.

Thin spears are usually caused by weak plants that have been cropped too early, cut too heavily in previous years, or not fed sufficiently – or a bit of all three.

Slugs can attack emerging spears, but the main pest is likely to be the asparagus beetle. It is about 8mm long with yellow and black wing cases and appears from May onwards. Both it and its greyish black caterpillar-like larvae (12mm long) feed on spears and mature foliage. Both are easily seen and can be picked off by hand. Look out for the clusters of black eggs on foliage and pick these off, too. Clean up the bed in autumn to discourage over-wintering pests, and burn the cut foliage to kill any that may be lurking within it.

Asparagus pea

Tetragonolobus purpureus

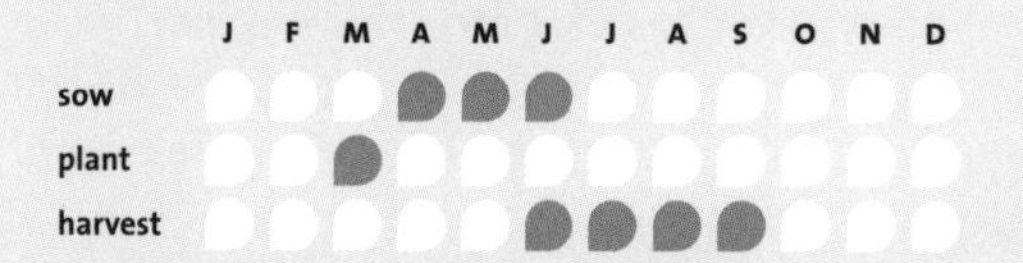

Never heard of it? No, I'm not surprised – this is one of those unusual veg that only gets a brief mention in some of the larger seed catalogues, and unless you grow your own you'd never see the edible pods because they aren't sold in greengrocers.

This is a novelty crop that's best grown in a tub on the patio. It doesn't need much room, since the semi-spreading plants only grow about 30cm (12in) high. They look quite attractive, with delicate foliage and small, dull-red, pea flowers followed by curious four-angled pods with a small green flange or 'wing' running down each edge. They are cooked whole – just like mangetouts – but they taste faintly of asparagus.

HOW TO GROW

Degree of difficulty: Easy, but needs a warm sheltered spot to succeed; picking is tedious.

Sow: Best in small individual pots on a windowsill indoors in April, or in a warm, sheltered spot outside from mid-May to mid-June where they are to crop.

Plant: Pot-grown plants outside without breaking up the rootball from late May; wait until warm weather.

Spacing: 20cm (8in) apart.

Routine care: Water sparingly; feed lightly while cropping with general-purpose liquid feed. Plants can be left unsupported but may look better on a patio given a few twiggy sticks or the sort of 'cage'-type support used for perennial plants in a border.

VARIETIES

No named varieties are available; this vegetable is a botanical species – *Tetragonolobus purpureus.*

HARVEST

Start picking pods when they are 2.5cm (1in) long, as they become woody if left to grow bigger. They are difficult to see because they hide among the foliage and are almost the same colour as the leaves. Check plants two to three times a week as pods are produced little and often from around late June until late August or early September. They are light croppers, so a few small handfuls are all you get at any time.

PROBLEMS

Plants can be killed by overwatering, especially early on, or by a cold, miserable summer, or simply by attempting to grow them in a windy spot. If they are not actually killed by these conditions, they may just sit still without growing. If space is available in an unheated greenhouse, they'll do better there.

Culinary tip – Cooking asparagus peas

The asparagus pea has a mild, delicate flavour which is easily ruined by overcooking; it's best eaten stir-fried or steamed gently and topped with a little melted butter.

Aubergine

Solanum melongena

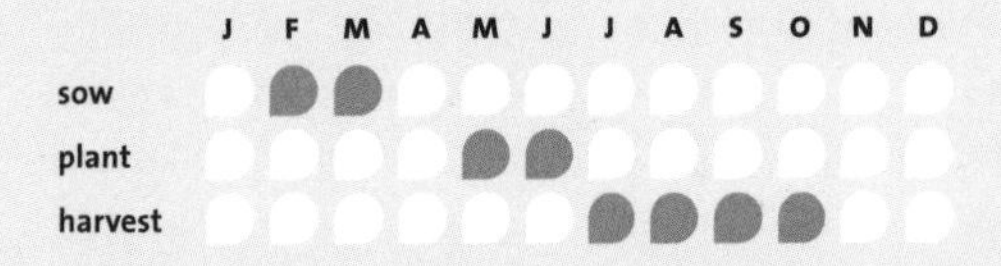

Closely related to tomatoes and peppers, plump purple aubergines have become a popular ingredient in Mediterranean cookery.

Relatively few people grow aubergines, since they really need a greenhouse to produce a worthwhile crop in our climate, but it's worth trying one or two plants in large pots or tubs on a warm, sunny, sheltered patio and hoping for a good summer. Unnamed aubergine plants are sold in garden centres around planting time – one aubergine tastes very much like another, so there's little to choose between named varieties anyway.

If you raise your own plants from seed, you could try the various 'fancy' varieties with pink, white or finger-like fruit (from specialist seed firms) which are fun and decorative and edible, though frankly the flavour of those I've tried is nothing special.

HOW TO GROW

Degree of difficulty: Easy to grow if you have the right conditions; little routine work – less than tomatoes.
Sow: February/March at 21–24°C (70–75°F), prick out into small pots and grow on at 16–18°C (60–65°F) until May, then harden off.
Plant: In a cold greenhouse in early May, outside late May/mid-June – but wait for warm weather.
Spacing: 60cm (2ft) apart in all directions.
Routine care: Water sparingly at first and increase as plants start carrying a crop, especially in hot weather. Feed weekly with liquid tomato feed once the first fruits have set. Support the main stems by tying them to 90cm (3ft) canes or trellis; stems are prickly but brittle and easily broken, so be careful when picking the fruit. Plants grow to roughly 90cm (3ft) high and 60cm (2ft) wide outdoors, and slightly bigger under glass.

GROWING IN CONTAINERS

Grow one plant per 30–40cm (12–15in) pot. Move the pots into a conservatory or porch towards the end of the season, when nights grow cool, to encourage any small fruit to reach picking size.

VARIETIES

1 **'Black Enorma'** – huge, fat, purple-black aubergines that slice into 'steaks'. Expect few per plant and they're slow to reach full size. Best grown in a greenhouse.
2 **'Moneymaker'** – reliable and high-yielding standard, purple, sausage-shaped aubergines, produced early, so this is one of the most suitable for growing outdoors.

HARVEST

Start picking aubergines from July as soon as they around 7.5cm (3in) upwards; use secateurs to cut the prickly stem 1cm (½in) beyond the fruit. If they are left on the plants too long they taste bitter and small hard seeds develop inside – these are distributed through the flesh and can't be removed. Plants crop until cold nights strike in autumn – usually late September outside and mid-October under glass.

PROBLEMS

Greenfly, especially early on and under glass, can set back plants so they stop growing. Treat infected plants with an organic spray or wipe off with a damp cloth.

Whitefly and red spider mite are best dealt with by biological control – introducing predators to the greenhouse. Low yields, slow growth and plants that are generally 'poor doers' are usually a sign that growing conditions are too cold, too wet or too windy.

Beans, broad

Vicia faba

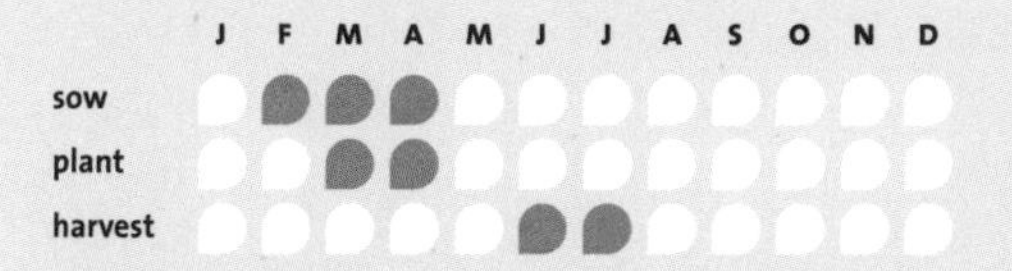

Broad beans are one of the traditional crops for large veg patches and allotments. The plants can be a tad difficult to accommodate in small gardens – you'll need quite a few for a worthwhile crop and they grow 90cm (3ft) tall and need lots of support, otherwise they flop and sprawl badly. But home-grown broad beans picked fresh from the garden and shelled straight away far outshine bought ones, or even frozen ones, for flavour, so they are well worth growing if you have room.

HOW TO GROW

Degree of difficulty: Need regular time and attention right up until cooking them; shelling broad beans takes time.
Sow: In pots or trays under cover in February/March, or in April in rows where you want plants to crop.
Plant: Indoor-raised plants once the weather improves in late March/April.
Spacing: 20cm (8in) apart, in double rows 20cm (8in) apart, with 45cm (18in) between these double rows.
Routine care: Water in – after that, water routinely if it's a dry season. No extra feeding is needed. Plants need support, so hammer in a 1.2m (4ft) post at the end of each row and run two horizontal strings along each side of the row of plants, at heights 30 and 60cm (12 and 24in) above the ground. As the plants grow taller the strings will hold them up without breaking the fragile stems.

GROWING IN ALLOTMENTS

When you have plenty of room, sow or plant broad beans 20cm (8in) apart in double rows 20cm (8in) apart as usual, but increase the spacing between the double rows to 60 or 90cm (2 or 3ft). The extra room means plants have a larger area from which to take up moisture from the soil, so they can cope in dry conditions without watering, and you can easily walk through the crop to tie up plants, pick pods, or do routine weeding and hoeing.

If you enjoy your beans, make several sowings to keep yourself supplied over a longer season, starting with early and late spring sowings, followed by a November sowing using a suitable variety. Don't be tempted to cover autumn-sown beans with cloches, since the protection encourages fungal disease – they need to rough it. But be aware that over-wintered crops will suffer large losses, due to predators and bad weather, so they are far less successful than late-spring-sown crops. In some years over-wintered beans can be a complete failure!

VARIETIES

1 **'Aquadulce'** – the old faithful, not the best for flavour, but rugged, weatherproof plants good for early sowing, and can also be sown in November to give extra-early crops in spring (in a good year).

2 **'Green Windsor'** – an old 'heritage' variety with truly outstanding flavour, but it's hard to find outside specialist organic seed suppliers nowadays. For spring sowing only.

3 **'The Sutton'** – a compact variety growing 30cm (12in) tall. Good for small or windy gardens as they need little support, this variety can be sown in November for early crops and, being short, you can protect early-spring-sown crops with cloches or fleece.

4 **'Witkiem Manita'** – very early, heavy-cropping broad bean with superb flavour. For spring sowing only.

HARVEST

As pods start to swell, open one to check on the progress of the beans inside. Start picking once the beans are the size of your thumbnail – they are most tender while small. Pods can be picked even younger and eaten like French beans, but you might regard this as a bit of an indulgent luxury!

To pick the pods, pull the pod gently back against the direction in which it is growing, twisting slightly and taking care not to 'skin' the stem of the plant. If pods are difficult to pick by hand, snip them off with secateurs instead. Expect to be picking most of the crop over six weeks from mid-June to the end of July.

PROBLEMS

Seeds sown early outside in spring are at the mercy of mice and pigeons who'll happily eat the lot, or cold wet weather that makes them rot; so unless you have well-drained ground and can protect seeds under cloches or fleece, sow later when conditions are better or raise your plants under cover first.

The black bean aphid (blackfly) is a regular pest of broad bean plants; they congregate round the plant tips and on young developing pods. Once the plants start to set pods, nip out the growing tips to remove the blackflies' favourite feeding sites.

Chocolate-coloured markings on leaves are caused by a fungal disease called chocolate spot; this can be prevented by adequate feeding, good drainage and good air circulation around the plants. Another fungus, rust, produces rusty brown markings on the leaves. It occurs late in the season and isn't a worry.

Notches chewed out of the leaves are a sign of pea and bean weevil. It's disfiguring, but it seldom affects yields.

Crafty tip

If you don't need the space to re-use for another crop after you have picked all your broad beans, instead of pulling the old plants out, cut them off a few centimetres above the ground. The stumps will very often re-sprout so they'll give you a light, late crop of beans for a few weeks towards the very end of the summer, when fresh broad beans aren't usually available at all.

Rescuing overgrown broad beans

If you forget to pick broad beans, or you go away and find they've grown huge and the beans are big and tough, don't let them go to waste. Shell them, then skin each bean (it's quite quick once you've learnt the knack – make an incision and then squeeze to 'pop' the bean out from its coat), throw the hard skins away and cook the small green 'bean heart' in the centre as usual. Delicious!

Beans, French

Phaseolus vulgaris

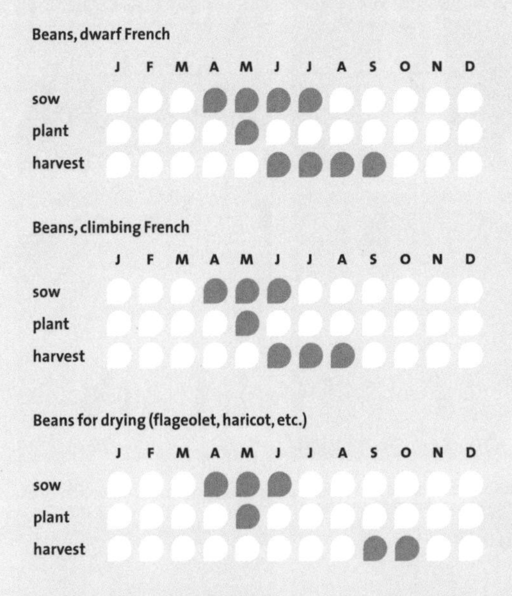

If you only grow a limited range of veg, you really should grow green beans. They are one of the most useful edible crops and one that gives a good return from a small space, besides making the best use of your time when you are busy.

They are quite fashionable too – French beans are developing a cult following in much the same way as tomatoes, and a real enthusiast can branch out into choice connoisseur's kinds, heritage varieties, coloured varieties (with purple or golden pods), and haricot-type beans for drying. Added to that, you can produce out-of-season beans from an unheated greenhouse, which must be far better than flying in beans from Kenya.

HOW TO GROW

Degree of difficulty: Not difficult, but French beans need better growing conditions than runner beans. Beginners are best advised to start with runners.

Sow: On a warm windowsill indoors in mid-April for an early batch of plants to put outdoors. Sow outdoors in May, June and early July where you want plants to crop.

Plant: Plants are not frost hardy, so don't plant outside until mid- or late May.

Spacing: Grow dwarf varieties 15cm (6in) apart with 25cm (10in) between rows; grow climbing varieties up supports 20cm (8in) apart in a single row.

Routine care: Water bean plants sparingly while they are young and in dull, cool weather, but more generously after they are well established and carrying a crop. The plants don't need extra feeding if the ground was well prepared before planting. Climbing varieties need support, so put up netting or trellis before planting, but they don't need tying up since they twine round supports by themselves.

VARIETIES

Dwarf French beans:

1 **'Golden Teepee'** – similar to 'Purple Teepee', but with bright golden pods.

2 **'Opera'** – good, modern, heavy-cropping variety of green beans. Reliable, well-flavoured and resistant to several bean diseases, so good for organic growing.

3 **'Purple Teepee'** – heavy-cropping plants with tasty, purple pods held well above the foliage, making them easy to see for picking.

4 **'Sonesta'** – an early variety which makes neat, compact plants producing heavy crops of tasty pale gold pods. Good for growing under glass out of season, in pots on a patio, or down the garden.

Climbing French beans:

5 **'Cobra'** – heavy cropper with long, slim, green, pencil pods and lilac flowers; attractive for the patio, and can be grown early under glass. Plants have a long cropping season.

6 **'Blue Lake'** – old, traditional variety with cylindrical, green pods. Dual-purpose; eat some as green beans and leave the large pods at the end of the season to dry out for haricots.

7 **'Hunter'** – huge crops of large, flat, but very tender, tasty, stringless pods for slicing, which are sometimes mistaken for runner beans. Good for growing in spring and late summer/autumn under glass as well as outdoors. Seeds can be hard to find outside specialist firms now, otherwise try 'Helda', which is similar.

Beans for drying:

8 **'Borlotto Lingua di Fuoco' (tongue of fire)** – traditional Italian climbing variety; pods are red-dappled when ripe, with maroon and white speckled beans inside. Dwarf borlotto beans are now also available, but you won't get anywhere like such a big crop since the plants are so much smaller.

9 **'Pea bean'** – old variety with small pods containing rounded, pea-sized, cream beans with maroon spots and 'yin-yang' pattern. These can be shelled and used fresh, like broad beans, or allowed to dry out in the pods for winter use. Delicious both ways but not always obtainable, so save seeds to sow each spring.

10 **'Soissons'** – climbing bean with green seeds which dry out as flageolets – an upmarket type of haricot known for their particularly fine, almost asparagus-like, flavour.

GROWING IN ALLOTMENTS

Climbing French beans are more productive than dwarf varieties, and have a longer cropping season, so grow these where you have space. Put up a row of beanpoles as for runner beans, or knock in two posts to support a 'fence' of bean netting. Space plants 25cm (10in) apart and allow 90cm (3ft) between rows.

GROWING IN CONTAINERS

Beans make fairly decorative plants in tubs or growing bags when space is short, and do well grown this way. Climbing beans need sheltere, so plant dwarf beans if conditions are rather exposed because, being close to the ground, they'll be more sheltered.

Space plants in tubs and containers 10cm (4in) apart in each direction. Plant out a single row in a deep trough or growing bag and let them scramble up netting or trellis against a wall, or plant ten plants round the edge of a free-standing 30–40cm to (12–15in) pot and place an obelisk or tall cone of canes in the centre for them to grow up.

GROWING OUT OF SEASON UNDER GLASS

Sow seeds of dwarf French beans on a warm windowsill indoors in March then grow on in a cold greenhouse for an early crop. Plant these in a well-prepared soil border, growing bags or large pots from mid- to late April, avoiding a cold spell, and you can be eating beans in May/June. Make a late sowing in mid- to late July for plants to grow in an unheated greenhouse for cropping in September/October.

Beans for drying

Varieties intended for drying, including haricots and flageolets, should not be picked green. Leave the pods to grow to full size, and later they'll start to turn yellow-brown as the beans inside ripen. Let them dry out naturally, but if damp weather or frost threatens, pick the pods to finish drying in shallow trays in a warm, dry place indoors. When completely dry, shell the pods, spread the bean seeds out to dry further then store in jars to use in winter soups, stews and casseroles, or as bean salads.

HARVEST

Green French beans can be picked as soon as the beans are big enough to bother with – they are most tender while small. (Depending on the variety this may be anything from 5–7.5cm (2–3in) long; climbing varieties normally have longer beans 13–15cm (5–6in) or more). Snap pods off the plants cleanly, along with part of the short stalk that joins them to the stem. Pick regularly – pods left too long on the plant develop fat beans inside.

Depending on sowing times, it's possible to have green beans outdoors from late June to the end of September, although you'll need to sow several crops of dwarf varieties to achieve continuity over this time.

Climbing French beans tend to crop from late June to the end of August.

PROBLEMS

Seeds can rot or young plants will rot at the roots if they are planted in cool, dull and damp growing conditions. Early in the season, or in a cold spell in summer, seeds come up best in small individual pots indoors with compost kept very much on the dry side. Don't put plants out until growing conditions are warmer and drier.

French or runners?

People tend to grow either climbing French beans or runners, but which you prefer is up to you. Fans of runners claim they have more flavour – and they certainly give heavier crops over a longer period than climbing French beans (the same runner bean plants will keep cropping all summer) and put up with worse weather. But fans of French beans (of which I'm one) say they have a finer flavour and don't go stringy or tough the way runners do. Many of these varieties will crop very well under glass to produce upmarket, slim, cylindrical 'pencil pods' or the large flat-podded types which can then be had out of season without any heating. You'll also get a reasonable return from dwarf varieties (which are handy when you don't want the bother of putting up supports). These plants do have shorter-cropping lives, though, so you need to sow them little and often throughout the season.

Beans, runner

Phaseolus coccineus

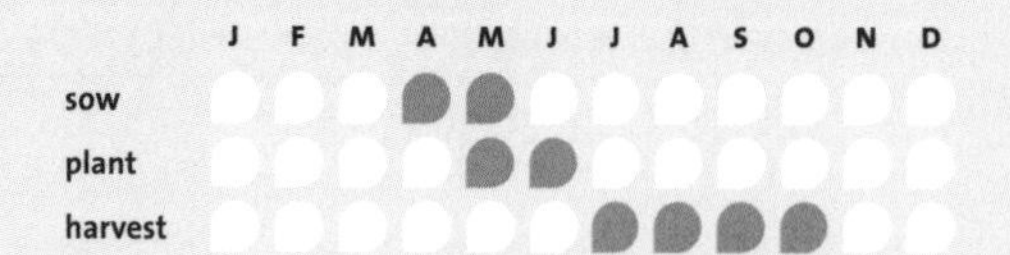

The 'People's Bean', the runner bean is a great allotment favourite due to its easy-going nature and reliably heavy crops, whatever the weather. Runner beans are also popular on the showbench – certain varieties will give you spectacularly long pods without any special attention.

HOW TO GROW

Degree of difficulty: Easy, and once your beanpoles are up, little work apart from regular picking.
Sow: Mid-April in pots on a windowsill indoors, or the end of April/mid-May outdoors where the plants are to crop.
Plant: Mid- to late May or early June.
Spacing: 23cm (9in) apart; grow a single row along the foot of trellis or netting against a fence or wall in a small garden, a double row on a long, tent-like construction of bean poles, or plant round the edge of a free-standing wigwam in a flower bed or decorative potager. The canes or poles will need to be 2.4m (8ft) tall to allow 30cm (12in) or so to be pushed into the ground.
Routine care: Water plants in and water thoroughly in prolonged dry spells. Extra feeding won't be needed for plants grown in well-prepared veg patches, but in containers or intensive veg beds you should apply liquid tomato feed weekly while plants are carrying crops.

GROWING IN ALLOTMENTS

In winter, dig a large trench 45cm (18in) deep and 90cm (3ft) wide and put into this anything you'd normally put on your compost heap. Towards the end of March, fill what's left of the trench with soil; this will produce a deep, rich bed ideal for runner beans, which are notoriously 'greedy' plants. Sometime before your plants are ready to go in, put up a traditional beanpole framework over the trench.

Put a 1.8m (6ft) post in at each end of the row, run a strong wire or length of slender timber between them and lean pairs of 2.4m (8ft) tall beanpoles up against this so they straddle the trench, pushing the bottom of each pole into the ground for stability. Tie them together at the top and fix firmly to the horizontal wire or timber so they can't slip sideways. Place the pairs of poles 25 or 30cm (10 or 12in) apart, and at planting time, when danger of frost is past, plant one bean plant alongside each so that it can twine its way up to the top of its own personal pole.

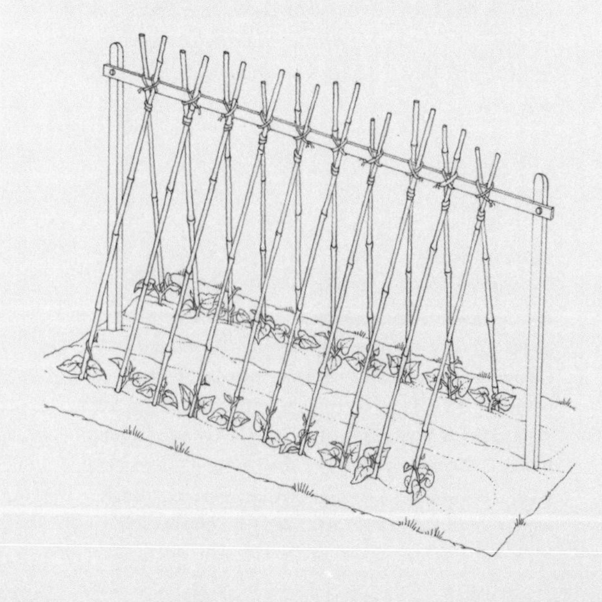

VARIETIES

1 **'Enorma'** – very long, even-sized straight pods which are good for the show bench but also tender and well-flavoured. Good for everyday eating.

2 **'Lady Di'** – long, slender, tasty, high-quality, stringless beans that remain seedless until quite well developed, and have a long growing season.

3 **'Sunset'** – very early variety with pretty pale pink flowers and rather short but well-flavoured beans. Good for a patio or decorative bed.

4 **'Wisley Magic'** – a modern variety with an 'old-fashioned runner bean flavour'; long, good-quality beans on heavy-cropping plants.

GROWING IN CONTAINERS

Runner beans grow well in large pots, growing bags or troughs in the same way as climbing French beans. Dwarf runner beans *are* available (there's a variety called 'Hestia') but these have never caught on in the same way as dwarf French beans, and though they might be worth considering if you have a particularly exposed garden, frankly they don't produce a very worthwhile crop.

HARVEST

Runner beans can be picked as soon as they are long enough to use. Start small, from 10cm (4in) long, although many large-podded varieties are still good to eat at 30–38cm (12–15in) long if they are young and have been grown quickly without a check. As a general rule, large old pods may be tough, tasteless and stringy, and once they have seeds inside (you can tell by the row of regular bulges along the sides of the pods) they aren't worth using.

PROBLEMS

In dry weather, runner bean flowers are prone to falling off without leaving a baby bean behind to grow; the standard remedy is to water the plants regularly and gently spray them with the hose daily to increase humidity, which helps flowers to set. But in prolonged dry weather, go one better: give the plants a thorough soaking then mulch them with garden compost or even old newspapers – anything to help hold moisture.

Birds can sometimes peck off the flowers.

Culinary tip – Quantity surveying

Summer gluts are an unfortunate side effect of growing runner beans. These beans can be difficult to give away, especially if the pods are big and tough or you are overwhelming the same few friends with your home-grown gifts. It takes time to prepare runners for freezing (and frankly they don't thaw out well, especially if stored for more than a few weeks) and the old-fashioned way of salting beans isn't to modern tastes. My advice is grow a lot fewer runner bean plants than you think you'll need and enjoy them more, knowing you'll never be swamped.

Beetroot

Beta vulgaris subsp. *vulgaris*

	J	F	M	A	M	J	J	A	S	O	N	D
sow	○	○	●	●	●	●	○	○	○	○	○	○
harvest	○	○	○	○	○	○	●	●	●	●	○	○

If you think beetroot only means the pickled stuff sold in jars, try growing your own. You'll be amazed just how good it can taste, and you'll discover several ways of using it creatively! The plant-chemicals responsible for its gaudy colour are claimed to be particularly good for you because they are powerful anti-oxidants. So if you hated it at school, try it again now that your palate is more sophisticated.

HOW TO GROW

Degree of difficulty: Needs some care and regular attention.

Sow: Early bolt-resistant varieties in late March and protect with cloches or fleece; sow other varieties mid-April to the end of June.

Spacing: Sow seeds a couple of centimetres apart in rows and thin seedlings out in several stages to 7.5cm (3in) apart for early or baby beet, or 15cm (6in) apart when you want full-sized roots. Each 'seed' is actually a case containing several seeds, hence the need to thin out later.

Routine care: Water in after sowing and water during dry spells to keep plants growing steadily; no extra feeding is needed. Keep plants well weeded, since weeds delay the crop and may smother plants and create gaps in your rows.

Culinary tip – Being creative with a large crop

Instead of boiling beetroot, which makes the kitchen look like it's had a visit from Count Dracula, try baking them instead. Wash well and trim off the tops and tiny thread roots, pile into roasting pans or baking trays and bake in a medium heat until the biggest roots are soft right through when prodded with a skewer. When cool they are easy to peel and less inclined to 'leak' everywhere.

Try beetroot baked whole (as for baked potato) and served hot or cold with sour cream. Use cubed, baked beetroot tossed in olive oil and a little fresh crushed garlic as a salad, or liquidize with chicken stock and seasoning to make beetroot soup.

Grated raw beetroot makes a great salad dressed with olive oil and black pepper or garlic; or mix it 50:50 with grated raw carrot.

VARIETIES

1 **'Burpee's Golden'** – a traditional, though unusual, globular golden beetroot with superb flavour. Good for picking small and grating raw for salads or cooking; the beet-like foliage can be steamed as a separate vegetable.

2 **'Cylindra'** – tall, tubular beetroots that grow above ground, growing up to 20cm (8in) tall and producing a huge crop from a single row, but no good for baby beet.

3 **'Detroit 6-Rubidus'** – reliable bolt-resistant, globular-rooted variety for early crops. Roots are non-fibrous even when allowed to grow large.

4 **'Egyptian Turnip Rooted'** – a superb heritage variety (obtainable from specialist catalogues) with very tasty, heart-shaped roots for early sowing; very bolt resistant and does not turn tough or fibrous, even when large, if left until the end of the season.

HARVEST

Start pulling baby beet when they are big enough to use – usually 2.5cm (1in) diameter. (Early varieties could be ready from mid-May.) Pull every third or fourth plant, leaving the rest with more space in which to continue growing. Use full-sized roots as needed during the summer. Roots tend to store best in the ground, but any still remaining in October should be lifted and kept under cover to use as soon as possible.

PROBLEMS

Beetroot is very prone to running to seed if sown too early, when growing conditions are cold or dry or if it gets a check to its growth. Use suitable varieties for early sowing, protect the plants with fleece if the weather deteriorates and keep them well watered in dry spells.

Woody roots are generally caused by dry growing conditions.

Beetroot leaves for salads

Young, tender beetroot leaves are a popular ingredient of mixed baby-leaf salads; you can pick a very few leaves from the centre of plants you are growing for roots, but don't overdo it. If you plan to use lots of leaves, sow a row of 'Bull's Blood' – you can keep picking over the same plants again and again all through the summer. And if you leave a few beetroot plants in the ground through the winter, these will start growing early next spring and produce a natural harvest of strong young leaves you can pick for salads before they eventually 'bolt'.

Broccoli, sprouting

Brassica oleracea Italica Group

Sprouting broccoli

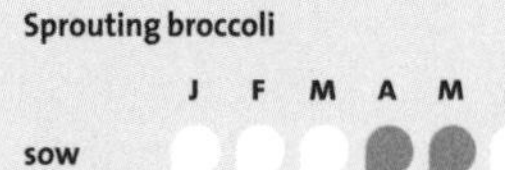

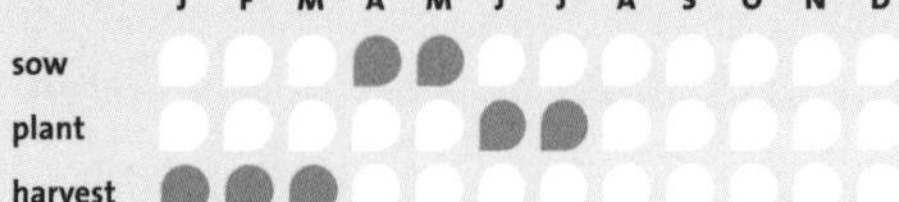

Summer purple sprouting broccoli

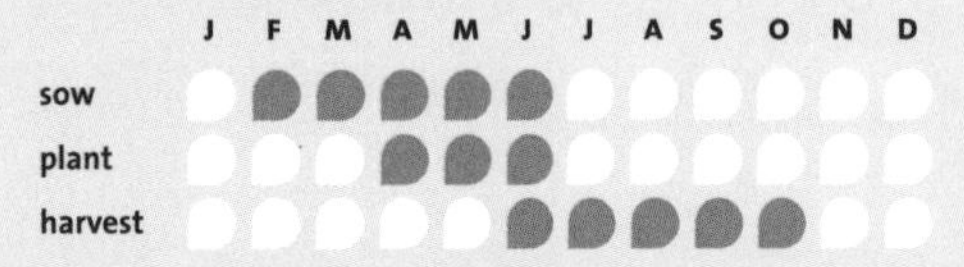

Sprouting broccoli is one of the most reliable, successful and delicious winter veg, and it has been justifiably growing in popularity for some time. It is not to be confused with calabrese, which is all green and a summer crop. Even if you can find sprouting broccoli on sale in the greengrocer's, it's far better picked fresh from the garden and cooked within minutes. Just a few plants are enough to supply something special to go with the Sunday roast at a time when there isn't much fresh food from the garden, and the occasional frost makes it all the better.

HOW TO GROW

Degree of difficulty: Easy and hardly any work involved.
Sow: Late April/early May in a seedbed and thin out to 7.5–10cm (3–4in) apart.
Plant: Transplant strong young plants to their final positions in June or early July.
Spacing: 45cm (18in) apart in each direction.
Routine care: Plant in very firm ground, water in, and keep watered in dry spells to keep plants growing without a check. To boost growth, give a general-purpose liquid or granular feed in early August with plenty of water to wash it in. In exposed areas it may be necessary to support plants by tying the main stems to stakes – do this before autumn gales set in.

GROWING IN ALLOTMENTS

Where you aren't short of room, space plants 60cm (2ft) apart with 60cm or 75cm (2ft or 2ft 6in) between rows, to allow plants better access to moisture when you can't water. This also allows easy access when picking in winter; when plants may be wet or frosty and not so nice to rummage through.

GROWING PURPLE SPROUTING IN SUMMER

Sprouting broccoli isn't a veg that freezes well, but now enthusiasts can enjoy fresh purple sprouting through the summer by growing one of the new summer-cropping varieties, such as 'Bordeaux' or 'Summer Purple'. Sow seeds under cover in February/March for the earliest crop in July, and sow again at intervals of three weeks until early June to provide plants that produce fairly quick, but short-lived, crops.

Summer-sprouting broccoli plants aren't frost-hardy, so they won't stand winter. But, in theory at least, a real fan growing several varieties and staggering their sowing times could have fresh purple sprouting broccoli from the garden 365 days a year.

VARIETIES

1 **'Claret'** – very late purple sprouting, for picking in April.
2 **'Late Purple Sprouting'** – takes over where 'Rudolph' leaves off, cropping in March and early April.
3 **'Rudolph'** – an extra-early, purple sprouting type, ready from just before Christmas until February from an early sowing, with large, plump, tasty shoots.
4 **'White Sprouting'** – greeny-white spears are ready from March to April.

HARVEST

As soon as the colour of the developing purple or white spears is visible (in late winter, from mid-Jan to the end of March, depending on your variety) cut the entire shoots 5–10cm (2–4in) long with a sharp knife. Once plants start 'sprouting', check them over at least twice a week so that no shoots are left to run to seed; the more immature shoots you cut, the more will grow, but once you let them flower they stop growing new young shoots.

PROBLEMS

Unlike most brassica crops, sprouting broccoli stays amazingly pest free, since the edible part forms in winter and early spring when there are no cabbage white butterflies about. Even if the plants themselves suffer a slight outbreak of caterpillars during the summer, it's usually not a problem. If it's a bad attack – enough to check the plants' growth – then yes, pick the pests off by hand or use a suitable organic spray.

Clubroot is the biggest worry. It is usually 'bought in' on young plants raised in infected ground. Roots of affected plants swell up and eventually rot, but before this their growth becomes badly stunted and they stop forming shoots. Once you have clubroot it stays in the ground for 20 years or so, during which time you can't grow other members of the cabbage family – there's no 'cure'. Make sure you don't get it by raising all your own plants, by growing plants in well-drained soil and by liming acid soil prior to planting. The disease is less common on chalky, alkaline soils and where drainage is good.

In a cold winter when food is scarce in the countryside, pigeons will often raid gardens and allotments, and brassica crops are among their favourites – they peck away leaves and reduce plants to bare stems and ribs, so they don't produce a crop. If pigeons look like being a problem, cover brassicas with fine-mesh crop-protection netting or even old net curtains to protect the plants. If you use bird netting, raise it well up above the plants on a support frame so that birds can't push their beaks through to peck the plants.

Brussels sprouts

Brassica oleracea Gemmifera Group

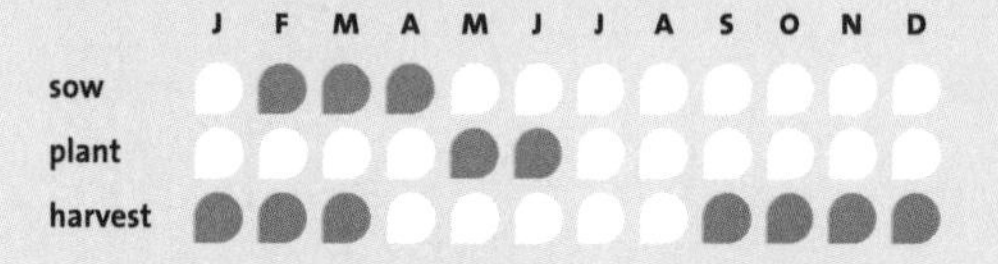

The veg that's every schoolboy's pet hate, and compulsory eating at Christmas dinner, has been turned into one of the latest healthy-eating sensations. This is due to the presence of cancer-busting mustard oils which is what gives sprouts their characteristic, slightly bitter flavour.

But for those who do enjoy them (and for whom Christmas dinner would be a pale shadow of itself without them – and that includes me) sprouts are worth growing. Go for a high-quality variety even if it's not all that heavy cropping, as you won't be able to buy anything like it in the shops and the flavour makes the effort worthwhile.

HOW TO GROW

Degree of difficulty: Very nearly as foolproof as sprouting broccoli and, like broccoli, needs little work.
Sow: At 13–16°C (55–60°F) indoors in February/March for early varieties; in an outdoor seedbed mid-March/April for later cropping kinds.
Plant: Early-cropping varieties in May, later varieties in June. Very firm soil is essential – dig it over the autumn before planting, not immediately beforehand, and consolidate by treading down well. Make planting holes with a big dibber and firm plants in after planting by treading the soil down all round them with your heel.
Spacing: 60cm (2ft) in all directions in small gardens or intensive beds where space is short.
Routine care: Water in well, and water in dry spells so plants don't suffer a check to their growth. Give a top-up feed of general-purpose organic fertiliser in early August and water in.

GROWING IN ALLOTMENTS

Space plants 75cm (2ft 6in) apart with 90cm (3ft) between rows. In exposed situations, support each plant by tying the main stem up to a stake – if plants are shaken loose by autumn gales the root disturbance results in 'blown' sprouts or no sprouts at all. Protection from pigeons is also essential on country allotments.

If you don't want to keep travelling to the allotment to pick a few sprouts at a time, especially over Christmas, either pull up a whole plant to bring home and 'plunge' the roots in a tub of compost in a cool place by the back door, or else cut the top section of stem and stand the base in a jar with a centimetre of water to keep sprouts fresh. They'll keep in perfect condition this way for a week or so.

VARIETIES

1 **'Falstaff'** – the red-cabbage version of a Brussels sprout with excellent, mild-tasting, tight, reddish-purple buttons.

2 **'Trafalgar'** – one of the new breed of sweeter-tasting sprouts that children are said to like more than old-fashioned kinds, which they find bitter. Heavy crops of tight buttons from December to March.

HARVEST

Start picking sprouts as soon as they are large enough to use, starting from thumbnail size; early varieties are ready late September onwards, later ones in succession from November/December and the latest are not ready until after Christmas, but will stand until the end of March. If you grow several varieties that overlap, make sure you use them in the right order. A frost is said to improve the flavour of sprouts (it certainly makes picking them more painful).

Sprout tops

When I was a lad, the end of the sprout season was always marked by the serving up of the loose, open, cabbage-like growth from the top of a stem of sprouts, which was also sold by greengrocers as 'sprout tops'. They've started to reappear in some small, independent greengrocers who obtain supplies locally, and delicious they are too – combining a mild sprout taste with a spring cabbage texture. Don't overlook this end-of-season bonus, which you can reap even if your sprouts 'blew' instead of forming proper tight buttons.

PROBLEMS

Brussels sprouts suffer the same set of problems as sprouting broccoli (see page 85).

Instead of forming tight buttons, 'blown' sprouts open out into small, flattened, 'exploded green rosettes'. It's usually due to growing in loose soil, or when plants suffer a check in growth or are shaken loose by gales. Don't underestimate how hard the perfect Brussels sprouts bed needs to be, and stake plants in exposed areas.

Cabbages

Brassica oleracea Capitata Group

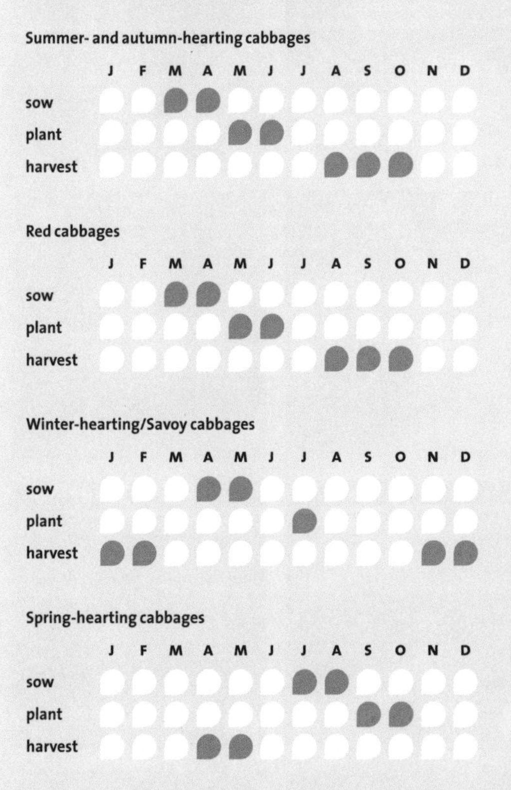

Cabbage has long since lost its '1950s school-dinner' image, and today it's seen as healthy, verging on the cool. But there's more than one sort of cabbage – there are crispy summer types, crinkly-textured Savoys, red cabbage and some outstanding heritage types which, when well-grown and imaginatively cooked, move the humble cottage-garden standby into a new league.

HOW TO GROW

Degree of difficulty: Repays a little care.

Sow: Summer- and autumn-hearting cabbage, including red cabbage, in March/April, and winter-hearting cabbages from mid-April to late May; sow spring cabbage mid-July to early August. Sow the earliest varieties under cover, the rest in an outdoor seedbed, then thin out seedlings to transplant later.

Plant: Summer and autumn cabbages, including red cabbage, in May/June; winter cabbage and Savoys in July, spring cabbage in late September/early October.

Spacing: Compact varieties 30cm (12in) apart in each direction; larger growing kinds allow 45cm (18in)gap.

Routine care: Plant in firm ground, water in well and keep watered in dry spells so that plants can grow continuously without a check. Apply a general-purpose or high-nitrogen feed during the summer to top up nutrients and encourage leafy growth, and water in well. Keep well up to date with weeding, but take care if using a hoe, as cabbages are shallow-rooted.

GROWING IN ALLOTMENTS

When you're growing on a larger scale with plenty of room, space cabbages slightly wider apart, 45cm (18in) for compact varieties and 60cm (2ft) for larger kinds, and allow 75cm (2ft 6in) between rows. Instead of cutting cabbage hearts to bring home, pull up the whole plant (you have nothing to lose since you only get one heart per plant) and wrap the roots in a plastic bag. Back home you can keep a cabbage fresh without filling the fridge if you store it in a cool place out of direct light – a garage or utility room is ideal – and keep the roots moist by standing them in a container with a centimetre of water, or a plastic bag with a little damp soil inside.

GROWING IN CONTAINERS

The 'big new thing' in cabbages are compact kinds for producing baby veg or growing at close spacings in small intensive beds. You can find compact red cabbage and Savoys as well as conventional cabbages suitable for these uses. New varieties are constantly coming on stream, so keep an eye on the seed catalogues to see what's available from year to year.

VARIETIES

Summer cabbage

1 **'Golden Acre'** – an old favourite, ball-headed variety which matures in early summer from a February/March sowing.

2 **'Hispi'** – well-flavoured and fast-growing with small conical hearts; a versatile variety suitable for sowing from early February under cover and from March to July outdoors to provide hearts throughout summer and autumn. It can also be sown under cover in October: plant seedlings in the border of a cold greenhouse when big enough (ready to cut April/May), or outdoors under cloches or fleece in spring for early crops. This variety is also suitable for growing as 'baby veg' – plant 23cm (9in) apart in each direction for small, individual-serving-sized hearts. If you only grow one cabbage, 'Hispi' is a good choice and has stood the test of time.

Red cabbage

3 **'Red Drumhead'** (pictured), **'Autoro'** and **'Maestro'** – all are good varieties; the latter two are especially suitable if you fancy entering the local flower show.

Spring cabbage

4 **'Wheeler's Imperial'** – another oldie. Being compact, it fits small gardens and as well as being sown in summer for a spring harvest, it can also be sown in February/March and harvested in autumn.

5 **'Offenham Flower of Spring'** – this can be sown in July and August to crop in April/May.

Savoy cabbage

6 **'Tundra'** – a green Savoy cabbage with firm, round hearts which mature in autumn from a late-spring sowing and stand all through the winter.

7 **'Winter King'** – a favourite Savoy with dark, crumpled leaves. Make several sowings in late spring and early summer to stagger the crop and extend the Savoy season from late summer into late winter.

Winter cabbage

8 **'Cuor di Bue' (Bullsheart)** – an old, full-flavoured variety with tight conical hearts in late summer and autumn. Seeds need searching out from specialist suppliers' catalogues.

9 **'January King'** – a fat-headed winter cabbage tinged with red, maturing in autumn and standing well through winter from a late-spring sowing.

HARVEST

It's easy to see when cabbages are ready for picking, as the centre tightens up and forms a solid 'heart' – which may be rounded or conical depending on the variety. Use a stout knife with a scalloped blade, or secateurs, to cut through the stem just below the heart and remove damaged outer leaves. Leave the rest on during transportation to protect the more delicate interior. As a rough guide, expect spring cabbages to be ready April/May, summer kinds August/September, autumn varieties September/October, red cabbage August to October, Savoys and winter cabbage November to February.

Recipe – Stuffed cabbage leaves

Use clean, whole leaves from the outer layer of the heart for wrapping several spoonfuls of a well-seasoned, herby, minced meat mixture. Sit the 'parcels' close together in a deep baking tray, and just cover them with stock. Bake in a moderate oven for 40–60 minutes or so until the meat is cooked and the cabbage leaves are tender.

PROBLEMS

Cabbage whites (large white and small white butterflies) can't resist any kind of brassicas, and their caterpillars eat the leaves and deposit frass (excrement) in the cabbage hearts, making them unfit to eat. Cover growing crops securely with fine insect-proof mesh throughout the growing season (not fleece, which overheats in summer). Remove caterpillars regularly by hand, or use an organic pesticide.

Small slugs can work their way into the heart of developing cabbages and perforate them, besides leaving their droppings behind. Use organic slug remedies before planting (see page 51) to reduce the population in your veg patch and take regular precautions during the growing season.

In winter, use netting held up over the plants on supports to foil pigeons, or leave insect mesh in place. Cabbage root fly lays its eggs at the base of the stem and the larvae eat the roots, causing the plant to wilt and eventually die. Small roofing-felt collars are available which can be slotted around the base of each cabbage stem to prevent egg laying. Crop protection mesh held above the plants also helps.

Clubroot can also occur, see broccoli (page 85).

Culinary tip – Getting the best from cabbage

The heart of a garden-fresh, organically grown cabbage has a crisp texture and sweet semi-nutty flavour, making it ideal to shred or grate for use raw in salads or homemade coleslaw. There's no need to grow solid, white, winter cabbage especially for this job; most summer and autumn cabbages are fine, as are some mild-flavoured winter kinds and Savoys that produce firm hearts.

When you're cooking cabbage, don't ruin it by boiling it in lots of water, which is what always puts people off eating it. Instead, drop lightly shredded cabbage into a centimetre of fast-boiling water or chicken stock, or steam it instead, and cook for the minimum amount of time so that it retains some 'bite' and its crispy, crunchy texture.

Calabrese

Brassica oleracea Italica Group

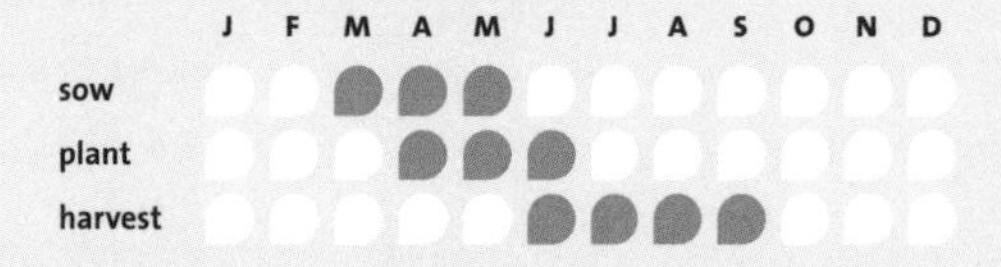

Calabrese, also sometimes called green sprouting broccoli, is one of the most popular green veg, known for its healthy properties. It's also very easily grown, and anyone with a walk-in polytunnel on their allotment can have good-quality heads for much of the year without needing any heat. Calabrese plants yield one large, green, central head, but if you leave them in the ground after cutting this off they start to produce secondary 'broccoli spears' a few weeks later; so you have two crops in the space for one.

HOW TO GROW

Degree of difficulty: Needs some regular care.
Sow: Under cover in early March, or in a seedbed outside from mid- or late March until the end May.
Plant: April to June; transplant outdoors May/June.
Spacing: 30cm (12in) apart in each direction
Routine care: Water and weed regularly, it's vital to keep plants growing steadily. 'Top up' with a general-purpose or high-nitrogen feed halfway through the growing season. Protect against caterpillars.

GROWING YEAR-ROUND IN POLYTUNNELS

By sowing little and often, top-quality calabrese can be grown and picked virtually year round in an unheated polytunnel, in all but the midwinter months. It's probably *the* most successful outdoor veg grown this way, thriving in the warm 'buoyant' atmosphere and quickly producing large, tender heads.

Sow a few seeds roughly every six weeks from March until August and plant a short row each time there's room and when you have young plants ready. Ventilate the tunnel on hot days in summer, but fix insect-proof mesh securely across open doorways to screen out cabbage whites and prevent your crop being riddled with caterpillars. Then, in September and October, sow a short row of seeds, well spaced out where you want plants to grow (don't transplant them), then thin out to 30cm (12in) apart. Leave to grow undisturbed through the winter. Autumn-sown crops will produce heads ready to pick March/June.

VARIETIES

1 **'Crown and Sceptre'** – long-lasting plants that produce a large central head in summer followed by several pickings of smaller shoots which can be used as broccoli spears in late summer and autumn.
2 **'Chevalier'** – medium-sized heads followed by small broccoli spears, starting September onwards.
3 **'Trixie'** – green heads on compact plants that are resistant to clubroot.

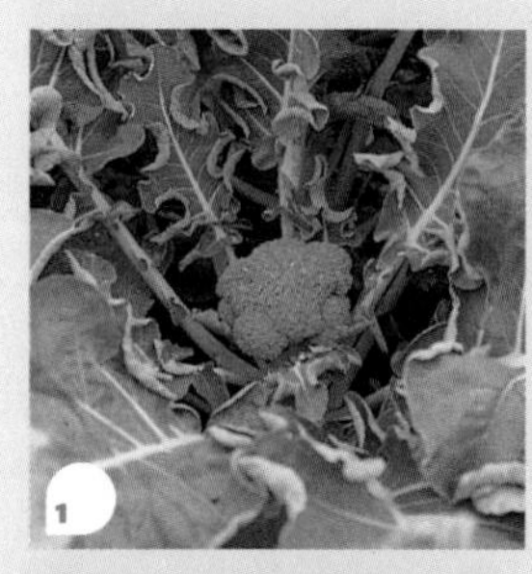

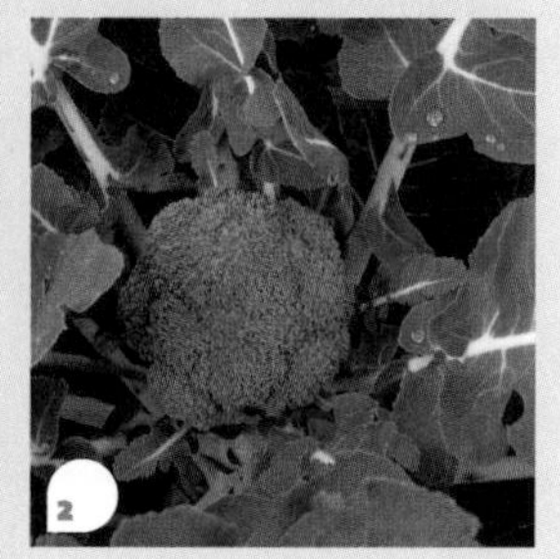

HARVEST

June to late September. Cut the whole head off, complete with a couple of centimetres of stalk. Use straight away or keep in the fridge for up to three days.

PROBLEMS

As for cabbages.

Carrots

Daucus carota sativus

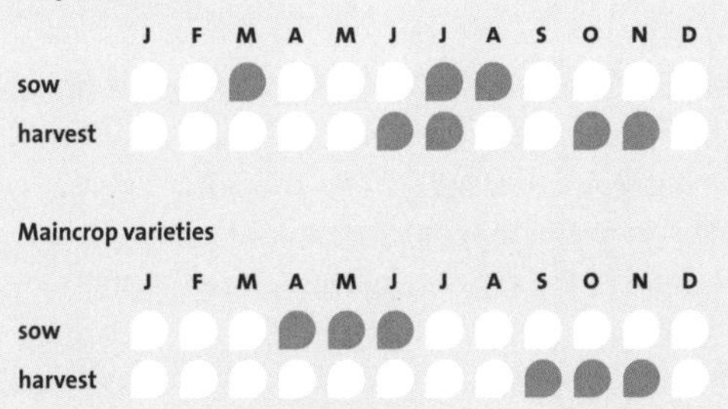

Unless you have an allotment and you use an awful lot of carrots, then you might not want to bother with maincrop varieties, but it's well worth growing your own when you want organic carrots (which really do taste far better), baby carrots – the sort sold in bunches with the leaves on, for salads – and unusual, particularly coloured, varieties. With any of these you'll save yourself a fortune on shop prices.

HOW TO GROW

Degree of difficulty: Need attention to detail and regular care.

Sow: Thinly in shallow drills in soil that hasn't been manured in the last 12 months, and cultivate the ground well so it looks finely sieved. Sow early varieties under cloches or fleece in March; maincrop varieties in the open from April to the end of June, then for a few quick, late crops sow an early variety in July and early August and cover with fleece in the autumn to extend the growing season.

Spacing: Thinning carrots releases an aroma that attracts carrot fly, so it is always best to sow the seeds thinly to start with and avoid pulling out the seedlings later. Allow 30cm (12in) between rows.

Routine care: Water well during dry spells and weed regularly to keep the roots growing steadily without a check. Even if you don't thin out the seedlings, you'll need to take precautions against carrot fly, which could ruin the crop.

Culinary tip – Making carrot juice

Juicing machines are increasingly popular with health-conscious veg growers and are a great way of using up surplus produce. Simply feed in well-washed veg after cutting them up small enough to fit into the hopper – there is no need to peel them first since much of the vitamins and nutrients are in the skin, but scrub root veg such as carrots to remove soil.

Carrot juice is rather bland and unexciting on its own, but a half-and-half mix of carrots and tomatoes, plus the juice of a lemon, makes a tasty 'liquid breakfast' or post-gym pick-me-up that's enjoyable at any time of day. Vary the flavour slightly by adding a stick of celery, slice of celeriac, an apple and/or a few lettuce leaves or cabbage heart. You can also add fresh herbs, but if you add much greenstuff the green colour added to the orange/red of the tomato and carrot turns the juice an unappetising khaki colour that few people fancy drinking.

VARIETIES

1 **'Autumn King'** – large, tapering carrots, ready in autumn but which keep well when left in the ground over winter, as long as the soil is fairly well drained.

2 **'Healthmaster'** – deep reddish-orange flesh which contains a third more beta-carotene than other carrots; good for using raw in salads or for juicing.

3 **'Nantes'** – an excellent sweet-tasting, early variety with long, slightly tapering roots, produced in June from a February sowing under polytunnels, in a cold greenhouse border or outside under fleece. Can also be sown from March to June for normal use, and in July or even early August for protecting under fleece for late autumn crops. Good for cooking or using raw in salads. Improved forms of 'Nantes', such as 'Nantes 2', are identical.

4 **'Purple Haze'** – an unusual purple-skinned carrot with orange flesh inside, best used raw and scrubbed rather than peeled to retain the antioxidants in the coloured skin. Ready in summer and autumn.

5 **'Yellowstone'** – bright yellow carrots with a very sweet flavour, best for use raw in salads.

6 **'Flyway'** (pictured), **'Sytan'** and **'Resistafly'** are three of several carrot-fly-resistant varieties that are available and which appear periodically in seed catalogues. However, the range is limited and they often prove to be less exciting eating than the varieties you'd have chosen for reasons of flavour or colour. Resistant varieties are not 100 per cent effective, and small outbreaks of carrot fly may still occur. That said, you can improve the odds if you alternate rows of fly-resistant carrots with onions or spring onions.

HARVEST

Start pulling baby or salad carrots as soon as the first ones are big enough to use, leaving the rest to keep growing (around June/July). It's a good idea to pull alternate carrots to thin out the rows (the flies are less trouble later in the season), leaving the rest more room to keep growing. As a general rule, maincrop carrots 'keep' best in the ground, but any that are left once seriously rainy weather starts in autumn are liable to produce fine, white, root hairs all over and start to re-sprout. Or, at worst, they'll rot, so dig up the remainder then, brush off the dirt and let them dry off. Don't wash them – they store better left 'dirty'. Late-sown early varieties will be ready October/ early November.

STORAGE

Keep carrots in a cool, dark, dry place, like the garage or shed, and use as soon as possible. Check them over regularly and remove any that are starting to rot before infection spreads to the rest. Traditional techniques for storing in 'clamps' – piles of roots embalmed in soil or sand – are tricky to get right, take a long time to construct and aren't all that successful. Paper sacks are so easy.

Recipe – Carrot-slaw

For a fresh, healthy carrot salad that tastes delicious, finely grate fresh, organically grown carrots, add chopped spring onions, grated apple or firm cabbage heart (optional), a sprinkling of whole pine nuts and dress with lemon juice.

PROBLEMS

Carrot fly is almost inevitable, unless you take precautions. The best solution is to keep crops of carrots securely covered with fine, insect-proof mesh throughout their lives. As belt and braces, grow rows of onions around areas of carrots (pictured above) as well as protecting them with mesh, to disguise the scent that attracts carrot flies in the first place.

You'll nearly always have one or two fanged roots, but when a whole crop is badly affected it's usually due to heavy doses of manure or incompletely rotted organic matter having been used on the soil within a few months of sowing. Only sow carrots on ground that's had organic matter applied one or two winters beforehand. Stony soil can produce the same effect, so if your earth is full of stones, make a raised bed out of old floorboards and fill it with sieved soil mixed with old potting compost so that the carrots have a good root run.

Cauliflowers

Brassica oleracea Botrytis Group

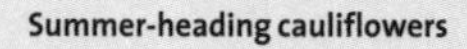

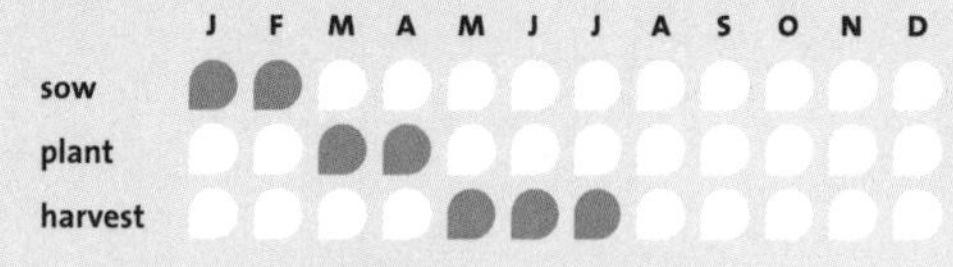

Autumn-heading cauliflowers

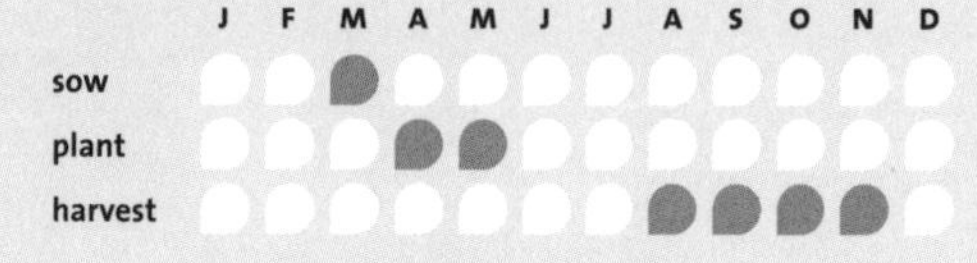

Winter-heading cauliflowers

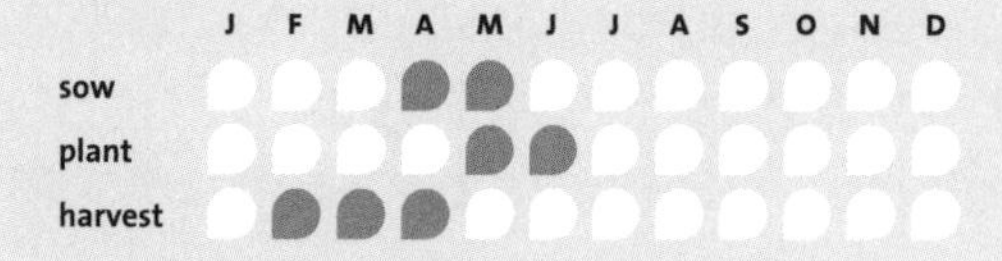

The plain, white, old-favourite cauliflower may have rather gone out of fashion, but don't be in too much of a hurry to write off caulis – purple and lime-green varieties are some of our top 'designer' veg nowadays. Cauliflowers have a delicious buttery flavour and cost a fortune if you buy them in the shops. True, they aren't all that easy to grow, but if you have space to spare it's worth risking a few. (You'll get better results as you gain experience.) Anyway, I love cauliflower cheese made with the bog-standard type.

HOW TO GROW

Degree of difficulty: Challenging; need attention to detail, but they are not time consuming.

Sow: Summer-heading varieties in January/February at 16°C (60°F) indoors; autumn varieties March indoors or in a seedbed under cover (coldframe, greenhouse border or under fleece outdoors) and winter varieties April–May in a seedbed outdoors.

Plant: When young plants are 7.5–10cm (3–4in) tall with good root systems; plant summer-maturing varieties March–April, autumn-maturing varieties April–May and winter-maturing varieties May–June.

Spacing: 60cm (2ft) apart each way for summer and autumn varieties, allow 75cm (2½ft) between winter varieties.

Routine care: Really good soil preparation is vital and firm soil is essential. Water plants in well, and keep well weeded and watered through the summer. 'Top up' with a general-purpose or high-nitrogen feed around July–August. Keep a watch on plants approaching maturity and when the first signs of small curds appear in the centres of the plants, bend a few outer leaves over for protection from weather – snapping their midribs so they stay in place. This is especially important for white varieties as they easily discolour.

GROWING BABY VEG

Small, one-man-sized mini-cauliflowers are increasingly popular, both for eating fresh and for freezing. Special 'baby veg' or 'patio varieties' are needed for this – the range regularly changes in seed catalogues, so see what's available each year. If you can get them, look for 'Igloo' or 'Avalanche' and plant them closer together than for normal caulis – approx 30–45cm (12–18in) apart.

VARIETIES

1 **'All Year Round'** – if you only want to buy one packet of seeds to keep yourself supplied with traditional white caulis all season, this is the one to go for. Sow in October for plants to grow through the winter under a polytunnel or fleece to give you spring crops.
2 **'Autumn Giant'** – one of the most reliable-heading cauliflowers, with firm white heads. Ready September–October.
3 **'Purple Cape'** – a reliable winter variety with large, deep purple heads of superb flavour. Not ready until March, so it takes nearly a year to grow, but the result is well worth it. Wonderful flavour, and keeps its colour fairly well on cooking.
4 **'Romanesco'** – lime-green caulis that form a series of pinnacles instead of a smooth flat head; good for dividing into florets to use raw as crudités or to steam as individual 'shoots' (they keep their colour when cooked). Several different 'Romanesco' varieties are available, including 'Veronica' and 'Celio', and all are very similar. Ready in late summer/autumn.
5 **'Trevi'** – bright jade-green heads of conventional cauliflower shape and superb flavour and keeps its colour on cooking. Sow in May, eat in September.
6 **'Violet Queen'** – medium-sized, mauve heads ready in late summer/early autumn; turns green when it's cooked.

HARVEST

Wait until developing curds form a dense, tightly packed hemisphere that looks full size and cut it off just beneath the base of the head (include the small collar of bract-like outer leaves as these will offer protection in transit). But don't wait until too late, otherwise individual florets will start to shoot out and spoil the shape of the head. Once that happens they quickly open out into flowers and you've lost your cauli.

Summer-heading varieties should be ready May to July, autumn varieties will crop August to November, and winter caulis between February and April.

PROBLEMS

As for cabbages.

Sometimes sheer bad luck due to unfavourable weather produces small or ruined heads, but often this is a sign of poor cultivation leading to poor or checked growth. Suspect inadequate soil preparation with insufficient organic matter, drying out and/or shortage of feed during the growing season, or weeds being allowed to smother the growing plants at some stage. Don't give up; make a mental note to do better next time!

Celeriac

Apium graveolens var. *rapaceum*

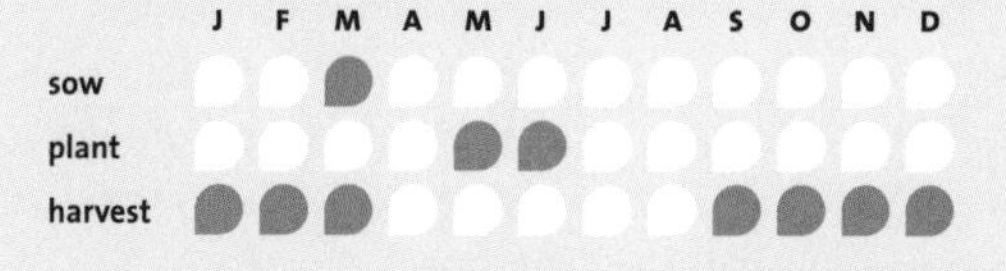

Unlike its troublesome cousin celery, celeriac is well worth growing. Also known as turnip-rooted celery, it is an unusual vegetable with dark green, celery-like leaves on short, stiff stalks that grow from the very top of a swollen 'root' that sits half out of the ground. In the shops celeriac sells for over £1 a piece, so even a small patch is a good investment. While it needs similar growing conditions to celery, it's more reliable, so you stand a better chance of achieving useable results with halfway reasonable care.

HOW TO GROW

Degree of difficulty: Needs attention to detail.

Sow: As for celery: in early March on a windowsill indoors at 16–21°C (60–70°F). Prick out seedlings into small pots or cellular trays as soon as they are big enough to handle and continue growing in warmth (13°C/55°F minimum) on a windowsill or in a heated propagator in a greenhouse. Keep well watered and in adequate light, but not strong direct sun.

Plant: Harden off carefully, then plant into well-prepared soil containing plenty of well-rotted organic matter when the last frost is safely past, in late May or early June. The rich soil is the key to fat roots – on poor, dry earth they will fail to develop.

Spacing: 30cm (12in) apart in all directions, when growing in intensive beds in small gardens; in allotments or where there's more space, increase this to 45cm (18in) apart.

Routine care: Water in well after planting, liquid feed regularly (every two weeks is not too often, using general-purpose feed) and keep plants well watered at all times; celeriac needs to grow without a check. Weed regularly until plants are big enough to cover the ground and shade out weeds naturally.

VARIETIES

'Monarch' – a reliable modern variety that produces high-quality roots.

HARVEST

Start using celeriac as soon as the first roots are big enough – dig the whole plant up carefully. The season generally runs from September until Christmas, although in mild winters plants can be left in the ground until March without harm – they 'keep' better this way than when lifted and trimmed of leaves.

To prepare for cooking once dug up, rest the plant on a hard surface, such as a path outside or a large chopping board and, using a long strong knife, slice off the top of the plant, including all the leaves. Roughly cut off the irregular-shaped roots at the base and remove the worst of the dirt. Indoors, pare off the knobbly skin thickly. Slice the fat 'root' or cut it into cubes. If you aren't cooking it straight away, stir a little lemon juice through to prevent it discolouring.

PROBLEMS

Stunted plants that fail to make decent-sized roots are usually due to poor growing conditions, which may mean not enough organic matter in the soil, not enough feeding and watering, or just a very bad growing season which 'checked' the plants and stopped them growing properly. Small roots with hollow or fluffy-fibrous centres are usually caused by much the same things, but even if they aren't perfect affected roots can usually be used for soup or in stews. Celery fly (see celery, page 101) may also attack.

Cooking with celeriac

Cubes or slices of celeriac are good roasted alone or with a mixture of winter root veg; it is also delicious steamed and served with a sauce, or cooked in winter casseroles, stews and soups. It's positively brilliant sliced and laid in layers alternating with potatoes and baked in a casserole dish with a creamy cheese sauce. But the classic French celeriac recipe is for remoulade, which is basically raw celeriac grated in a mustard mayonnaise sauce.

Celery

Apium graveolens var. *dulce*

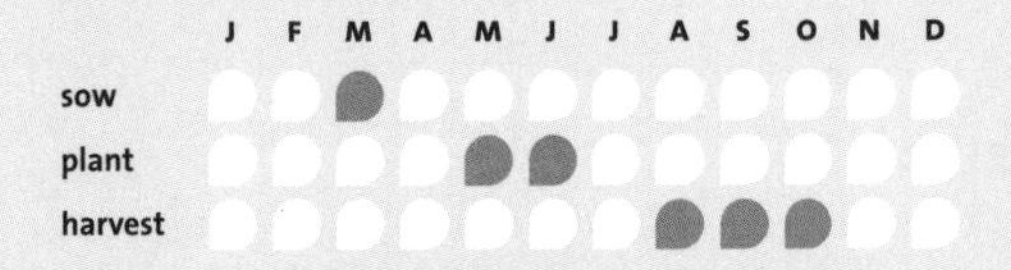

Celery is a surprisingly useful vegetable for cooks, since lots of recipes call for a stick of celery to be chopped in or added to stock, and it brings a crunchy texture to salads and is a great snack when accompanied by cheese. But it has to be said it is a bit of a fiddle and one of the most difficult crops to grow well. Commercially, celery is grown in the Fens, where the highly organic soil and damp conditions suit it down to the ground, but in gardens it takes a lot of effort and enormous watering to 'get right'.

This crop must be grown throughout its life without even the tiniest check, and if you don't spend much time on the veg patch or down the allotment, that's not easy. Don't let me put you off having a go, but if you aren't prepared to go the whole hog, then don't bother. Seed catalogues list several varieties divided into two groups: self-blanching and trenching types. My advice is stick to self-blanching varieties – old-fashioned trench types take much more work and have even more potential for going wrong. That said, I do like a stick of celery...

HOW TO GROW

Degree of difficulty: Decidedly challenging, needs veg-growing experience and a lot of attention to detail.

Sow: Early March on a windowsill indoors at 16–21°C (60–70°F). Prick out seedlings into small pots or multi-compartmented trays as soon as they are large enough to handle, and continue growing in warmth (13°C/55°F minimum) on a windowsill or in a heated propagator in a greenhouse. Keep well watered and in adequate light, but not strong direct sun. Harden off carefully before planting out.

Plant: Good soil preparation is essential – celery is only worth growing on rich, fertile soil that's had a lot of well-rotted organic matter worked into it beforehand. Plant when the last frost is safely past, in late May or early June.

Spacing: 23cm (9in) apart in all directions (such close spacing is vital since the plants need to shade each others' stems to help them blanch). You can grow them in blocks, with an edging of raised boards to prevent the stems of the plants at the edges being turned green by the light.

Routine care: Water in well after planting and keep plants well watered at all times – they must never go even slightly short of moisture, let alone actually dry out. Liquid feed regularly from planting time onwards, using a general-purpose or high-nitrogen feed, to keep plants 'pushing on'. Keep well weeded until plants cover the ground enough to shade out weeds for themselves, and take strict precautions against slugs at all times, starting several weeks before planting.

GROWING CELERY THE TRADITIONAL WAY

Dig a trench 45cm (18in) wide and 30cm (12in) deep where the celery is to be grown. Spread a thick layer of well-rotted garden compost or manure in the base and then return the soil to within 10cm (4in) of the top. Arrange the rest of the soil evenly in ridges down either side of the trench. Sow the seeds of trenching celery inside at 16°C (60°F) and prick them out into multi-compartmented trays of multi-purpose compost as soon as they are large enough to handle. Harden off and plant out in early June. Space the plants 15cm (6in) apart in a single row down the centre of the trench. Water thoroughly and never allow them to dry out subsequently (that does not mean you should drown them).

Remove sideshoots as soon as they are seen, and any weeds. When the plants are 30cm (12in) high, wrap corrugated cardboard or several layers of newspaper around the stems and tie it in place with string. The leaves should just stick out of the top. Scatter a general-purpose fertiliser around the crop and pull 7.5cm (3in) of soil up around the cardboard from the ridges at the side – a technique known as 'earthing up'. You will need two more earthings up at roughly three-week intervals, at which point the entire stem will be covered with soil and only the leaves will be

protruding from what is now a mound of soil. Harvest the stalks between October and February.

Traditional celery is hardier than the self-blanching sort, though straw spread over the foliage will offer welcome protection in really cold weather. And if you do all this and produce a fine crop? Have a drink, then go and lie down. You've earned it.

VARIETIES

'Victoria' – an F1 hybrid, and by far the most reliable and easy to grow successfully. Personally, I wouldn't bother with any other.

HARVEST

Start cutting the first celery when the plant is large enough to have 'sticks' virtually the same size as ones you buy in the shops. Self-blanching celery is ready from August to October. It's not hardy, though, so try to use it all before the first proper frost.

PROBLEMS

If small slugs can possibly get themselves in between the stems they will hole up inside, feeding on soft central stems. These then become scarred or, worse, the open wounds allow bacterial disease in, causing the internal stems to rot. Either way, affected celery is unuseable. Avoidance is the only answer; starting before the crop is planted.

Celery fly is another predator on celery; it has the appearance of leaf miner – blotches and ribbon-shaped tunnels in leaves. Pinch off and destroy any badly infected leaves. Carrot fly may also attack (see page 95).

Chicory

Cichorium intybus

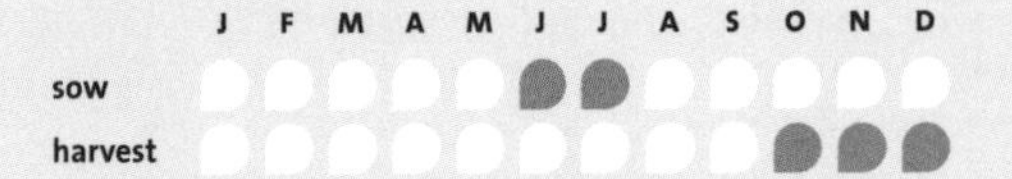

For most people, chicory means the slender, pointed, tightly packed yellow and white leaf vegetables sold in winter for braising or slicing up into salads, but there are several other types available, including radicchio (which people often don't realise is a type of chicory).

At home there are also others you can grow as winter veg under cover. None of them are, frankly, in the front rank of must-grow veg, but if you like varied and colourful winter salads or an unusual steamed or stewed leafy vegetable to go with cheese sauce, then do give them a go.

HOW TO GROW

Degree of difficulty: Fairly easy; hearting varieties don't need much work, but forcing varieties take a bit more time and effort.
Sow: June/July, where you want plants to grow – thin seedlings out, but don't transplant them.
Spacing: 30cm (12in) apart in both directions.
Routine care: Water in dry spells, and keep well weeded. Slugs and snails are less of a problem than they are with lettuce, as they dislike the bitter taste of chicory.

FORCING CHICORY

In November, cut the tops of witloof chicory plants down to 7.5cm (3in) above the ground, then dig up the roots and store them in buckets of nearly-dry sand or soil in a cool shed or garage.

Take 3–4 roots at a time to force, so as to stagger your crop; 'plant' them right end uppermost, in pots of moist compost, so the tops of the roots are barely buried below the surface. Keep in a warm, dark place – the cupboard under the stairs is ideal. Check occasionally and water lightly if need be, but don't overdo it.

After a few weeks you should see 'buds' pushing up. When these are 7.5–10cm (3–4in) tall, cut them off at the base to use – don't wait until they start to 'blow' or 'bolt'. The same roots will usually produce several more chicons over the next few weeks before they need replacing with a new batch from the shed.

If you have hearting chicory that fails to heart naturally and you need some for Christmas, it's also worth lifting a few roots to force in this way – it often works.

VARIETIES

1 **'Rossa di Treviso'** – a slightly different type of radicchio producing slender 7.5–10cm (3–4in) tall, flame-shaped hearts with few outer leaves. No forcing is needed.

2 **'Rossa di Verona'** – an Italian radicchio variety with the familiar, tightly packed, mahogany-red, solid, round, fist-sized hearts with few outer leaves. No forcing is needed.

3 **'Sugar Loaf'** – a green, leafy, upright heart with few outer leaves, something like a cos lettuce in shape, but with substantial leaves that have the typical, semi-bitter chicory 'bite'. Very reliable to grow and good in polytunnels, where mature heads stand for some time in winter without rotting or bolting.

4 **'Witloof Zoom'** – the standard variety for forcing; plants grow up tall and flower in summer, the roots are then dug up and forced in winter.

HARVEST

Hearting chicories are ready when they form a heart (no surprise there), which happens mostly from October to Christmas. Do not pull out any plants that don't do so on cue, since they may heart later – any time until the end of March, when they run to seed instead. Forcing chicory should be dug up in November and stored until later in the winter when you want to start forcing the roots to produce the familiar 'chicons'.

PROBLEMS

Hearting varieties failing to heart up may be due to their being sown too late or because of poor growing conditions that mean plants have not been able to make enough growth. Radicchio is not 100 per cent reliable, so bad luck comes into it too – the green 'Sugar Loaf', however, hearts virtually 100 per cent of the time, given fair growing conditions.

A common problem is that red varieties stay green. This is easily solved, since red radicchio starts off green and only changes colour towards autumn, when the nights start growing cold. If it hearts up but happens to stay green, you can still eat it.

Culinary tip – 'Neutralising' bitter chicory

Chicory does have a slightly bitter taste, which some people don't like very much. However, you can minimize it by cutting the hearts a few days before you need them and storing them in a cool, dark place, such as the salad drawer of the fridge. If you don't have time, slice the heart up thinly and leave the pieces to soak in ice-cold water for a couple of hours – this also makes them beautifully crisp.

Another alternative is to serve chicory with something very slightly sweet tasting – try braised witloof chicory with a rich creamy cheese sauce made without too much salt, or raw radicchio in a salad with 'Little Gem' lettuce hearts and tomatoes.

Chillies

Capsicum annuum

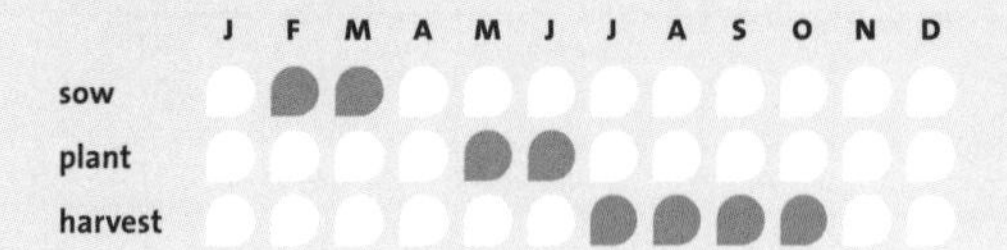

Chillies are the fiery-hot little brothers of the better-known sweet peppers. Their strength varies considerably between varieties, so read the small print in the seed catalogue or on the back of the packet before buying. Mild varieties have a rich, fruity flavour that spices up all sorts of food, especially casseroles, sauces and stews enriched with tomato. Mid-strength chillies are great for giving curries and chilli dishes their oomph, but to my mind the very hottest are for macho types intent on testing their tastebuds to destruction.

Chillies are an up and coming crop that's popular with foodies and they are decorative to grow on the patio (you'll sometimes find varieties that produce a mixture of coloured fruits, which are especially attractive), though for the biggest crop of ripe fruit grow the plants in a greenhouse or conservatory. There even are a few dwarf varieties suitable for a windowsill indoors.

HOW TO GROW

Degree of difficulty: Given mild, sheltered conditions they are easy to grow and little work.

Sow: February to March in warmth 21–24°C (70–75°F), prick out into 7.5cm (3in) pots when big enough to handle.

Plant: Under glass (unheated) from mid-May to mid-June; outside (warm sheltered spot essential) from mid- to late June – wait for a warm spell.

Spacing: Allow 30cm (12in) between compact varieties, 45cm (18in) between normal varieties and 60cm (2ft) between rows.

Routine care: Water sparingly. Avoid over-watering, especially in cool conditions and at the start of the season while plants are small – increase the supply gradually as they grow bigger and in warmer weather. Two to three weeks after planting, start feeding weekly with liquid tomato feed. Support plants by tying the main stems to canes pushed in alongside, since stems are brittle and easily broken. No pruning or trimming is needed.

EXTEND THE SEASON

In autumn, when the nights start turning cool, chilli plants stop growing, no new fruit forms and existing green fruit may not ripen. So move container-grown plants into a greenhouse or conservatory to extend the season, or pull the whole plant up, hang it upside down in a shed or not-too-hot kitchen to continue ripening, and let the chillies dry out naturally on the plant. When completely dry, pick and store in airtight jars.

VARIETIES

1 **'Apache'** – a compact variety that's ideal for pots, small containers and windowsills, with green/red medium-strength fruits.
2 **'Jalapeño'** – short, tapering, medium-strength green/red chillies with a characteristically slightly cracked skin. Often used green on pizzas.
3 **'Joe's Long Cayenne'** – very long (20–30cm/8–10in) slim, green chillies, ripening red with a rich, fruity yet fairly mild flavour. A huge crop per plant.
4 **'Thai Dragon'** – large crops of slender 7.5cm (3in) long green/red fruits. Extremely hot.

HARVEST

Pick green chillies as soon as they are large enough to use (usually from July to October), or leave them on the plants to reach full size and turn red, since this is when the full flavour and strength develops. To pick, use secateurs to snip off a whole chilli plus part of the short green stalk attaching it to the plant.

GROWING IN CONTAINERS

Grow one chilli plant per 20–30cm (8–12in) pot, or three in a large tub (45cm–18in plus).

STORAGE

Dry surplus red chillies. Grind whole chillies in an old coffee grinder to make your own strong chilli powder. Remove the seeds first for a milder mix or leave them in for maximum strength. Mild varieties make paprika, or use appropriately named varieties to make cayenne.

PROBLEMS

Greenfly are mostly seen soon after planting. Wipe them off by hand or use an organic spray as soon as you see them, or plants are badly set back.

A cold, dull or windy situation or a poor summer can mean the fruit fails to develop.

Antidote

If you taste an unwisely hot chilli, don't drink water to de-tingle your mouth – it just makes it worse. Dairy products are the best antidote. Yoghurt, cream cheese, a creamy dip, or a glass of full-cream milk works much better; even a piece of bread and butter or a mouthful of rice smothered in butter will soothe, due to the fat content.

Chinese cabbage

Brassica rapa Pekinensis Group

	J	F	M	A	M	J	J	A	S	O	N	D
sow	○	○	○	○	○	●	●	●	○	○	○	○
harvest	○	○	○	○	○	○	○	●	●	●	○	○

Chinese cabbage are tall plants rather like cos lettuce in shape, but with thick, fleshy, central veins down the middle of their thick, pale green leaves. They are not a regular sight in veg gardens, though they do appear in greengrocers. Consider Chinese cabbage an optional extra rather than a must-have veg, but it is fun to grow and handy for filling the gaps that appear in plots around midsummer once you start harvesting earlier crops. It is useful to have some 'on tap' for salads and stir fries during its shortish season, in late summer and early autumn.

HOW TO GROW

Degree of difficulty: Challenging. Needs regular care and attention to detail.
Sow: Mid-June to mid-August, then thin seedlings out – don't try to transplant them.
Spacing: 30cm (12in) apart in each direction.
Routine care: Keep very well watered and weed regularly and liquid feed using a general-purpose feed occasionally; it's essential to keep the plants growing steadily without a check. Slug control is vital.

HARVEST

Start cutting as soon as the first heads begin to form hearts; there's no need to wait for them to reach full size. Cut the biggest plants first and leave the rest to keep growing – they catch up fast. Most modern varieties 'stand' for quite some time without bolting once they are ready; even if a few of the outer leaves deteriorate a bit, the heart usually stays in perfect condition. Expect to be cutting from mid-August until early October.

VARIETIES

1 **'Green Rocket'** – tall, slender, vase-shaped heads, quick to mature.
2 **'Wong Bok'** – very large, oval heads, light green outside with lighter hearts.

PROBLEMS

The tender tasty flesh is very attractive to slugs and snails, and once they find their way inside the Chinese cabbage plants they just spend their time eating through the centre of the hearts, which completely spoils them. To keep them at bay, use a range of organic remedies, starting several weeks before sowing, which will reduce the population.

Plants may run to seed without forming proper hearts if the weather turns bad or growing conditions are poor, particularly if plants run short of water or are overrun by weeds; but the most common reason is sowing too early.

Courgettes, marrows and summer squashes

Cucurbita pepo

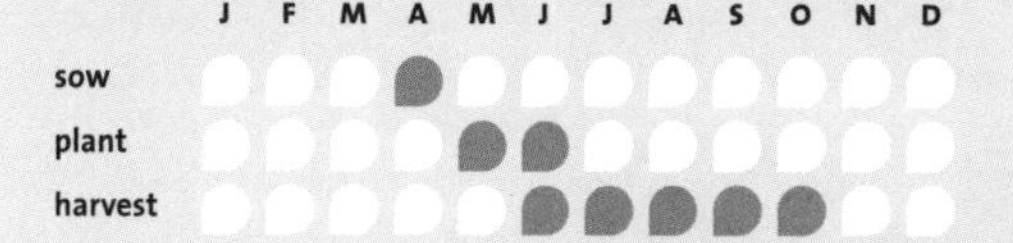

Courgettes have totally taken over from their big brother, the vegetable marrow – star of many a 1950s village produce show – and, thanks to the arrival of modern bush varieties, nature's original 'baby vegetables' are prolific and easy to grow.

One plant is enough for a couple or a small family, and though it's compact enough to fit in a corner of a veg patch, you could also grow it on top of the compost heap or in a tub on the patio to save space. The one big essential is to keep cutting your crop, even if you end up giving some way – otherwise instead of tender tasty courgettes you just end up with one or two big, fat marrows.

Grow and use summer squashes, which are to all practical purposes courgettes with peculiar shapes, in exactly the same way as courgettes.

HOW TO GROW

Degree of difficulty: Easy and little work.
Sow: Singly in small pots on a windowsill indoors mid- to late April. Harden young plants off well before planting.
Plant: Outside, shortly after the last frost, from late May to mid-June. (Courgette plants and their relatives are not hardy.) They need rich, fertile soil with plenty of organic matter. For very best results, prepare a pit filled with pure garden compost in winter and mound up soil over the top, then plant on top of this so plants have a rich, moist root-run.
Spacing: 60cm (2ft) apart in each direction. Trailing varieties, including summer squashes, need more room; allow 90cm–1.2m (3–4ft) in each direction.
Routine care: Water young plants carefully, but increase the quantity as they grow. For best results, apply liquid tomato feed every week while plants are carrying a crop.

GROWING IN ALLOTMENTS

Give the plants wider spacing when there is a lot of room but no ready supply of water – 60–90cm (2–3ft) apart with 90cm–1.2m (3–4ft) between rows. Lay black plastic sheet or an organic paper mulch down over the soil between the rows of plants to smother weeds and reduce routine hoeing, as well as to seal moisture in the ground.

Courgettes are a good crop for growing in walk-in polytunnels; they can be planted up to a month earlier than outdoors, but if a cold night is forecast early in the season, give them extra protection by covering them with fleece overnight. Since pollinating insects will not be able to reach the flowers when they are grown under cover, it's advisable to look out for a parthenocarpic variety, which will set fruit without being pollinated. You could also plant courgettes under cover in late August to extend the cropping into the autumn, until cold nights halt production.

VARIETIES

1 **'Clarion'** – unusual, pale green, courgettes with a very mild flavour, good for using raw cut into strips as crudités, grated or thinly sliced for veg salads, or for halving, de-seeding and stuffing.

2 **'Defender'** – an early, heavy-cropping variety producing high-quality, dark green courgettes. Plants have an open centre which makes it easy to find and pick the fruit. A very neat, compact, bush variety, resistant to the virus that often ruins courgette crops.

3 **'Gold Rush'** – bright gold courgettes with a very mild flavour, attractive for growing in tubs on a patio or in a decorative potager.

4 **'Parthenon'** – a parthenocarpic variety which reliably sets fruit without being pollinated even in dull, cool conditions – making it ideal for growing under cloches, fleece or in a greenhouse or tunnel for out-of-season production. Good-quality, dark green fruits. A compact bush variety.

5 **'Patty Pan'** – summer squash similar to the above and equally good quality, but in pale, eau-de-nil green.

6 **'Sunburst'** – a summer squash with small, tasty, golden, pie-shaped fruits with scalloped edges, picked at golf-ball size to use as for courgettes. Also good for stuffing whole.

7 **'Tiger Cross'** – if you are going to grow marrows, don't just let courgettes grow over-size, grow a proper marrow variety. This is a superb, compact, heavy-cropping and virus-resistant bush type, producing large crops of tender, verdant, light-and-dark green striped marrows, best picked at 15–25cm (6–9in) long.

HARVEST

Pick courgettes and summer squashes as soon as they are big enough to use, from 7.5cm (3in) to 15–20cm (6–8in) long, depending on how you prefer them; there's no need to wait until the flower has fallen off the end of the fruit. Cut individual fruits with a sharp knife, slicing carefully through the narrow stalk which attaches it to the plant. Take care not to slash the plant or damage other developing courgettes with the tip of the knife – it's easily done. Remove any rotting, misshapen or damaged courgettes at the same time, and discard. Expect to be picking from the same plants from June to October.

PROBLEMS

Newly planted young plants that fail to thrive and eventually die off are often found to have lost most of their roots when they are dug up. This is usually due to poor growing conditions and/or overwatering, but it could very occasionally be due to disease organisms in the manure or garden compost used before planting. To be on the safe side, start again with new plants in another spot.

Cucumber mosaic virus appears on affected plants as yellow-flecks at first, then leaves turn increasingly yellow-patterned and crinkled and plants become stunted and produce fewer new flowers and fruits, then eventually none at all. The disease is spread by aphids, but also by knives used for cutting crops, and is widespread on a range of garden plants so it is difficult to avoid other than by choosing resistant varieties, where possible. Affected plants are best pulled out and disposed of, but since courgettes are so fast growing it's worth putting in new seeds and starting again even in July/early August, especially if you can protect late crops so they continue into the autumn.

Powdery mildew is a small grey/white talcum-powder-like deposit on the foliage, which often appears in the autumn as nights start turning cool, especially following a dry summer. Plants usually continue cropping lightly. Try to improve vigour by liquid feeding and watering generously and remove the worst affected leaves by hand.

If fruits fail to develop it is almost always due to weather conditions – the warmer it is the less likely the problem. Hand pollination can help. The failure of female flowers to appear is usually also due to cold weather.

Culinary tip – Using courgettes in salad

If you have enough to spare, try finely grating raw courgettes to use as a salad. Mix in with French dressing, pine nuts and a few fresh, torn basil leaves (this is especially good with the mild-tasting gold or pale green varieties).

If you have a serious surplus late in the season, and also of other Mediterranean-type crops, make some ratatouille, which freezes far better than plain courgettes.

A bit of history

In wartime, vegetable marrows (which, ironically, are a fruit, since they contain seeds!) were often grown over the roofs of Anderson shelters in back gardens, and though they were never exciting vegetables, they were popular as they produced huge veg so easily at a time when food was short. They were hollowed out and stuffed, not with savoury mince, which would have been in short supply, but with breadcrumbs and leftovers flavoured with herbs. These marrows were served with gravy as 'mock goose', as a reminder of better times. Mature fruits at the end of the season were made into marrow and ginger jam, which was the closest you could get to marmalade, since oranges were largely unavailable and sugar was in short supply.

Marrows

The cultivation of marrows is exactly the same as for courgettes, but only two or three fruits should be grown on each plant. Many spurn them, but I love them. When the marrow is of a sufficient size, cut it off the plant, peel off the skin, slice it in half and remove the pippy centre before stuffing it with savoury mince. Wrap it in foil and cook in the oven until the flesh is tender. Serve with white sauce. Delish!

Cucumbers

Cucumis sativus

Greenhouse cucumbers

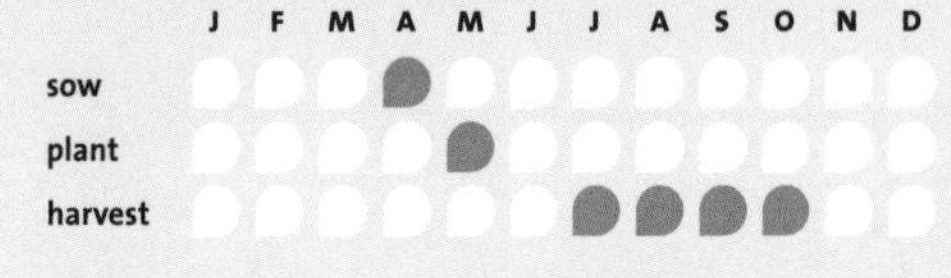

Outdoor cucumbers

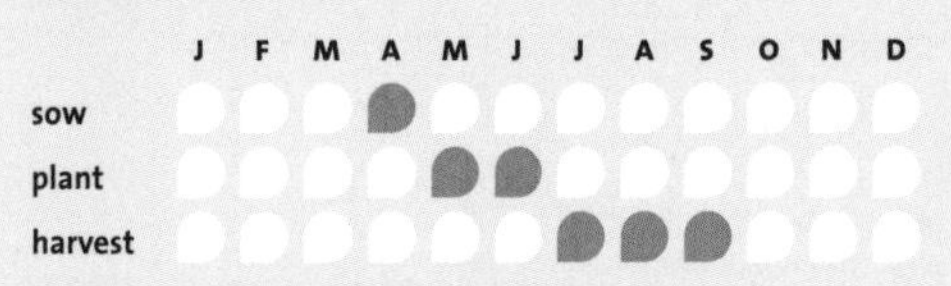

If you think cucumbers are uninteresting, flaccid, watery things, try growing your own; once you've tasted the difference you'll soon be converted. Whether you have a greenhouse or grow outdoor varieties down the garden, they are well worth it – and one plant is all you need.

HOW TO GROW

Degree of difficulty: Need time and attention to detail.

Sow: Seeds singly in small pots on a windowsill indoors in early April (for plants that will be grown in a cold greenhouse), or mid- to late April for outdoor plants.

Plant: In mid- to late May after hardening off the plants in a cold greenhouse. Outdoors late May to mid-June. Plants are not hardy and also hate cold, wet, windy conditions, so a warm, sunny, sheltered spot is essential.

Spacing: 45cm (18in) apart.

Routine care: Water sparingly at first and increase gradually as growth takes off – on hot days in summer when carrying a crop they may need large amounts of water. Feed regularly with general-purpose, liquid feed. It's essential to keep plants growing steadily without a check otherwise developing cucumbers may abort. Supports are needed, too, since this is a climbing crop. Outdoors you can allow cucumbers to scramble up trellis, over netting or over a wigwam of canes; plants grown in free-standing pots will need an obelisk to climb up. Some tying in will be needed, though the plants hang on to some extent with their own tendrils. In a greenhouse, fix netting in front of the glass or give each plant its own cane to grow up – it will then need regular tying in and training.

Training and pruning: Under glass, I'd recommend growing all-female hybrid varieties. Training is easy. Tie the main stem carefully to a vertical bamboo cane as it grows taller. Remove all sideshoots from the bottom 60cm (2ft) of the plant, then allow subsequent sideshoots to develop. By the time each of these has grown about 15–20cm (6–8in) long, you should see a baby cucumber with a flower at the tip. Nip out the end of the shoot, one or two leaves beyond the developing fruit. (This diverts all the plant's efforts into growing cucumbers and prevents the shoots getting too big and the plants becoming tangled together.) You'll need to check over greenhouse cucumber plants twice a week – nip out the tips of sideshoots and tie up the main stems to keep them under control. They grow very fast.

If growing varieties other than all-female types, you'll need to prevent the female flowers being pollinated by the males, otherwise the fruit becomes bitter, full of hard seeds and inedible. Under glass, put insect-proof screens over the ventilators to keep out pollinating insects. Outdoors, remove all of the male flowers every few days as new ones grow (they are easily recognised as they don't have a baby cucumber growing at the back of them).

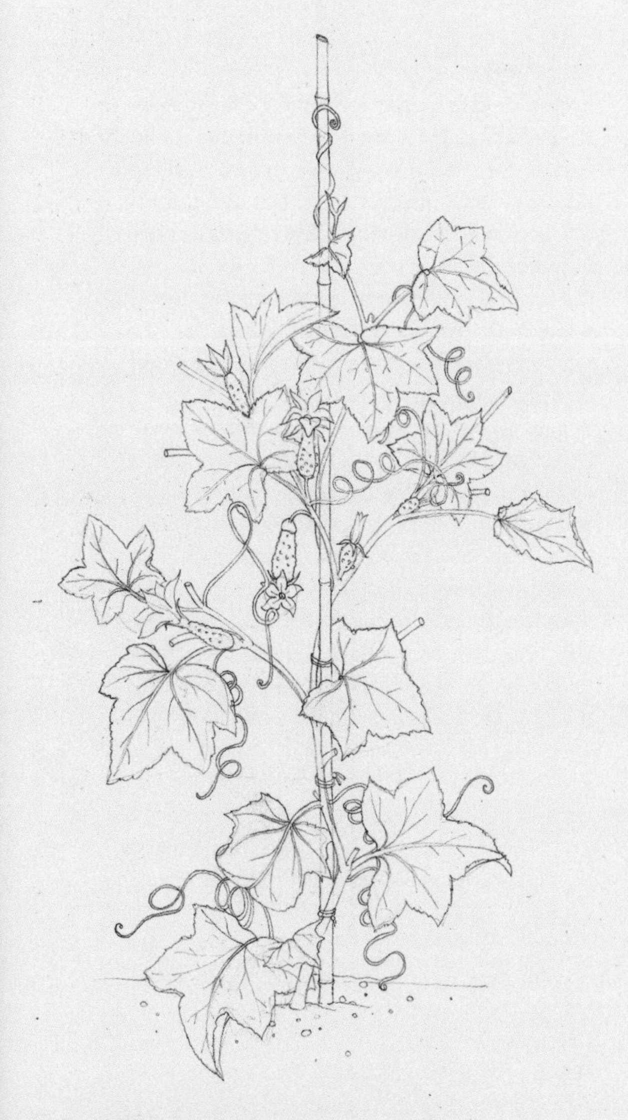

To train cucumbers, tie the main stem to a cane as it grows and let the top sideshoots develop from 60cm (2ft) up. Keep nipping out the ends of the shoots one or two leaves beyond each developing fruit.

INDOOR AND OUTDOOR CUCUMBERS

Outdoor cucumbers have improved a lot in the last 20 years; they are no longer just the short, fat, prickly ridge cucumbers that were the only sort our grandfathers could get. You can still buy seeds of those if you want, but modern varieties are just like normal cucumbers, long, green and smooth.

Outdoor cucumbers don't *have* to be grown outdoors, they can also be grown in a greenhouse or walk-in polytunnel, and in a poor summer they will perform a lot better under cover than outdoors. If you grow them inside in a good year they'll do incredibly well.

Greenhouse varieties, though, really *must* be grown under cover. Avoid varieties intended for growing in heated greenhouses, as they won't like cool nights without any background heat and, frankly, it's not worth going to the expense of heating. But do go for all-female F1 varieties as they do so much better and are less effort. There's very little to choose taste-wise between different varieties, so if you don't grow your own plants from seed you'll do just as well with unnamed plants from a garden centre.

A good recent development is mini-cucumbers, which are only about 15cm (6in) long, and handy as you regularly get lots of small fruits instead of a few very long ones occasionally. New varieties can be found in the seed catalogues; the range on offer changes almost every year.

Culinary tip – Versatile cucumbers

Vegetable juice: Add cucumber to other veg in a juicer; it's especially good with tomato and carrot. Peel green cucumbers thinly first (using a potato peeler), otherwise the green skin turns red tomato juice an unappetizing khaki colour.

Cucumber raita: Grate raw cucumber, add natural Greek yoghurt and flavour with garlic or chopped spring onions. This is good with curries or as a simple summer salad.

Cosmetic use: Plain cucumber juice chilled in the fridge makes a refreshing summer facial toner, or soak into cotton-wool pads and rest on the eyelids while relaxing. Great for busy gardening writers and indeed anyone else who spends hours in front of a computer screen!

VARIETIES

1 **'Crystal Lemon'** – an outdoor variety with small, round, yellow, lemon-sized fruits and an incredible flavour. Plants are very prolific, so one is enough.

2 **'Long White'** – an outdoor variety with short, cream-skinned fruits and a mild flavour; good for people who find most cucumbers indigestible. Also ideal for anyone with a juicer, since you can push a whole fruit through the machine without peeling it first, producing a neutral-coloured juice that won't alter the colour of the finished product if it's mixed with tomato juice.

3 **'Flamingo'** – an F1 hybrid producing slender, long fruits on strong plants. An all-female variety for greenhouse growing.

4 **'Carmen'** – very productive and high-quality, all-female F1 hybrid greenhouse variety. Resistant to powdery mildew and several other diseases.

HARVEST

The cucumbers are ready as soon as their shape fills out all the way along and they look big enough. Use secateurs to snip through the narrow stem connecting the cucumber to the plant – don't pull or twist them off or you'll drag the plant down or break off long lengths of stem that should carry the next lot of cucumbers.

PROBLEMS

Powdery mildew may appear at the end of the season. Keep plants going by regular watering and feeding and remove the worst affected leaves. It's worth choosing disease-resistant new varieties when buying seeds.

Cucumber mosaic virus is less often seen on cucumbers than on courgettes (see page 109), but it can happen, so choose virus-resistant varieties where possible. Destroy affected plants.

Young plants are particularly prone to root rot caused by overwatering or organisms in the manure or garden compost used for improving the soil prior to planting, so water sparingly until plants take off and the weather is warmer. In hot, sunny weather, and especially under glass, large plants may need a great deal of water when carrying crops.

Red spider mite affects crops under glass, spinning minute webs and sucking the sap until leaves become bleached and straw-coloured. A damp atmosphere helps to discourage attacks and biological control using the predator *Phytosieulus persimilis* is also effective.

Dandelions

Taraxacum officinale

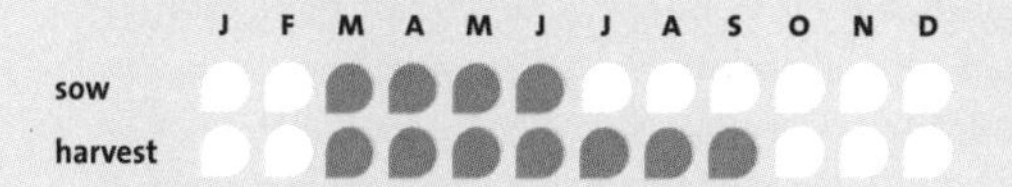

If you've spent a lifetime trying to eliminate dandelions from the lawn, let me persuade you to try cultivating one or two plants – they make excellent leaves for a mixed salad with lettuce and sprouting seeds and are easily 'forced' out of season when there's very little fresh salad stuff around. Dandelions are tasty and packed with vitamins and minerals. Don't risk picking wild leaves from the countryside, though, as they may have been sprayed, or affected by traffic fumes or other pollution, and cultivated plants will be more tender.

HOW TO GROW

Degree of difficulty: Easy, and virtually no work.
Sow: Use seeds of cultivated dandelions or some you've saved from plants that set seed in the garden. Sow thinly in shallow drills where you want the plants to crop any time from early spring to early summer, then thin out the seedlings when they are big enough to handle.
Plant: As an alternative to sowing seeds, carefully dig up small dandelion seedlings you may find when weeding the garden and transplant them into a row in your salad bed. (Yes, I know it goes against the grain, but you can get your own back when you eat them.)
Spacing: 15cm (6in) apart.
Routine care: Water in the seedlings until they are well established; water and liquid feed regularly to encourage strong, tender growth. Remove flower heads regularly to keep the plants concentrating on leaf production and so they don't set seed – which they quickly will – or you'll have a weed problem. Cut plants back to within a couple of centimetres of the ground periodically through the summer to remove tough old leaves and encourage a flush of fresh young growth.

BLANCHING DANDELIONS

Plants can also be 'blanched' to provide large, tender, pale-coloured leaves. Sit a large, upturned flowerpot, with the holes blocked, over a well-established plant after removing any emerging flower stalks. Take anti-slug precautions and check inside the pot daily, removing snails and slugs which congregate in the cool, shady confines and will spoil your salad leaves. After a week, remove the pot and cut the younger blanched leaves. The same plant can be blanched again as often as you like, but just let it recover for a few weeks between times. For out-of-season leaves in winter, dig up a few roots, pot them up and 'force' them in a warm cupboard indoors in much the same way as chicory.

VARIETIES

'Pissenlit a Coeur' – the only cultivated variety I've met, only available from organic suppliers' catalogues. This stays much leafier than wild dandelions (at least, in the first year) so you have more salad and fewer flowers.

HARVEST

Choose small leaves from the very centre of the rosette for salads. You can pick from March to September, though they are at their best in spring when plants are growing most vigorously and before they start flowering in earnest.

Edible flowers

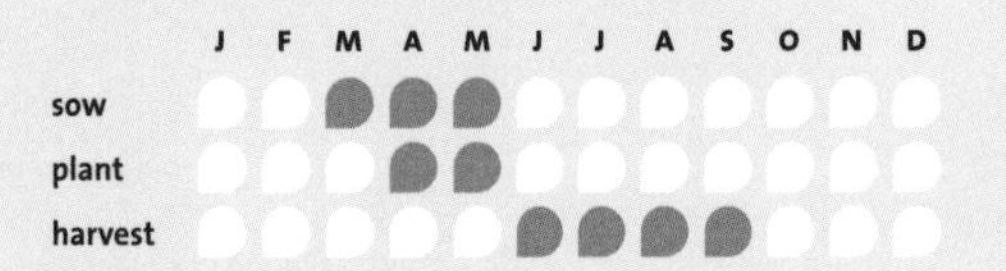

Edible flowers are a good way of introducing cheery touches of colour to a salad patch, and they're invaluable for sprucing up a potted veg garden on a patio or a decorative potager. In the kitchen they are very useful for geeing up salads, and even if you don't actually eat them they make great creative garnishes or buffet table decorations when you're pushing the boat out for 'foodies'. Scatter flowers or petals over a pine table or a plain white cloth.

HOW TO GROW

Degree of difficulty: Quick and easy to look after.
Sow: Mixed or individual varieties in spring, from March to May, either in pots indoors or in rows where you want the plants to grow.
Plant: Late April or May. If you don't grow your own from seed, you can usually buy young plants of the individual varieties of your choice from a herb farm or nursery.
Spacing: 10–15cm (4–6in) apart.
Routine care: Water carefully and use liquid tomato feed once a month when growing in containers. Remove dead heads regularly so that plants keep producing new flower buds.

Recipe – Rose petal jelly

Rose petals are good for scattering on tables, or used as garnishes for fruit salads or on the dessert table at a buffet, but they also make a very delicate-flavoured and delicious jelly for polite afternoon tea parties. What's more, they are something you can gather in fair quantities when you grow roses in the garden. Towards the end of the first flush of the season, roughly mid-June, is probably the best time to attempt this.

You need: 85g (3½oz) rose petals (all red, or a mixture of red and deep pink are probably the best colour for this job), one lemon and 1.2 kg (2lb 6oz) preserving sugar. Bring the petals to the boil in just under ½ litre (1 pint) of water then, when they are cooked and feel soft when prodded with a fork, add the sugar and lemon juice. Boil up the mixture three times until it starts to set when a blob is put on a cold plate. It shouldn't set too firm – it's meant to be rather runny. Make small quantities and keep in the fridge.

VARIETIES

1 **Borage** (*Borago officinalis*) – blue flowers with 'beaks'; a very attractive, all-white variety 'Alba' is also sometimes available. Use whole individual flowers after removing the stalks.

2 **Calendula or pot marigold** (*Calendula officinalis*) – available in red, orange, cream, yellow or nearly white varieties, as well as the traditional orange. Use only the petals, which you'll need to pull off the hard green calyx.

3 **Chives** (*Allium schoenoprasum*) – whole spherical chive heads can be used for decoration, but remove petals to use on their own in salads. They have a mild onion taste.

4 **Heartsease and violas** (*Viola tricolor* and *Viola* species) – wild heartsease and very small-flowered varieties of viola can be used as edible flowers; use whole heads without their stalks. Larger, cultivated pansy flowers can be used for decoration, but they are a bit too bulky for eating.

5 **Nasturtium** (*Tropaeolum majus*) – often recommended for use as edible flowers, but frankly they are a tad too peppery so I'd use them for decoration instead.

6 **Pinks** (*Dianthus* species) – remove individual petals, otherwise use whole heads for decoration only.

7 **Polyanthus** (*Primula* hybrids) Pull the flowers away from the calyces before sprinkling over salads. They have a nutty flavour.

HARVEST

Pick fresh, healthy, unblemished flowers shortly after they have opened; don't use old ones that are starting to go over. Snip them off neatly just beneath the flower head with scissors or snips. If you aren't using them straight away, lay them in shallow water in a cool place out of direct sun to keep them fresh.

PROBLEMS

Aphids and other pests can really ruin the look of these on a plate, so use organic remedies or else look flowers over carefully before picking to avoid 'infested' ones. A quick rinse under the tap will usually freshen them up.

Endive

Cichorium endivia

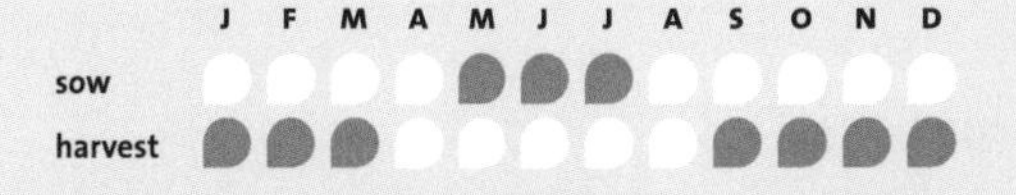

Endive is an unusual vegetable, certainly, and if you see it in greengrocers' shops it looks like a green, frilly mophead. In both flavour and use it's very much like chicory, except that the leaves are loose and shaggy instead of forming nice neat hearts. It's not to everyone's taste, and even if you do like it, you don't need a lot of it, but it can be quite useful when you are trying to extend the home-grown veg season through the winter – and a sandwich stuffed with endive and cheese can be mouthwatering.

HOW TO GROW

Degree of difficulty: Needs some attention to detail but not a lot of time.

Sow: In May, June and July, according to the variety (check the small print on the seed packet). Sow thinly in shallow drills where you want the plants to grow; they don't transplant well.

Spacing: Thin seedlings out in several stages so the plants end up 30cm (12in) apart. Allow 30cm (12in) between rows.

Routine care: Water sparingly at first, but increase the supply slightly as plants mature to keep them growing steadily. Plants remaining in the ground as winter approaches won't need watering, since natural rainfall should be adequate.

BLANCHING

The plants do not form solid hearts like chicory, but by the time they are three months old they should have a dense head of shaggy, loose, frilly foliage making the classic mop-head shape. Choose a few of the biggest plants to blanch. Stand a heavy bucket or large clay flowerpot upside down over the top, making sure that drainage holes through which light could enter are bunged up. Leave the pot in place for 3 weeks. Alternatively, with very large endive plants, simply bunch up the outer leaves round the heart and use raffia or soft string to tie them in place – you won't have such a pale yellow heart, but it's easier and there are fewer problems with worms or slugs getting in. When one batch is ready, start to blanch a few more.

VARIETIES

'Moss Curled' – known in France as 'frisée'. Large heads of finely divided, narrow, branching, curly leaves, which are tender and tasty when blanched. This is more to British tastes than the endive varieties with broad ribbon-like leaves often found on the continent.

HARVEST

When the blanching time is up, cut the whole plant by sliding a long knife underneath it to sever the stem just above the ground, and trim off the tough outer leaves. The soft, tender, yellowish leaves at the centre of the rosette are the part that's used. By staggering sowings you can eat endive from September until early spring.

Unblanched plants will usually 'stand' quite well until late February, if the weather stays kind, but you can help by covering them with cloches or fleece for protection until you are ready to blanch them.

PROBLEMS

Slugs and snails like to hide inside blanching pots, where they ruin the tender heads underneath; check regularly and remove them, but also take anti-slug and snail precautions throughout the growing season around the endive plants to reduce the population.

Culinary tip – What to do with endive

Use tender, blanched endive leaves raw in green salads, after detaching them from the head – they are best mixed with lettuce to dilute what is still a slightly bitter taste. Or you can braise whole heads and serve them in the same way as braised chicory, with cheese sauce, or you can adapt any recipe for cooked chicory (see page 103).

Florence fennel

Foeniculum vulgare var. *azoricum*

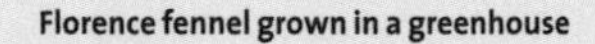
Florence fennel grown in a greenhouse

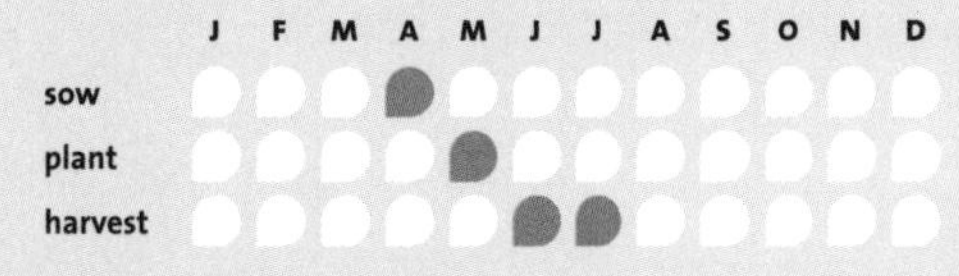

Florence fennel grown outdoors

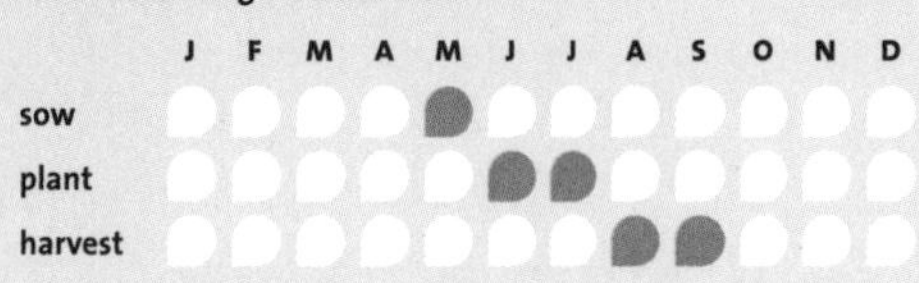

Not to be confused with the herb fennel, Florence fennel is the one with the swollen, white, bulbous base that's something like an aniseed-flavoured celery heart. It is used raw in salads or cooked as a vegetable (try it with roasted veg in olive oil in the oven). Again, it's an unusual crop that's fun to grow in small quantities and which is most reliable when grown under cover, unless it's a very good summer.

HOW TO GROW

Degree of difficulty: Challenging and needs attention to detail but not much time.

Sow: On a windowsill indoors in April for crops to be grown under cover, and May for outdoor crops. Sow three seeds each in small pots and weed out the weakest two (if they all come up), to leave one strong, undisturbed seedling.

Plant: After hardening off carefully, plant in the soil border of a greenhouse or walk-in polytunnel in May, or outside in a sunny, sheltered spot at the end of June or early July once the weather has settled down to a prolonged warm spell. You could try putting one or two plants in a large tub on a sheltered patio; they are very decorative with feathery deep green foliage.

Spacing: 23–30cm (9–12in) apart in each direction.

Routine care: Water carefully – they need little water at first but must not dry out. As they grow and the weather warms up, increase watering and ensure the soil stays moist at all times. Feed with a general-purpose feed. It's essential to grow plants fast and without a check or the 'bulbs' will be too tough and the plants will bolt prematurely. Cover outdoor-grown plants with fleece at night if it's chilly, even in June.

VARIETIES

There's nothing between varieties, flavour-wise, but F1 hybrids are the most reliable to grow. Seed firms offer much variety, which changes from year to year. A good one is 'Victorio', which is a fast-growing F1 hybrid variety with short, squat 'bulbs'.

HARVEST

The 'bulbs' (they are really swollen leaf bases) can be cut as soon as they are big enough to use – start cutting when they are the size of a tennis ball. Don't delay until they are as big as ones you get in the shops, or by the time you've used one or two the rest will have bolted. Plants planted in May under cover will be ready mid-June to mid-July. Plants put outdoors in late June or early July will be ready late August or September. To harvest, pull the whole plant up, trim off the root and lower leaves and cut the foliage back to a few centimetres above the top of the 'bulb'.

PROBLEMS

Bolting is the main one; any little problem sets them off and once this happens they are useless for eating, although you can still use the leaves in the same way as the herb fennel. Dry soil, poor growing conditions, lack of organic matter, old age, sudden swings in temperature…if you do pull it off, you've done well.

Culinary tip – Using florence fennel

Salads: Strip off the tough, very outer layers (you can braise these). Slice the heart thinly for salads; fennel is great with celery, spring onions and slices of apple or avocado in a lemon juice and olive oil dressing. Or if you have lots of 'bulbs' to use, braise or roast halved hearts and serve with parmesan shavings and good olive oil as a light lunch.

Garlic *Allium sativum*

	J	F	M	A	M	J	J	A	S	O	N	D
plant	○	●	●	○	○	○	○	○	○	○	●	○
harvest	○	○	○	○	○	●	●	●	○	○	○	○

A member of the onion and shallot tribe, garlic is as easy to cultivate as growing onions from sets, and the more you grow the more you use. (Which is great for your health if modern medical opinion is to be believed, even if it's not so clever where your social life is concerned!) But don't grow garlic from bulbs you've bought from the greengrocer's; get named varieties from a garden centre or seed firm, using virus-free cloves if at all possible, and start with new stock each season – don't save your own.

HOW TO GROW

Degree of difficulty: Easy, and doesn't take up much time.

Plant: Whole garlic cloves ideally in November or, as second best, February/March. Autumn-planted garlic is ready earlier and usually makes larger 'heads', since it has a longer growing season. To plant, separate the cloves from the head of garlic (if whole heads are delivered), choose the biggest for planting and discard the tiny central few. Push the cloves into the soil upright, so the tip is a couple of centimetres below the surface.

Spacing: 15cm (6in) apart in each direction.

Routine care: Check occasionally and push back any cloves the birds have pulled up; water both autumn- and spring-planted garlic in dry spells in summer.

VARIETIES

1 **Elephant garlic** – not a true garlic at all (it's more closely related to leeks) with a thick stem and a huge underground 'head' up to 10cm (4in) across, which is made up of a very few enormous cloves. The flavour is mild; brilliant for roasting and good for anyone a tad dubious about garlic. The big drawback is the price of the cloves for planting, which are also often in short supply, so order early. Use within four months.

Several Isle of Wight strains are available (garlic grows well there) and there's little to choose between them for flavour, though some are slightly stronger than others.

2 **'Lautrec Wight'** – pink cloves with a white skin and stores well until March.

3 **'Purple Wight'** – a good early variety, ready for use 'wet' from mid-June, with attractive purple-tinged cloves. Use over the summer; it's not a keeper.

4 **'Solent Wight'** – one of the most suitable for the British climate and is ready from mid-July.

5 **'Purple Modovan'** – a striking vintage variety with mauve-tinged skins, and very pungent aroma. Use within four months.

HARVEST

Real garlic lovers can't wait until the harvest in late summer, so they pull garlic from June onwards – as soon as reasonable-sized heads have formed underground but the plants are still green and leafy. 'Wet' garlic, as this is known, will not keep for long, so only pull what you need straight away and use it within a few days. Leave the majority of the crop in the ground; in July (autumn-planted) or August (spring-planted) the foliage turns yellow as it starts to dry off naturally. When it has turned brown and raffia-like, dig up the plants and let them finish drying off on the ground in the sun

STORING

Once dry, store in a cool, dry place under cover. Some varieties don't keep well – about four months is top whack – but long-keeping kinds can be kept through the winter until March or April before they go soft or start sprouting. Keep them in the house, which is warmer, rather than a cooler shed or garage where they can start growing again more readily.

Antidote

Garlic does tend to linger like an aura for some time after you've eaten it, so if you enjoy garlic but don't want to run the risk of anti-social breath, chew a little fresh parsley to neutralize the fumes.

PROBLEMS

Rusty-red spots on the foliage are a sign of rust disease. In bad attacks the leaves turn yellow and die prematurely, resulting in a shorter than usual growing season and smaller 'heads' of garlic. It's often worse on poorly drained ground or where a lot of nitrogen fertiliser has been used, but particularly in soil where members of the onion family have been grown before, or debris from previously infected onions, leeks, etc, has been left behind or composted and returned to it. There is no cure, but try to prevent it by improving cultivation techniques. Destroy affected foliage and move onions, leeks and garlic to a new site every year. Don't grow garlic and its relatives on the same ground for at least 4–5 years afterwards, to reduce the risk of more outbreaks.

As with onions and leeks, bolting sometimes just happens. It's thought to be brought on by certain sequences of temperature; something that's outside our control. If some plants bolt, use those first, but it's nothing to worry about; some varieties bolt easily but it doesn't affect their ability to produce a good bulb, as the flower spike comes out through the centre of a group of perfectly good cloves. In the USA garlic flower stalks, or 'scapes', are simply sliced up and used as extra garlic. Give it a try!

Make a garlic plait

Choose a long-keeping variety and, when it's harvested, leave the dead leaves attached to the heads. Plait them together, starting with three heads at the base (1) and adding in another layer of three with their own dead leaves in turn (2). This will extend the plait so it's solid garlic from bottom to top. Finish off by looping the foliage over (3) and fixing it with wire or strong string, so you have a loop to hang it up on a hook. It looks good in the kitchen, but if the atmosphere is steamy the garlic won't keep as long as it would in a drier atmosphere. It makes a good decoration for the conservatory, once the smell has worn off!

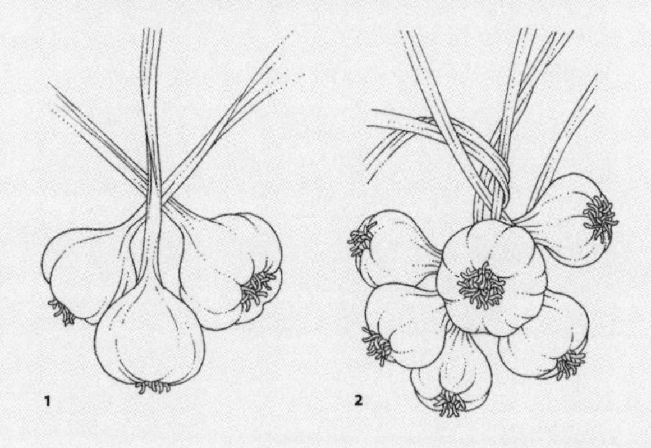

Culinary tip: Using garlic

Garlic tends to be regarded as a herb or a flavouring, but if you've grown large quantities, try it as a cooked vegetable.

Roast garlic: Large individual cloves of garlic, such as 'elephant garlic', are brilliant roasted in olive oil in a shallow dish alone, with other vegetables or around a joint of meat.

Garlic used as a veg: Drench a whole head of good-sized garlic (in its skin) with olive oil and roast until soft, so it pops out of its skin when squeezed between your fingers. Serve warm in butter as a separate vegetable, add to beans such as haricots for a hot or cold bean salad, or add cooked and skinned cloves to a vegetable stew such as ratatouille.

Chicken stuffed with garlic: Part-cook two handfuls of peeled garlic cloves by sweating them in a little butter or olive oil, then stuff them into a chicken. Roast as usual and use the garlic as an accompanying vegetable, or add it whole to the gravy to make an intensely garlicky sauce.

Kale

Brassica oleracea Acephala Group

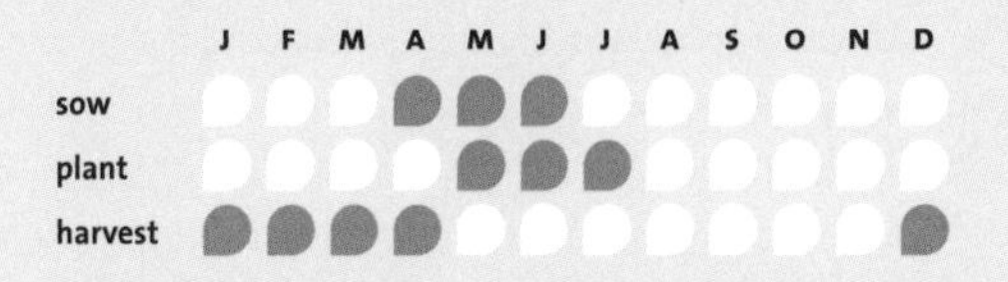

Kale was always considered the old cottage gardener's standby; it was one of the few veg that could be relied on to withstand a harsh winter because it could be picked when there was virtually no other fresh veg in the garden. When the advent of the freezer made summer veg available all year round, kale was largely abandoned. But just lately it has made a mighty comeback as a 'foodie favourite' – look for the less rugged but tastier varieties.

HOW TO GROW

Degree of difficulty: Easy and needs little time.
Sow: In pots or sedtrays in April/May/June.
Plant: Into beds in May/June/July.
Spacing: 45cm (18in) apart in each direction.
Routine care: Plant in firm ground and water in; keep watered in dry spells through the summer.

VARIETIES

1 **'Black Tuscany'** ('Nero di Toscana') – this is the kale made famous by the River Café some years ago, which was responsible for bringing kale back to the nation's attention. Rather upright plants (can be spaced 30cm/12in apart) with long, narrow, deep green leaves with a bobbly texture and fab flavour. It can be picked from August to mid-March.
2 **'Pentland Brig'** – probably the most tender and best-flavoured curly kale; pick leaves as usual in winter, then in spring plants produce succulent shoots that are delicious used as broccoli spears.
3 **'Redbor'** – deep bronzy-purple curly kale with crinkly leaves; tender young leaves are sometimes torn up in green salads in summer, then picked as usual all winter. It's good to eat and popular with flower arrangers!

HARVEST

You can start picking a few leaves as soon as they are big enough to make it worthwhile, but don't take too many at once. It's more common to wait until winter when the crop is fully grown. April-sown kale is ready just before Christmas; May- and June-sown plants follow on afterwards. The original plants make a new spurt of growth in spring and produce a quick crop of succulent young leaves before running to seed, so don't be in too much of a hurry to pull them out.

Culinary tip – Cooking kale

Today it's 'not done' to boil veg to death, and kale in particular responds to gentle treatment. Try steaming it until tender instead, or use it in stir fries – they don't have to be Chinese, you can make a medley of stir-fried winter veg instead, such as celeriac, carrot and onion, then add 5cm (2in) squares of kale just before everything else is cooked and finish with a little chicken stock, so it reduces and makes a nice gravy.

PROBLEMS

Caterpillars are less of a problem with kale than on many brassica crops, but watch out anyway, since an infestation will spoil the leaves. Remove any you see by hand or use an organic pesticide and keep the plants covered with fine insect-proof mesh if it's a real problem.

Clubroot produces a large bulbous base at the bottom of each plant and causes stunted growth, and in some cases plants will often die off before they are able to produce a crop. The disease is best avoided by raising your own plants instead of buying them in, and by growing brassica crops on a new piece of land instead of using the same patch all the time. If you are unlucky enough to get clubroot, there's no cure – all you can do is avoid replanting brassicas on that spot for 20 years and destroy all affected plants, including stalks and roots. Don't put them on the compost heap, though, or you will just spread clubroot as you spread the compost. This disease is prevalent on poorly drained, acid soil but is unlikely to be a problem on chalky soil and well-drained earth.

Kohl rabi

Brassica oleracea Gongylodes Group

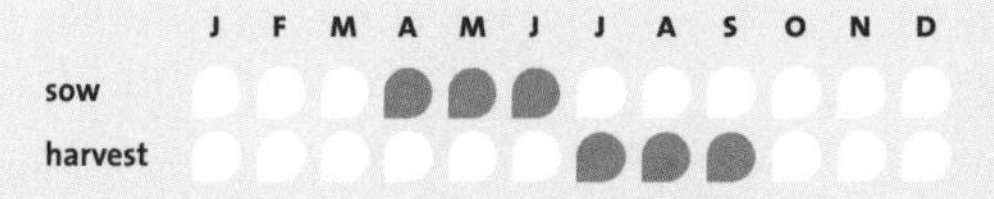

One of the most extraordinary-looking vegetables around (it always makes me think of a tennis ball with leaves), kohl rabi has never really caught on in Britain, so it's rarely seen in shops here, though it's more popular *sur la continent*. But if you are a confirmed veg fan, or if you simply fancy trying something 'different', give it a go – you'll be surprised just how good it is. Used raw it tastes rather like a cross between celery and apple, while when cooked it's more like a very mild cabbage/turnip cross – but better. Two things you must do are grow it fast and pick it small – otherwise you might just as well be eating a tennis ball.

HOW TO GROW

Degree of difficulty: Needs some attention to detail, but doesn't take much time.
Sow: April–end of June where you want plants to grow, then thin out.
Spacing: 15cm (6in) apart in each direction.
Routine care: Good, fertile soil with plenty of organic matter is essential; keep well watered throughout the life of the crop – it's vital it never has a check.

GROWING AS BABY VEG UNDER COVER

Since kohl rabi takes up little room and it does best in very good conditions, it's worth sowing a row in a greenhouse border, cold frame or polytunnel in spring, around early April, and again in August. Sow seeds slightly thicker than usual and thin out to 10cm (4in) apart to produce a fast crop of early and late 'baby veg'. They are delicious grated raw in coleslaw or very thinly sliced for fresh 'vegetable crisps'. Or top and tail, steam or boil them whole to eat with a sauce.

VARIETIES

1 **'Blusta'** – vivid purple globes with matching stems and richly coloured leaves. It has an excellent flavour and remains tender without turning woody.
2 **'Logo'** – a readily available variety with white globes. Fast-growing and slow to bolt.
3 **'Supershmelze'** – an unusual variety that needs hunting out. It is capable of growing to almost the size of a child's head whilst remaining tender and without bolting or splitting. Green globes with crisp white flesh.

HARVEST

Start pulling up whole plants when the first 'globes' reach golf-ball size (these are best for using raw); use the rest of the crop before they reach tennis-ball-size. The season is from July to late September. To prepare kohl rabi for the kitchen, top and tail then peel thinly.

PROBLEMS

As a member of the cabbage family, kohl rabi can suffer from clubroot, but this is rare since it grows so fast. The thick leaves don't attract caterpillars either.

A woody texture, splitting or bolting usually suggests poor growing conditions or that the plants ran short of water, causing a check to growth.

Leeks

Allium porrum

An invaluable winter veg that's hard to do without – a couple of rows down the garden is as good as an outdoor larder, and leeks are a great allotment crop. By growing two or three different varieties you can have continuous supplies right through autumn, winter and early spring, but the latest trend is to grow slim pencil leeks in your salad bed as baby veg, to use in summer.

HOW TO GROW

Degree of difficulty: Easy and little work.
Sow: Mid-March–April and thin out to 2.5cm (1in) apart.
Plant: Transplant strong 15–20cm (6–8in) high seedlings in June.
Spacing: 15cm (6in) apart; leave 30cm (12in) between rows.
Routine care: Plant into fairly firm soil, make an individual hole 7.5–10cm (3–4in) deep for each seedling with a big dibber or a short, thick piece of cane and drop a seedling in. When you've planted the whole row, water them in well so that the planting holes fill with water – don't firm soil round the roots as you normally would when putting in veg plants.

GROWING BABY LEEKS

It's essential to use the right varieties for this job; the favourite is 'King Richard', which gives a naturally early crop of long, slender leeks. Prepare the ground well – in the same way as you would for a seedbed – in a sunny, sheltered place. Sow thinly, making several sowings at three-weekly intervals from mid-March to late April, with another batch in early June. Thin the seedlings out to 2.5cm (1in) apart, but don't transplant. Keep them well watered and use general-purpose liquid feed to keep them growing steadily without a check. Start pulling baby leeks when they are barely more than pencil-thick and resemble spring onions. You'll be using them from June until September. If you have space in a walk-in polytunnel, it's also worth making a sowing in early March to give you a late May crop of baby leeks.

VARIETIES

1 **'Apollo'** – a new F1 hybrid variety. High quality and with good rust resistance. Good for winter use.
2 **'Neptune'** – a new, high-yielding, rust-resistant midwinter variety.
3 **'King Richard'** – a good early leek for use from September to Christmas, with long, slender stems. The best for growing as baby 'pencil' leeks.
4 **'Musselburgh'** – a really reliable old faithful that stands hard winters. Ready December to March.

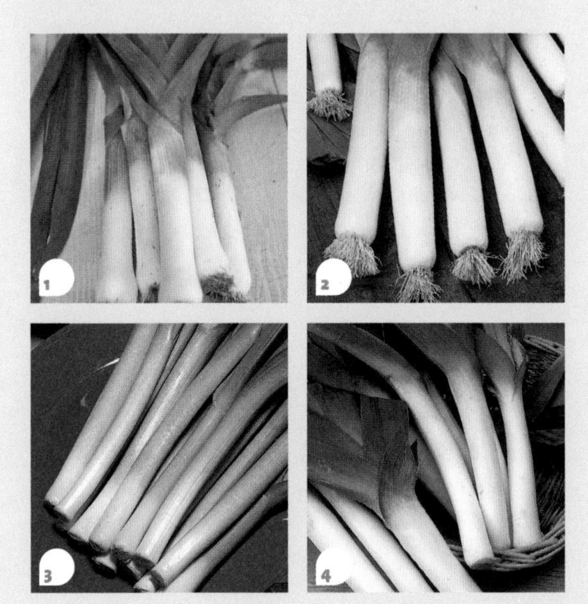

HARVEST

Start lifting a few leeks at a time to use fresh when you need them, as soon as they reach usable size. Use early varieties first as they won't 'stand' long in winter, and move on to later varieties as soon as the earlies are finished. Early varieties are ready between September and December, and late varieties from December to March; there are also some very late varieties for use in April and May.

PROBLEMS

Just like onions and garlic, leeks are prone to bolting. Some years are worse than others, depending on how temperatures fluctuate. Nip emerging flower heads off leeks that are starting to bolt and use such plants as soon as possible, even if they are still small and not really ready, to prevent them going to waste. Leeks sown very early indoors (January–February, which is often suggested for raising plants for very early crops or for showing), tend to produce more 'bolters' than those sown outside in late spring. Bolting can sometimes occur if plants dry out badly or growing conditions are poor.

The same rust disease affects all members of the onion family. Infected plants develop rusty-orange dots and leaves eventually turn yellow, so plants make little or no more growth. There's no cure. If an outbreak strikes, salvage what you can by cutting leeks to use early. Dispose of leaves and trimmings by burning, if possible – don't put them on the compost heap – and grow leeks and onions on another piece of land. Rotate crops so you don't grow members of the onion family on the same ground more than once in 4–5 years. Look out for rust-resistant new varieties coming along.

Useful bit of kit – a big dibber

Gardeners are usually familiar with the sort of dibber used in the greenhouse for pricking out seedlings, which is about the size of a pencil, but a bigger version about 30cm (12in) long and 2.5cm (1in) or so wide is very handy for planting leeks and brassica seedlings. The point pushes down into the hard ground quite easily, and by working the handle round in a circle it leaves a conical hole that's just right for the job.

Old allotment gardeners often used to make their own big dibber by taking a broken spade or fork handle, shortening it to about 30cm (12in) long and whittling the tip to a point. You can also buy beautifully 'turned' dibbers, which are being sold by makers of reproduction antique gardening tools. These are a joy to own and use – they have leather straps so they can be hung up in the shed. You can get by without – just improvise and cut a thick piece of broom handle to use instead – but junk shops sometimes have the originals, and they are kinder on the hands than a simple bit of broom handle.

Lamb's lettuce (corn salad or mache)

Valerianella locusta

Lamb's lettuce is a little-known but tasty and very reliable rosette-shaped salad plant that's quickly and easily grown through the winter. It's best grown under cover, and it's ideal for windowsill cultivation.

HOW TO GROW

Degree of difficulty: Easy and needs little time.
Sow: Thinly September to March; sow September to October in finely raked, well-drained soil in a sheltered but open situation where you want plants to crop and protect with fleece. Alternatively, sow at intervals from September to March in cold frames or in a soil border in a cold greenhouse or polytunnel, or on windowsills indoors. No need to thin out unless seedlings are badly overcrowded.
Spacing: 2.5cm (1in) apart with rows 15cm (6in) apart to allow for hoeing.
Routine care: Water sparingly, but don't allow plants to dry out badly.

GROWING ON WINDOWSILLS INDOORS

Fill a half-size seed tray or shallow windowsill trough with seed compost, sprinkle seeds thinly and evenly all over the surface so they are spaced about 1cm (½in) apart. Barely cover them with a thin layer of compost or, better still, use horticultural vermiculite – which keeps the plants clean so there's less cleaning to do when you cut the plants later. Water in gently and water whenever the compost starts drying at the surface; take care not to over or under water. As soon as the first seedlings are big enough to use, start pulling them out to thin the remaining crop, which will carry on growing. Snip off the roots and rinse.

VARIETIES

1 **'Verte de Cambrai'** – an old French variety with small, deep green rosettes and an excellent flavour. Needs hunting for in specialist catalogues.
2 **'Vit'** – modern, fast-growing and high-yielding variety more easily found in mainstream seed catalogues. 'Cavallo' is very similar.

HARVEST

Use any small seedlings removed when thinning out as baby salads. Mature plants are ready within a few weeks of sowing but will 'stand' several weeks longer. The plants should be cut when they are 2.5cm (1in) across, and the roots and any yellowing lower leaves snipped off. The remaining rosettes can be used in salads. The season lasts from October to April.

PROBLEMS

Greenfly sometimes occurs under cover in winter, but the crop is generally trouble free.

Lettuces

Lactuca sativa

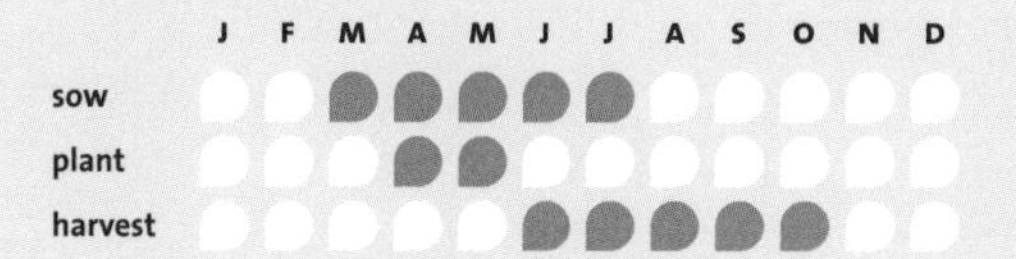

The best-known and most widely used of all salad crops is now available in a huge range of types, including cos, loose-leaved and oak-leaved, red-leaved and frilly, as well as the familiar, round, floppy sort as grown by Mr McGregor and scoffed by Peter Rabbit. Many gardeners fall into the common trap of assuming lettuce is foolproof and leave it to grow itself, but it repays a bit of care to 'get right'.

HOW TO GROW

Degree of difficulty: Needs some attention to detail, but does not need a lot of time.
Sow: Early March to mid-July; early to mid-March sowings can be made in pots in a propagator, a windowsill indoors or a cold greenhouse, then pricked out into pots or 'plugs' for planting out when the weather has improved sufficiently. Seeds sown between mid-March and the end of May can be sown in a seedbed outdoors and transplanted when big enough. Seeds sown after the beginning of June don't transplant well as the weather is usually too hot, so sow those thinly in rows where you want the plants to grow and thin the seedlings out to the correct spacing. In a mild area you can make a late sowing at the end of July or even very early August for a late crop, but you'll need to protect the young plants with fleece once the evenings start growing chilly in autumn – and choose a variety that's recommended for this job.
Important: Always sow short lengths of row at fortnightly intervals to make sure of a succession of steadily maturing lettuces, rather than a sudden glut.
Plant: As soon as seedlings are big enough, usually 3–4 weeks after sowing.
Spacing: Depending on the eventual size of the variety, from 10–30cm (4–12in) apart.
Routine care: Plant in rich, fertile, loose soil containing plenty of organic matter; water in after planting and keep the soil moist throughout the growing season. Weed regularly until the lettuces cover the ground – they don't like competition and are easily smothered by weeds. Take precautions against slugs and snails.

GROWING LETTUCES OUT OF SEASON

Summer lettuces are by far the easiest to grow, but if you don't mind more effort and risking less reliable results, it's also possible to grow lettuce for cutting in winter and early spring. They aren't easy, and there are other more reliable salad leaves you could grow instead, which are almost guaranteed to give you a good crop, but if you want lettuce…

Choose the right varieties; your usual summer lettuce won't do at all. Suitable sorts for cutting in winter and early spring are nowadays mostly found in organic suppliers' catalogues.

Winter lettuce outdoors: 'Winter Density' (cos), 'Rouge D'Hiver' (red cos).

Winter lettuces in an unheated greenhouse, cold frame or polytunnel: 'Valdor' (round-headed cabbage lettuce) 'May King' (large, semi-frilly, red-tinged cabbage lettuce).

Sow late summer/autumn (but see individual instructions on each variety). Prick out into small pots and plant out later, disturbing the roots as little as possible, then water in.

Outdoor crops need a sheltered spot and well-drained ground. Cover plants with fleece for protection against cold, pests and pigeons, but even so they often 'disappear' over the winter.

Under cover, water sparingly when needed until the lettuces grow big enough to touch each other. Try not to water again after that stage, due to the risk of dampness on lower leaves causing mildew which can quickly kill the plants at this time of year. Ventilate at every opportunity to reduce the risk of mildew.

VARIETIES

1 **'Cocarde'** – red, oak-leaved lettuce with a wonderful flavour; stands well without bolting. Not widely available, so look in specialist catalogues.

2 **'Little Gem'** – a deservedly popular, fast-growing dwarf cos with crisp, sweet hearts; unlike regular cos varieties the plants do not need tying up to form a heart. Grow 10cm (4in) apart. This one is good for salad beds in small gardens and is my favourite lettuce of all time.

3 **'Lobjoits Green Cos'** – an old cos lettuce with a superb flavour; tall, narrow plants need tying round with raffia when half grown to encourage a dense, tender, pale heart to form. Well worth the effort. Cos lettuce has lately been 'discovered' by foodies, and occasionally appears in shops with fancy prices, so it pays to grow your own.

4 **'Lollo Rossa'** – beautiful, deep red, frilly, Italian loose-leaf lettuce which does not form a heart. Can be cut whole or used as cut-and-come-again lettuce. Superb for garnishing and sandwiches.

5 **'Marvel of Four Seasons'** – Stunning-looking, yellow and red lettuce with shaggy-frilly leaves that aren't quite oak-leaved. It has an outstanding flavour and brilliant quality. The same variety can be sown over much of the year and grown outside or under cover fairly successfully. Needs hunting down in specialist catalogues.

6 **'Tom Thumb'** – a Victorian variety that's a mini version of the traditional round cabbage lettuce, but with a wonderful flavour and just the right size for 1–2 people. Reliable and easy, it can also be grown under protection.

7 **'Webb's Wonderful'** – large, very crisp, slightly shaggy-looking lettuce, which traditionally has the outer leaves trimmed off leaving only the large solid heart, which is then known as Iceberg lettuce. Not the easiest to grow well.

HARVEST

Start cutting hearting varieties as soon as the first lettuce in the row forms a small heart – don't wait until they are all ready, or most of them will bolt before you can use them. Non-hearting varieties, such as the many frilly and oak-leaved kinds, can be cut as soon as they are big enough to use. To cut a lettuce, slide a long-bladed knife underneath the plant parallel with the ground and slice through the stem; trim off the lower leaves so you only take the useable ones inside.

'Cut-and-come-again' varieties are intended to be left in the ground to keep growing; with these you just need to cut off a few of the tender leaves from near the centre and don't over-pick from any one plant at a time. The same plants will keep growing for several months until they eventually bolt. By growing several varieties and making successional sowings during the season you can have fresh lettuce from mid-June to mid-October.

PROBLEMS

Slugs and snails are a constant problem since lettuce plants are so tasty and tender at all stages. Start taking precautions several weeks before planting, so as to reduce the population in the area, and continue throughout the summer – it pays to grow lettuce in a special salad patch that's given extra protection.

Greenfly is not a regular problem, but when it happens affected lettuce may be so infested that nobody fancies eating them, however well washed. Keep a close watch and use organic pesticide when necessary, or keep plants covered with fine insect mesh from planting time onwards.

Bolting is almost always caused by erratic growing conditions – temperature- or moisture-related. Keep the plants growing evenly and do not sow in cold, wet earth or hot and dry spells.

Aside from slugs, the biggest problem facing home growers is failing to water, weed or thin out seedlings in time. If time is tight, consider buying lettuce plants, or use pelleted seeds and sow 5cm (2in) apart so there's less thinning out to do; and above all *grow less lettuce, but grow it better.*

Mushrooms

Agaricus bisporus and others

Growing your own mushrooms is a fascinating experience, and one that is hugely educational for children. Seed catalogues frequently feature exotic mushrooms supplied as dowel implants for growing on logs outdoors or on straw wads under cover, and you might also see mushroom spawn sold for introducing to deep trays filled with well-rotted manure – which is most successfully grown in a shed. Feel free to have a go, but from experience I can tell you that a certain amount of luck is needed, even when you follow the instructions to the letter.

The most foolproof way of growing everyday cultivated mushrooms is indoors, from a kit. You may have to hunt around for one; they are sometimes sold in garden centres, sometimes in mail-order seed catalogues and occasionally at greengrocers' shops towards Christmas.

HOW TO GROW

Degree of difficulty: Needs attention to detail but doesn't take much time.

When to start: A mushroom kit can be started off at any time of year, but most people prefer to grow mushrooms indoors in winter, since there's relatively little else happening in the veg garden at that time and with central heating on it's easy to maintain the even temperature required.

Routine care: Follow the instructions carefully; the usual idea is to cut the kit open and water it well, then put it in a dark place at a warm, steady temperature. When you start to see the white threads of spawn 'running' over the surface, cover the top with the 'casing material' provided – this is a mixture of sterilised soil and chalk particles to adjust the pH. After few weeks, the mushrooms will start to erupt.

EXOTIC MUSHROOM LOGS

Pre-spawned logs are sometimes available, otherwise you'll need to insert the supplied dowels evenly into a freshly cut hazel log – choose one about 10cm (4in) in diameter and 45cm (18in) long and drill holes in it to insert the dowels. Do this at any time of year, but spring or autumn are usually best. Plant one end of the log in a cool, shady place so that over half of it is above ground, and keep it well watered. With luck, mushrooms will appear and you can crop on and off for several years. Water in dry spells during this time.

If you manage to get one going, it's worth planting another freshly cut log nearby, as the spawn often spreads. That way you can start a second one for free.

HARVEST

Pick mushrooms daily at the stage you prefer – button, cup or open – by twisting the entire stalk out; don't use a knife to cut it, as this leaves a short stub underground which rots, encouraging unwanted fungal organisms. Use a little of the casing material to fill in where you've removed mushroom stalks to deter mushroom fly.

PROBLEMS

Mushroom kits usually advise on any likely problems and their remedies, but the top two are:

If no mushrooms appear, or they stop appearing too soon, this usually means the kit has dried out or got far too wet, so always water with care.

Mushroom flies are easily attracted to moist compost (these are the same sort that can be found round pot plants indoors). Use net curtain or fine insect-proof mesh rather than spraying.

And afterwards...

The same kit should keep cropping for six weeks or so, and yield roughly enough mushrooms to repay the price of the kit, if you'd bought at shop prices. Still, it's fun for the kids. When no more mushrooms come up, spread the contents of the kit over a previously well-manured part of your veg patch, and if the weather is kind you may get a second crop.

Onions

Allium cepa

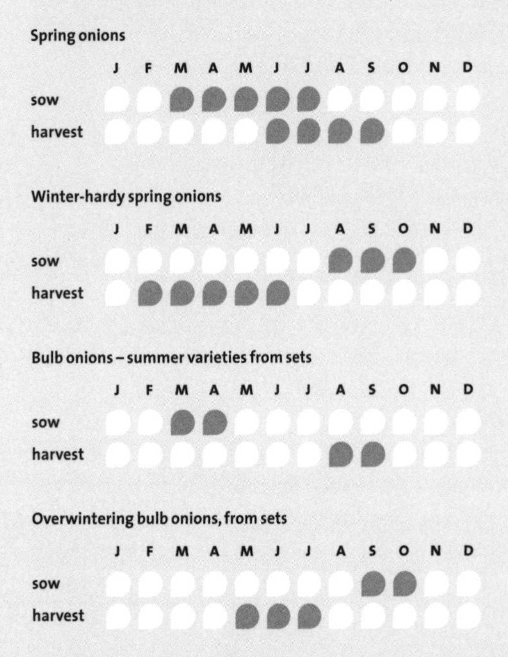

You might not think there's much point in growing onions, as they are cheap enough in the shops, especially if you buy a sackful from a market; and one variety tastes pretty much like another. And up to a point I'd agree – but I do find onion growing particularly satisfying, and over-wintering onions are a good bet since they occupy space when it's not needed for anything else. What's more, they are ready to use from May to July, when good-quality onions are expensive and hard to find in the shops. Then there are all the red onions, good for using raw in salads, which again are pricey to buy and not always available when you want them – so having your own on tap is very handy.

But the ones you should not be without are spring onions; they are tremendous in salads and can be cropped over most of the year by staggering your sowings. Best of all, any you don't use as spring onions turn into superb-quality red or white bulb onions, so you have the best of both worlds. The one thing you must do with onions of any kind is keep on top of the weeding.

HOW TO GROW

Degree of difficulty: Easy and need little time.
Sow: A short row of spring onions every 3–4 weeks from early March to late July, then use an overwintering spring onion variety to sow August to mid-October. Sow thinly and thin out if seedlings are overcrowded. For bulb onions, you can sow maincrop (summer) onions in March, and over-wintering onions in August, but frankly it's very much easier to grow both from onions sets (small bulbs) instead. Onion seed is relatively short lived and needs to be sown fresh.
Plant: Maincrop (summer) onion sets in March/April, and overwintering onion sets in September/October.
Spacing: Thin spring onions out to 2.5cm (1in) apart, and allow 15cm (6in) between rows – just enough to run a hoe between. Plant onion sets 10cm (4in) apart, in rows 20cm (8in) apart.
Routine care: Keep plants watered in dry spells so they don't suffer a check to their growth, which may encourage bulb onions to bolt. Weed regularly as the plants don't survive being smothered.

THREE-IN-ONE SPRING ONIONS

If you initially sowed spring onion seeds too thickly, you can have three slightly different forms of onion from the same row. When the seedlings come up and reach the grassy stage, thin them out by pulling up alternate young plants and use those as baby spring onions, or as chives. As the rest grow, rather than pulling out whole rows from one end to the other, keep thinning out alternate plants to use as spring onions. The remaining plants will keep growing and any that are still left after a few months will produce perfectly good bulbs that are mild enough to use chopped in salads, to put in the juicer with tomatoes and carrots to give your vegetable juice more 'bite', or for normal cooking.

VARIETIES

Spring onions

1 **'Overwintering White Lisbon'** – the out-of-season version, very hardy and reliable for autumn sowing in the open (protect with fleece ideally) but best in the soil border of an unheated greenhouse, polytunnel or under a cold frame.

2 **'North Holland Blood Red'** ('Redmate') – superb, reliable, purplish-red spring onion for spring sowing, which if left to keep growing produces good-quality red onions that keep until March.

3 **'White Lisbon'** – the classic, reliable, white-skinned spring onion for spring/summer sowing; if left to keep growing it produces white, silver-skin-type onions about the size of ping-pong balls which are good sliced raw in salads, or used for cooking or pickling.

Maincrop onion sets

4 **'Red Baron'** – superb red onion with outstanding flavour, mild enough to use raw in salads – it slices into attractive red and white rings – and keeps well through winter.

5 **'Sturon'** – reliable, fairly bolt resistant, with a good flavour and keeps well through the winter.

Overwintering onion sets

Various varieties available in garden centres each autumn, with little to choose between them in terms of taste. The autumn catalogues of the major seed firms also supply several varieties.

6 **'Electric'** – good-quality, well-flavoured red onions.

7 **'Radar'** – goldish-skinned onion; good for cold areas or bad weather.

8 **'Senshyu'** – a Japanese variety having semi-flattened bulbs with yellowy brown skin.

GROWING GIANT ONIONS

Giant onions were originally bred for the show bench, but despite what you hear about exhibition veg not being fit to eat, onions are the exception. Varieties that are naturally huge will always grow bigger than usual, even when under normal cultivation without resorting to the showman's secret methods. The eating quality is excellent, you get a lot of onion per row and, what's more, one big onion takes a lot less preparation time in the kitchen than several small ones, so you win all round. And it has to be said, you'll impress family and visiting friends no end with your gardening prowess.

Seeds of show onions are generally sown on Boxing Day (what an excuse to escape the house!) indoors or in a greenhouse, and the plants are pricked out and grown on until planting out time in late April. The soil needs to be very rich and the plants spaced 30cm (12in) apart each way. Apply a general-purpose feed in June. Pull off damaged outer skins three weeks before the show, so that the new outer skin has time to colour up; later, bend the tops over to encourage ripening.

Dig up the plants two or three days before the show, then prepare them the day before – cutting off the roots and leaves, and tying down the remaining 5cm (2in) of stem with soft twine so that it looks neat. Good varieties are 'The Kelsae', which is sold as seed or young plants; 'Robinson's Mammoth Improved', available as seeds from the firm specialising in exhibition veg; 'Showmaster'; and also 'Stuttgarter Giant', which are both available as sets.

HARVEST

Spring onions are ready as soon as they are big enough to use; spring-sown crops are ready June/July until September, overwintering spring onions sown in September/October are ready between February and May/June. Maincrop onions are ready in August/early September. Bend the tops over once the leaves start naturally turning yellow or brown to assist ripening; lift the bulbs when completely ripe and leave them on the ground in the sun to finish drying before storing in shallow trays in an airy shed. Overwintering onions are ready to use fresh from the ground from the time the first few reach useable size around mid-May, then keep pulling them as needed until late July.

STORAGE

Most varieties will keep in good condition to use from September until February. Keep them in the light to avoid sprouting (unlike potatoes, which need to be kept in the dark). Overwintering onions don't keep for more than a month or so after you take them out of the ground, so don't attempt to store them for long.

PROBLEMS

As with garlic and leeks, bolting is a problem in some seasons, and this is thought to be due to particular sequences of fluctuating temperatures. Choose a bolt-resistant variety and, where possible, buy heat-treated sets, which reduces the risk of bolting.

Mildew is a white or grey fungal growth which is disfiguring and debilitating. Cut off affected foliage and do not store affected bulbs.

White rot is quite common in onions, including spring onions, and it also affects leeks, although less so than other members of the onion family. Affected plants initially have foliage that turns yellow, then they develop white cotton-woolly clumps near the base in which small, black blobs can eventually be seen. It rapidly spreads between neighbouring plants, and the fungal organism remains in the soil for many years, affecting other onions. There is no cure; destroy affected plants and foliage – don't put it on the compost heap – and grow onions elsewhere. It's good practice to move onions and their relatives to a new site every year to minimise disease risk.

Useful bit of kit – an onion hoe

Weeding is one of the most important routine jobs involved in cultivating onions, but since the plants are shallow rooted they are easily disturbed if you try weeding with a hand fork, while a long handled hoe is very high risk as it is so easy to chop into part-grown plants. Most growers prefer to use a short-handled onion hoe, the sort with an elegant swan neck. This lets you get down close to the crop and work carefully between plants, skimming weeds off at the surface so there's no root disturbance and, being more precise, there's less risk of you damaging your onions.

Oriental leaves

Brassica species

Chinese mustard greens

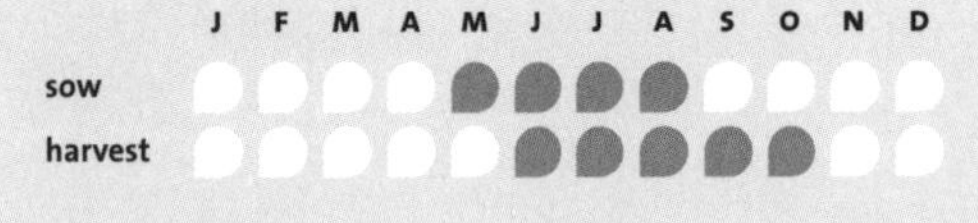

Mizuma

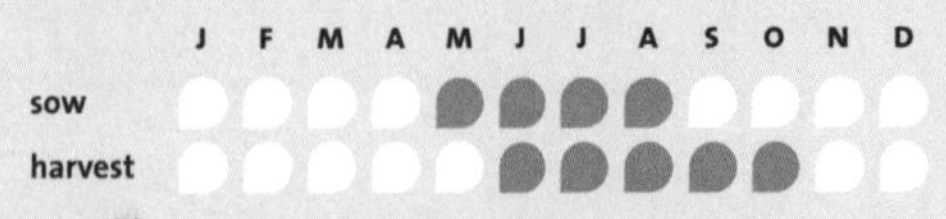

Pak choi

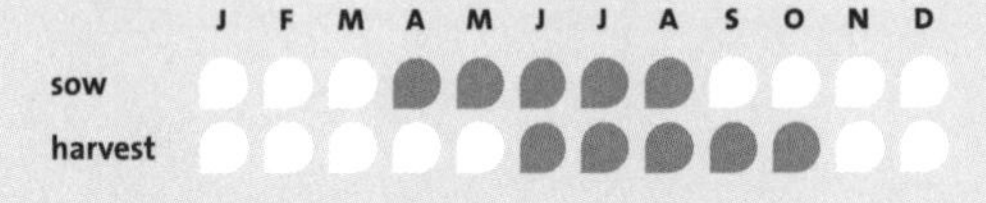

Several relatively little-known leafy veg come into the category of 'oriental leaves'. Some are used as baby salad leaves and are often included in packets of mixed salad leaf seeds. Others, such as red mustard and mizuna, are known mainly to enthusiasts as useful winter 'greens' for salads and stir fries (particularly for growing in polytunnels) and for the rest of us, let's just say they tend to be rather acquired tastes. Only one – pak choi – has a rather wider following as a superb baby vegetable, with a thick, white, bulbous stem and spoon-shaped leaves growing out almost horizontally. This is normally eaten steamed or braised whole.

HOW TO GROW

Degree of difficulty: Not difficult, need little time.
Sow: May to August outside, and September under cover for winter use. Sow pak choi in mid-June to mid-August in the open (as it bolts if sown outside too early) or in April or late August under cover for out-of-season crops. Sow oriental leaves where you want crops to grow and thin out to avoid checks in growth.
Spacing: Thin out pak choi and red mustard to 15–20cm (6–8in) apart, mizuma to 30cm (12in) apart.
Routine care: Keep plants watered and well weeded, and protect against slugs and snails.

VARIETIES

1. **Chinese mustard** (*Brassica juncea*) – mature leaves are really rather hot, but young leaves are pleasantly spicy.
2. **Chinese mustard 'Red Giant'** (red mustard) – attractive bronzy-red leaved form of Mizuma (Japanese greens) (*Brassica rapa* subsp. *nipposinica* var. *laciniata*) makes a shaggy green mophead 30cm (12in) across that's very winter hardy. July-sown plants stand until well after Christmas given the protection of fleece, or when grown under a polytunnel.
3. **Pak Choi 'Mei Quing'** (*Brassica rapa* subsp. *chinensis*) – fast-growing, tender, tasty, pale green pak choi with darker leaves.

1

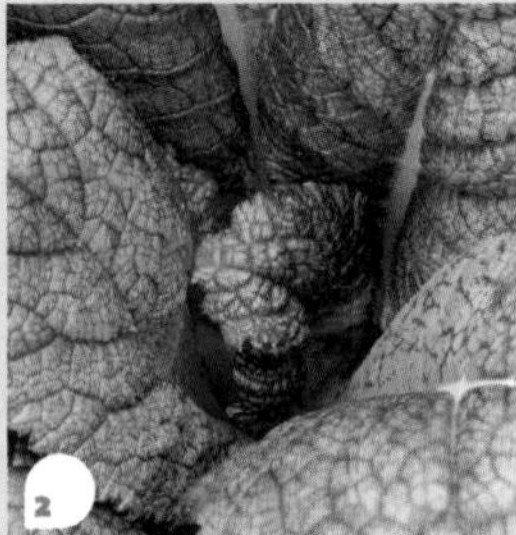
2

3

HARVEST

Start picking Chinese mustard plants as soon as they look big enough, usually from June to October; take out alternate plants as you thin the rows and use those young plants in salads while the rest keep growing.

Pick tender central leaves of mizuma on a cut-and-come-again basis from June to the following March (later-sown plants usually stand well through the winter), leaving the plants to continue growing.

Start pulling whole pak choi plants as soon as the swollen bases of the first plants reach about 2.5cm (1in) in diameter; don't leave them to grow too big or they become coarse and less appetizing. You should be cutting from August to October. (See also salad leaves, page 158.)

PROBLEMS

Slugs and snails are a regular threat; by reducing the leaves to ribbons they put you off using the crop. Start taking precautions several weeks before sowing to reduce the population and continue all the time crops are in the ground.

Flea beetles make small holes barely bigger than pinpricks in leaves, often seen along with tiny beetles that jump like fleas, especially in dry weather in spring and early summer. Attacks rarely kill plants, but the damaged foliage puts you off eating them. Help plants to get well established by good soil improvement and generous watering in dry weather, and cover with fine insect-proof mesh to prevent insects damaging the foliage.

Parsnips

Pastinaca sativa

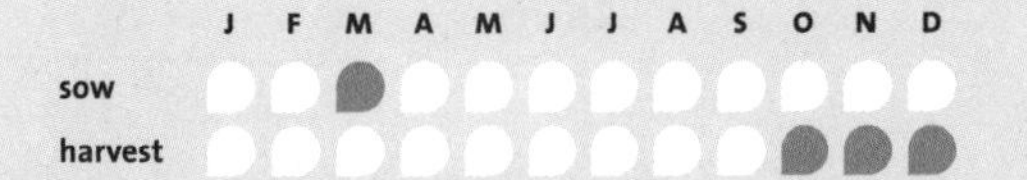

Parsnips are the sort of traditional winter root veg that are mostly grown on allotments nowadays, where there's plenty of room for a crop that occupies the ground for a long time. (And if you're going to grow them, it's worth doing so organically, since – just like organic carrots – the flavour is so much better.) They really are tremendously tasty when roasted with potatoes alongside the Sunday joint.

You might be put off growing them by the length of time they are in the ground – around nine months – but baby parsnips are a different story entirely. These veg are spaced close together and dug up while they are still small, in summer, so they don't need a lot of room – and they vacate the ground in time to plant a crop of winter brassicas, such as kale, sprouting broccoli or spring cabbage, afterwards. Not only is this a far better use of space in a smallish plot, but you're growing quite a high-price crop that's usually flown in from Australia – a tremendous saving in food miles.

HOW TO GROW

Degree of difficulty: Needs some attention to detail, but not much time. Can cope with a modicum of shade, but not deep gloom.

Sow: March, in drills where the crop is to grow, then thin out the seedlings – don't transplant them. Choose deep, rich, fertile, stone-free soil where manure was not used the previous winter, and wait for a mild spell if the soil is especially cold and wet. Parsnip seeds have a very short life, so always buy fresh seeds each year and sow them the same spring. It's not worth saving old seed in half-used packets since very few (usually none at all) will come up the next year, however well you store them.

Spacing: Sow seeds thinly so they rest about 2.5cm (1in) apart along the row, and thin the resulting seedlings to 15cm (6in) in several stages. Space the rows 30cm (1ft) apart.

Routine care: Water in dry spells and weed regularly.

GROWING BABY PARSNIPS

Choose special varieties listed as suitable for baby veg production in seed catalogues; some have whole pages of 'baby veg'. These fill out to the right shape while still small – normal varieties are not suitable for this job since their roots grow long and thin first, and don't start plumping up till much later in the season.

New baby parsnip varieties are coming along all the time, but try 'Dagger' or 'Arrow'.

Sow seeds as usual but thin them to 5cm (2in) apart and start digging up roots as you need them from midsummer onwards, without waiting for the leaves to die down.

VARIETIES

1 **'Avonresister'**– a deservedly popular variety that's very reliable, being resistant to canker, is less likely than most to bolt and produces good crops even on poor ground. The conical roots are smaller than average, so it is best grown 7.5–10cm (3–4in) apart.

2 **'Tender and True'** – an old favourite that's reckoned to be the best for flavour, with long roots with small cores in the centre.

HARVEST

Dig up roots as required from October (after the foliage starts turning yellow, which shows that the crop has finished growing) until around December. Parsnips 'keep' best in the ground, unless it's very wet and claggy or there's a particularly wet winter, when they may start to rot – in which case lift them and store in a frost-free shed. Parsnip fanciers reckon the roots are improved after a frost or two.

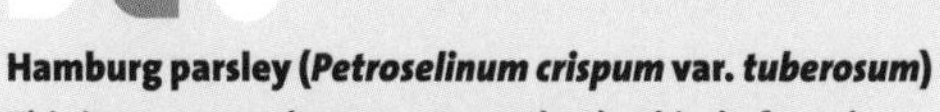

Hamburg parsley (*Petroselinum crispum* var. *tuberosum*)
This is an unusual gourmet veg that's a kind of parsley grown for its edible root; which looks like a short, conical parsnip and has a flavour that's halfway between the two.

Grow and cook as for parsnip (it also roasts well). Young roots can be pulled in summer to thin the rows and used straight away, leaving the rest to grow larger for winter use. It stands well in the ground.

PROBLEMS

Roots affected by canker develop brown 'scabby' skins, especially round the top of the roots, which eats into the flesh. A lot of the damaged areas can be cut off, but with badly cankered parsnips there's nothing fit to eat. Incidence is thought to be worse if there's too much organic matter in the soil or in a poor growing season with too much or too little rain. There is no cure, only prevention by growing resistant varieties.

Plants run to seed so the tap root, if it develops at all, turns tough with a particularly hard core and becomes unfit to eat. Poor growing conditions, sowing too early, or irregular water supplies can trigger bolting. Pull out any plants seen to be starting to bolt, leaving others to keep growing.

Fanged/forked roots are caused by growing in ground with too much fresh manure or inadequately rotted organic matter. It can also occur in stony soil or ground that hasn't been well dug and raked before sowing, so the roots have difficulty 'boring down'. Affected roots aren't inedible, just very difficult to prepare for cooking. On stony soil, build a raised bed that can be filled with sieved earth mixed with old potting compost. Showmen (and women) grow their parsnips in lengths of plastic drainpipe filled with pure potting compost.

Recipe – Roast winter root veg
Peel a selection of root veg, including parsnip, carrot, swede and celeriac, as usual. Cut parsnips lengthways into quarters and remove any hard cores. Halve large pieces leaving triangular shapes, slice carrots into four lengthways, cut celeriac and or swede into 2.5cm (1in) cubes. Toss lightly in olive oil and roast in a baking tray in the oven until tender. Serve round a Sunday roast, sprinkle with parmesan shavings for supper, or put in the blender with chicken stock to make a delicious and hearty roast winter vegetable soup.

Peas

Pisum sativum

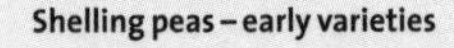

Shelling peas – early varieties

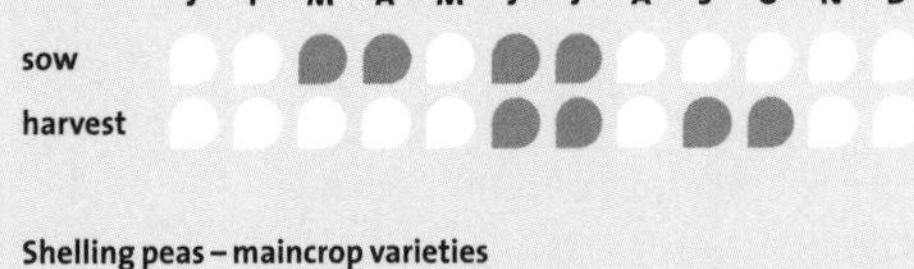

Shelling peas – maincrop varieties

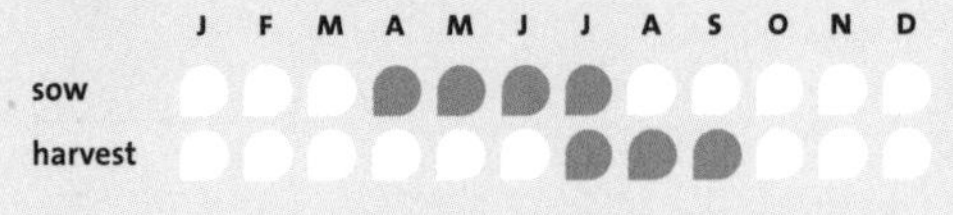

Mangetouts and snap peas

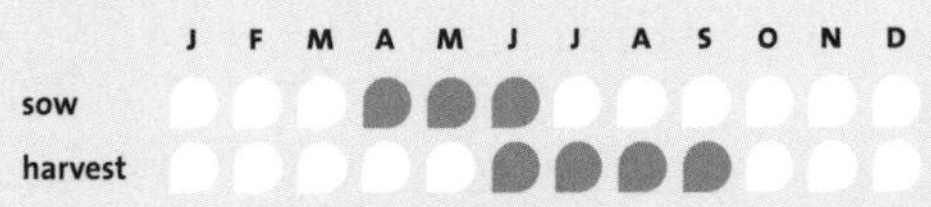

I do love fresh peas – the growing, the shelling and the eating of them raw in the garden – but if there's one vegetable that even top chefs agree is best frozen, it's peas. So unless you particularly enjoy growing them and can pick them at just the right state of perfection (mine always seem to come to maturity while I'm away on holiday), you may find your own home-grown crop a tad disappointing.

But the peas that are really worth growing are mangetout and snap peas. You get a big return from a short row of plants because you eat them pod-and-all, and they are great for busy people as there's no shelling or preparation to do – just steam or stir fry them whole. They are also unaffected by maggots, which often ruin whole crops of 'regular' peas.

HOW TO GROW

Degree of difficulty: Needs some attention to detail and a little time.

Sow: Two or three staggered rows of seed in a flat-bottomed drill about 20cm (8in) wide, so the seeds are roughly 7.5cm (3in) apart in each direction. Sow mangetouts and snap peas in April/May/June. If you want to grow the traditional types of peas that need shelling, sow early varieties in late March/early April, sow maincrop peas from mid-April to early June, and sow early varieties again from the end of June until mid-July for a late crop.

Plant: If buying plants, or raising them for early crops under cover, plant them in a double or triple staggered row with plants 10–15cm (4–6in) apart in each direction.

Spacing: Allow 45cm (18in) between rows of short-growing varieties and 90cm (3ft) between rows of tall-growing varieties, to allow access for picking and weeding.

Routine care: After sowing, push twiggy pea sticks in along the rows to support short-growing varieties. Tall varieties will need a stronger system, so put in a row of 1.8m (6ft) posts with horizontal wires holding up pea netting for them to scramble up. Keep the plants watered in dry weather and weed regularly.

Continuous cropping: In practice, most people at home don't even attempt to keep themselves constantly supplied with peas throughout the season; it's just pleasant to have one or two batches to look forward to some time over the summer. But with a little planning you could be cropping peas continuously all season. Most green 'shelling type' peas have quite a short season, so you'll need to grow separate early and late varieties and make several successional sowings; a tall heritage variety will give you a much longer cropping season and actually saves work, since you can get away with only one or two plants which need minimal support. Short varieties of snap peas are also virtually picked out after 3–4 weeks, but a tall variety of mangetout should crop for 6 weeks or more.

GROWING PEAS OUT OF SEASON

Shelling peas: Early peas sown outside are at the mercy of cold, wet soil which causes them to rot, and of rodents and pigeons in search of an easy meal, so there's a high failure rate.

You'll do very much better by raising plants under cover and putting them out later. Sow early varieties in trays of multi-purpose compost in a cold greenhouse or cool windowsill indoors in late February/early March and harden off before planting out in a mild spell about a month later. Cover young plants with fleece to protect from cold weather, pigeons, etc. Round-seeded varieties (which are marked as such in catalogues) are the hardiest and so best for very early sowings, and these can also be sown in October and November to overwinter outdoors and produce an early crop, though it's not too reliable a technique and protection with fleece is advisable in bad weather. Wrinkle-seeded varieties usually produce the best flavoured peas but are sown later. Take your pick!

Eat-all peas: If you have an unheated greenhouse or walk-in polytunnel with space in the border soil, it's also worth growing mangetouts or snap peas there. Raise the earliest plants in the same way as above and plant them once the weather turns milder, usually from late February to early March. Sow a few rows under cover where the plants are to crop in mid-March – they'll be weeks earlier than outdoor sown plants.

It's also worth sowing a wide row of 'Oregon Sugar Pod' mangetout peas into the ground under a polytunnel in late September to early October; plants will then overwinter and even if they look rather ropey during the worst weather they usually survive to produce a very early crop that's most welcome next spring.

'EAT ALL' PEAS

Mangetouts have large, flat pods and should be picked when they are about 5cm (2in) long; if left to grow much larger they develop a hard, stringy edge which needs to be stripped off when preparing pods for cooking. Large pods turn tough, but rarely develop peas inside that are worth shelling for – pick and discard old, tough pods that have been overlooked to encourage the plants to keep producing more flowers and young pods.

Snap peas have small, cylindrical pods which are best used when they are about an 4cm (1½in) long and cooked whole like French beans. If left on the plant, normal green peas develop inside so they can be shelled and used as usual instead of wasted, though their flavour isn't as good as varieties intended for shelling. In both cases, nip the pointed tip and the stalk off each end of the pods to prepare for cooking.

HARVEST

Mangetouts and snap peas are ready to start picking once the first pods reach usable size – expect to be picking from late June through July, August and early September. Check the progress of shelling peas by popping open one or two of the biggest pods occasionally to see how big the peas inside are – use them while young and tender (the same size as frozen peas, or smaller) and don't let them grow big and tough or they lose their flavour and turn starchy. Expect to be picking early varieties sown in March/April in June/July, maincrop varieties sown in April/June from July to mid September, and early varieties sown in June/July in late September to mid-October, given some protection.

Culinary tip – Marrowfat peas

If while you're picking you miss a few peas of a wrinkle-seeded shelling variety, such as 'Kelvedon Wonder', don't worry about them being left on the plant; they dry in their pods and will produce marrowfat peas. These are perfect for drying and storing for winter use in soups and stews.

VARIETIES

1. **'Alderman'** – traditional, tall, shelling variety 1.8m (6ft) high, rather late to start cropping as a result, but stays productive over a much longer cropping season than dwarf peas, which saves a lot of replanting and putting up supports. Sow March to mid-June.
2. **'Carouby de Mausanne'** – a similar mangetout and can be grown the same way as 'Oregon' (see number 6) but with outstanding flavour, though seeds are far less widely available.
3. **'Feltham First'** – round-seeded shelling variety suitable for the very earliest sowings in February. Can also be sown in October/November to overwinter outdoors. Dwarf plants, 45cm (18in) high, need little support.
4. **'Hurst Green Shaft'** – a delicious second-early or maincrop variety that I've been growing for years. Heavy cropping over a long period and 75cm (2ft 6in) tall.
5. **'Kelvedon Wonder'** – wrinkle-seeded shelling variety on dwarf plants 45cm (18in) tall with good flavour. This is a second-early variety that can be sown March to the end of June; it's ideal if you only want to buy one packet of peas.
6. **'Oregon Sugar Pod'** – tall mangetout reaching 1m (3ft 6in) tall, with a fairly long cropping season, suitable for growing outdoors from March–June, or early/late under cover.
7. **'Sugar Ann'** – tall sugar snap pea 1.5m (5ft) high, which takes longer to reach cropping stage but continues picking longer than dwarf varieties. Needs much more support.
8. **'Sugar Snap'** – early dwarf snap pea 75cm (2ft 6in) tall that needs little support. Good sown early under cover.
9. **'Waverex'** – if you want 'petit pois' – those tiny, tender peas that can be eaten fresh or frozen – this is the variety for you. Short plants that need little or no staking at 45cm (18in).

PROBLEMS

Plants failing to appear is almost always due to predation by mice or other rodents, or else to cold, wet soil at sowing time.

Maggots found on your crop are the larvae of the pea moth, which tunnel into the pods of 'shelling peas' to eat the seeds and deposit their mess inside the pods. Sometimes only a few pods are affected, but often all or most of a batch of plants are struck at the same time, making the entire crop inedible. Early and late-sown crops often escape attention, so stick to those; otherwise cover summer crops with fine insect mesh when in flower. Alternatively, grow mangetout and snap peas, which don't seem to be affected.

If young leaves and tips of shoots appear covered with grey/white talc, you've got powdery mildew, which can eventually spread to cover whole plants. It's usually worst when the soil is dry and in a sheltered place, and is especially common on plants grown under cover – older plants coming to the end of their cropping lives are often affected. It's best avoided by keeping plants well watered and ensuring good air circulation.

Pea weevil chomps irregular baby bite-sized notches out of the margins of leaves – you won't always see the culprit but they are 6mm (¼in) long buff-brown beetles with six legs and a short pair of forward-facing antennae. Damage looks worse than it is, and all but badly infested young seedlings nearly always grow out of it without problems.

If foot and root rot strikes, young plants or fairly recently emerged seedlings will turn yellow and shrivel up. On investigation the roots are found to have turned black and died off. This may sometimes be due to overwatering poorly established small plants in cold, dull weather, but may be due to an organism in the soil. To solve the problem, put new seeds or plants in a different piece of ground and rotate crops annually so you don't grow the same crops in the same patch several years running.

Peppers

Capsicum annuum

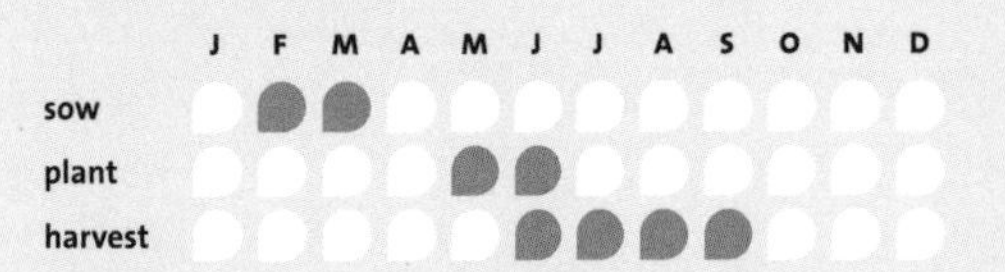

Also known as sweet peppers, bell peppers or capsicums, these are Mediterranean favourites that have taken off over here too. Though usually grown in an unheated greenhouse, they will produce a small crop when grown outside in a warm, sheltered spot, such as a sunny patio or courtyard garden, and they also do well in growing bags or tubs.

HOW TO GROW

Degree of difficulty: Not difficult given good growing conditions, and not time consuming.

Sow: In February or early March in a heated propagator or on a windowsill indoors at a temperature of 21–27°C (70–80°F). Prick out the seedlings into individual small pots when large enough to handle, and grow on at 16–19°C (60–65°F). Harden off carefully before planting out – they are frost-tender and can't stand cold.

Plant: In an unheated greenhouse late May to early June; outdoors early to mid-June.

Spacing: 45cm (18in) apart.

Routine care: Water in after planting, then water sparingly until plants are growing strongly and starting to flower or bear fruit. Feed weekly with liquid tomato feed after the first flower opens. Support plants by tying the main stem to a cane.

MULTI-COLOURED PEPPERS

There are varieties available that produce purple, chocolate-brown, orange or yellow peppers when ripe, instead of the usual red. Some seed firms also sell mixed colours; the range tends to vary year on year, so see what's available each season. Coloured peppers can be used just the same way as usual, though a mixture of colours looks especially attractive sliced raw in a salad bowl. All the peppers start out green and change colour as they ripen, often going through several shades before they are completely ripe and at their final colour. So don't be in a rush to pick them!

VARIETIES

1 **'Bell Boy'** – a standard commercial variety, heavy cropping and reliable, often sold in garden centres as ready-to-plant pepper plants. The blocky fruits have thick walls and start green, later ripening to red, just like the peppers you buy in the greengrocers.

2 **'Big Banana'** – long, tapering peppers to 25cm (10in), good grilled. Green, ripening through yellow to red.

3 **'Gypsy'** – narrow, tapering peppers 7.5–10cm (3–4in) long and 6cm (2½in) wide, with rather thin walls; starts off green, turning orange then red. Available from seed catalogues.

4 **'Redskin'** – rather compact, bushy plants, good for growing in containers on a patio, with large numbers of oblong peppers which start green and ripen to red.

Recipe – Stuffed peppers

Halve and de-seed one pepper per person; brush with olive oil and grill under a moderate heat until almost soft but so the shape still acts as a 'shell'. Fill with a mixture of cooked rice, pine nuts and sweetcorn, then sprinkle with grated cheese and grill lightly until warmed through and browned on top. Alternatively, stuff peppers with cooked mince fried with a hint of garlic and chilli, top with slices of fried potato, cover with grated cheese and grill. Makes a good light lunch or supper dish.

HARVEST

Start picking green peppers as soon as they are big enough to use – they don't have to be as big as the ones you buy in the shops – but if you want red peppers, leave green peppers to reach full size and ripen. This does take a lot longer and the plants won't produce any more peppers until the first batch have ripened and been picked. To pick, cut through the stem connecting the fruit to the plant with secateurs; don't tug or twist since the branches are brittle and you risk breaking them, which will reduce your cropping potential. Expect to be picking from late June to mid-September from plants grown outside. Cool evenings will prevent any further growth, so if there are any very small peppers remaining on the plants at this time you can pick and use them green, however small.

PROBLEMS

Young plants grown under cover are most at risk from greenfly. A bad infestation can stunt growth and distort leaves so that the total crop is considerably reduced, and in a very severe attack young plants can be killed. Check plants regularly and wipe greenfly off with damp tissue, or use an organic insect spray.

Plants can fail to produce fruit if growing conditions are poor (lack of sun, cold, dull or windy weather), or overwatering can also cause crop failure. Peppers are warmth-loving plants, so in a poor summer try to move them into a conservatory or enclosed porch for shelter, or drape them with fleece at night and on cold or windy days.

Potatoes

Solanum tuberosum

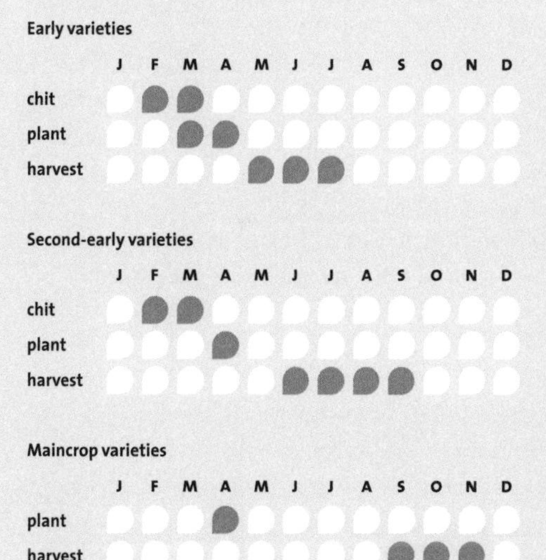

Potatoes are the most popular and versatile of veg, so much so that for many people a meal isn't complete without them. But in spite of their long history (Sir Walter Raleigh is reputed to have introduced them to this country in Elizabethan times), spuds are fast becoming a 'cult veg' on almost the same scale as tomatoes, now that such a huge range of varieties is available for home-growers to try. Besides bakers, chippers, mashers and salad spuds, there are also novelty potatoes, gourmet exhibition and heirloom varieties and coloured potatoes – not to mention an ever-expanding selection of disease-resistant spuds that are ideal for organic growing. Incredible though it would have seemed even a few years ago, displays of potatoes are now some of the most popular exhibits at flower shows.

The various potato varieties fall into three groups: earlies, second-earlies and maincrop, depending on how early or late in the season their tubers 'bulk up' ready for harvesting.

HOW TO GROW

Degree of difficulty: Easy; little work needed apart from regular earthing up.

Plant: Earlies in late March/early April; second-earlies a week later and maincrop potatoes a week after that.

Spacing: Plant tubers of seed potatoes 13cm (5in) deep; early potatoes should go 30cm (12in) apart in rows 60cm (2ft) apart; plant second-earlies and maincrop potatoes 38cm (15in) apart in rows 75cm (2ft 6in) apart.

Routine care: Hoe shallowly to keep down weeds until the potato shoots are 15cm (6in) high, then earth up the plants. (If frost threatens, you can earth up as soon as the shoots appear above ground.) To do this, use a draw hoe to pull soil up into ridges along the rows to cover the emerging shoots to about half their height, with gullies between them. Shortly after doing this potato foliage will completely cover the ground and no more weeding or hoeing will be needed, since weeds are literally smothered out. Except in a long dry summer, you shouldn't need to water potatoes.

SEED POTATOES

Potatoes are grown from 'seed potatoes', which are small, stored tubers that have been especially grown for the job and are certified disease-free – don't grow potatoes from the greengrocer that have sprouted in the cupboard under the sink. The ideal seed potato is the size of a hen's egg, but bigger or smaller ones are sent out. Don't cut large ones in half; the open wound makes them rot. Buy seed potatoes in late winter or early spring – seed firms often send them out in January/February. Keep them in a cool, dry place until it's time to plant.

Early varieties need 'chitting' to start them into growth before they are planted. On arrival, sit the tubers up on end in an eggbox or seedtray, so that the 'rose' end (where there is a cluster of tiny 'eyes' or buds at one end of the potato) is uppermost. Keep them in good light but out of direct sunlight. When the shoots are 6mm–2.5cm (¼–1in) long and the weather is suitable, your chitted potatoes are ready for planting. It's worth chitting second-earlies as well as first-earlies if you can, but you don't need to chit maincrop spuds, since they have a much longer growing season anyway.

VARIETIES

Earlies ('new potatoes')

1 **'Duke of York'** – well-flavoured, pale yellow tubers that can be used as new potatoes or left to grow for use later in the season; a good choice if you only have room for one variety .

2 **'Foremost'** – classic firm, waxy, white, salad-type new potato. Good eaten hot or cold.

3 **'Mimi'** – a new variety with small, round, pink-flushed, marble-like tubers and very compact foliage, making it ideal for growing in pots on the patio or under cover. Great fun.

4 **'Pentland Javelin'** – reliable and delicious white-skinned, white-fleshed, waxy potato, ideal for potato salad, with good disease resistance.

5 **'Rocket'** – one of the very fastest-growing earlies; large crops of round white tubers with reasonable all-round disease resistance. Good for growing early under cover.

Second-earlies

6 **'Charlotte'** – superb salad potato with golden skin and firm, waxy, cream-coloured flesh with a great flavour.

7 **'Edzell Blue'** – a Victorian variety with purplish skin and very floury white flesh; a perfect masher and very tasty. Spuds tend to burst open when boiled, so steam them instead. The plants have white flowers.

8 **'Estima'** – hefty crops of large tubers that bulk up early, good when you want potatoes for baking in summer. Oval tubers with pale yellow flesh inside; plants do well even in dry summers.

9 **'International Kidney'** – this is the variety that's sold as the 'Jersey Royal' when it's grown in Jersey, and whether it's the sea air and warm island microclimate or what, it never tastes quite the same when grown in mainland Britain, though it's still a good second-early spud. Kidney-shaped tubers with a waxy texture.

10 **'Kestrel'** – another very handsome spud, with perfectly shaped, off-white, oval tubers with purple rims around the eyes. This is a favourite at village shows and an excellent all-purpose cooking variety, with fair resistance to slugs and disease.

11 **'Yukon Gold'** – good-looking, round, pale yellow potatoes with very tasty rich golden flesh. Good for baking, roasting and luxury chips.

Maincrop

12 **'Belle de Fontenay'** – an old French early-maincrop variety known for its gourmet flavour, with rather tubular, light yellow tubers. Steam whole in scrubbed skins to preserve the full flavour.

13 **'Golden Wonder'** – a late maincrop, so leave this in the ground until after your other maincrops have been lifted. It has russety skin and superb flavour, which keeps improving with storage. Best for baking, roasting and frying.

14 **'King Edward'** – the old favourite with red-variegated tubers, known for its superb flavour, with cream-coloured flesh. Good for roasting and baking, but needs good growing conditions to do well.

15 **'Mayan Gold'** – bred from plants with Peruvian Andes parentage. A very successful newcomer with long, slender, tubular golden tubers with firm, golden, nutty-flavoured flesh. Good for all uses except boiling, as they break up, and are great deep-fried when whole or roasted.

16 **'Pink Fir Apple'** – possibly the best-tasting salad potato ever. An old variety that can be hard to find but which is worth the hunt; a very late maincrop with long, slim, slightly irregular-tubular tubers, best left in the ground until November. Cook whole in scrubbed skins and eat hot or cold. Produces heavy crops and stores well right through winter.

17 **'Smile'** – stunning-looking, bright-pink-skinned potato with a white 'smiley lips' pattern; a favourite with children. A well-flavoured all-rounder with cream-coloured flesh and good for all uses except boiling.

Recipe – Rosemary potatoes

Scrub some whole new potatoes (waxy salad varieties are best for this), and slice them evenly. Brush the slices with olive oil and dust with dried, ground rosemary. (Dry your own at home and put it in a spice mill, small blender or a coffee grinder that you keep specially for grinding herbs or spices.) Lay the slices of potato on a large sheet of tin foil in a baking dish with a few knobs of butter, and fold the sides of the tin foil loosely over the top like a parcel. Bake in a medium oven until the flesh is tender and just starting to turn crispy at the edges, and eat hot. Rosemary potatoes are fragrant and delicious.

GROWING EXTRA-EARLY NEW POTATOES UNDER COVER

For really early potatoes, buy seed potatoes as early as possible (January, ideally) and chit them immediately. When the shoots are 6mm (¼in) high, wait for a spell of mild weather (around early March) and plant three tubers 13cm (5in) deep in a 38cm (15in) pot filled with a half-and-half mixture of John Innes No. 3 potting compost and multi-purpose compost, or plant six in a growing bag and keep it in a frost-free greenhouse or conservatory with plenty of light. Once all risk of frost is past, the container can be moved outside if you need the space for other things, but for the very earliest crops, continue growing them under cover. You could be picking your first new spuds by mid-May.

Early varieties with compact foliage, such as 'Mimi', are especially good for growing this way; plants with normal, tallish foliage benefit from some support to stop the stems flopping and breaking, which reduces yields – the sort sold for supporting bushy herbaceous plants are ideal.

HARVEST

Earlies can be ready from early June, second-earlies from mid-June or early July and maincrop potatoes from September into the autumn. To harvest the crop, dig up each plant individually with a fork, pushing it in 30cm (12in) or so away from the base of the plant and lifting the soil gently. Take a handful of stems at the base of the plant and gently pull – a lot of the potatoes will come up with the roots. Sift through the soil for the rest before moving on to the next plant. Earlies often produce enough fair-sized tubers for a meal even before the plants start to flower (flowering is a good indication that the crop is ready), so feel around gently in the soil close to the base of the plants with your fingers, and you can usually ferret a few out without pulling the plants up. Meanwhile, they'll keep growing.

After digging up your earlies, move on to the second earlies and lift one or two plants as you need them – the rest will just keep fattening until you dig them up. Maincrop spuds keep quite well in the soil even after the haulm (the technical name for the foliage) has died down at the end of the season, but they need lifting before the weather turns really wet otherwise the tubers can be attacked by little black keeled slugs. They may also start to grow again, which spoils them and adversely affects their keeping qualities.

STORAGE

Early and second-early potatoes keep well in the salad drawer of the fridge or a cool airy shed or garage for a week or so, so only lift small quantities as you can use them.

Store perfectly dry maincrop potatoes in sacks made of paper or hessian, or spread them out in 'stacking' trays, making sure there is room for air to circulate between the layers. Keep them in the dark to prevent sprouting (unlike onions, which need to be kept in the light).

PROBLEMS

Potatoes are often bored into by keeled slugs, which live underground. Once they've been damaged, affected tubers are then very prone to rotting. Slugs are far worse on ground that's recently had a lot of organic matter dug in, particularly if it wasn't very well rotted, so only plant potatoes on ground that had compost or manure dug in at the start of the previous season, so it has been largely broken down to humus by the time you need to plant. Another good remedy is to use the biological control for slugs (see page 51), since these beneficial eelworms that are released into the soil attack slugs even when they are underground. Using slug pellets on the soil surface has little or no effect on keeled slugs (and too much of an adverse effect on wildlife). Where slug damage is a regular problem, look for varieties with in-bred slug resistance.

Tubers affected by scab look unattractive; they will be covered in irregular, roundish, corky patches on the skin. The disease is more prevalent on light soil that dries out badly in summer, especially on chalky ground, so adding plenty of organic matter a season before planting does help. But scab looks worse than it is and is really only a disease that's skin-deep – once potatoes are peeled or skinned they remain perfectly safe to eat, with no sign of scab.

Potato blight is the most serious potato disease; it's the one that was responsible for the Irish Famine in the 1800s. Outbreaks are most likely during a wet summer, when affected plants start to develop brown patches on the leaves in August and the foliage soon yellows and dies off. (You can diagnose blight as against other possible causes, since in damp conditions you'll see white fungal rings around brown spots on the backs of the leaves.) Early potatoes are rarely affected, since the tubers have usually been lifted before blight strikes, but if we have a rainy July, start spraying with Bordeaux mixture to prevent the disease. If you see the first brown spots, spraying will slow the advance of the disease but once it gets hold there's no cure. Dig up affected crops straight away and use tubers that are sound as soon as you can, since they rot quickly in store. Don't put the remains of affected plants, even peelings or foliage, back on your compost heap.

Several different viruses affect potatoes, too, which can be seen when the edges of the leaves roll inwards, or leaves may develop yellow mosaic patterns. Viruses are often spread by aphids feeding on the leaves, but often occur when people have saved their own tubers to replant. Affected plants are stunted with low yields. There is no cure, so avoidance is the best policy.

How to have new potatoes for Christmas

Several seed firms now supply a small range of seed potatoes that have been carefully cold-stored to keep them dormant for late planting; this produces very late crops of baby new potatoes in autumn and over Christmas. These can be found in their autumn catalogues. It's essential to order these as soon as the catalogue arrives, around late June, as you'll only have a worthwhile crop if you can get them planted in July or early August.

Plant without chitting them first, in a warm, sheltered part of your veg or salad patch. Planting in large pots is even better, since the containers can be moved under cover to continue growing when cold weather starts in autumn. Crops are usually on the light side and results aren't guaranteed – they are very weather dependent. But all being well, you should start finding the first baby spuds from October, and you can continue using them straight from the ground in November and December, as long as the weather stays kind. Being new potatoes they won't keep for long once they're out of the ground, but if a long freezing spell threatens, get them up and into the salad drawer of your fridge.

Alternatively, keep back a few potatoes from your first early crop and put them in a biscuit tin of damp sand. Put the lid on and bury the tin at the end of the veg plot, marking the spot with a cane. You can unearth the tin on Christmas morning and, with any luck, you'll have a few 'new potatoes' as part of your festive fare.

Purslane

Portulaca oleracea var. *sativa*

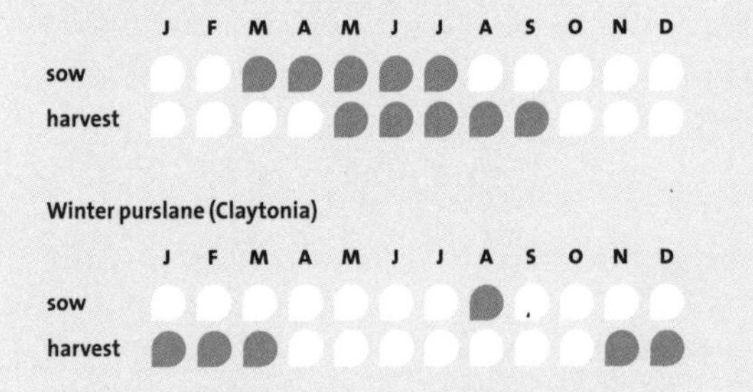

Salads, especially lettuce and baby spinach leaves, are notoriously difficult to grow well in a hot summer, but it is exactly that situation in which purslane thrives. It is compact enough to grow in patio tubs or indoors in a pot on the windowsill, so if you just fancy growing a little to try, it's ideal.

Purslane leaves are thick and succulent with a unique flavour and texture you'll either love or hate – add it to mixed green salads or eat it on its own with a French dressing. But, being a rather unusual crop, seeds are rarely available from the big seed firms – they usually have to be obtained from specialist suppliers.

HOW TO GROW

Degree of difficulty: Easy and no bother, given warm, well-drained conditions.
Sow: March to July.
Spacing: Thin seedlings out to approx 7.5cm (3in) apart; sow rows 30cm (12in) apart.
Routine care: Keep well weeded. Watering is rarely needed; it is a very drought-resistant crop.
Harvest: Start picking lightly as soon as plants are big enough, taking just the tips of shoots, about 2.5cm (1in) long. The same plants can be picked over regularly throughout the summer, even when they are trying to flower.

VARIETIES

1. **Golden Purslane** – a rare plant with pale gold leaves, more attractive on the patio but a slow, weak grower so you'll pick a lot less than with green purslane.
2. **Green Purslane** – the most readily available variety, making small, bushy green plants with ovalish leaves, about 15cm (6in) high with good flavour.
3. **Winter purslane, or miner's lettuce (*Claytonia perfoliata*)** – this is an unusual British native plant producing small rosette-shaped plants with thick, fleshy, succulent stems and leaves, which makes a handy and easily grown winter salad leaf. Sow seeds in August in a tub or trough on a warm, sheltered patio, or in the soil border of an unheated greenhouse, cold frame or polytunnel. Pick a few leaves as you need them from November until March.

1

2

3

PROBLEMS

Slow growth might occur in cold, wet conditions. A hot, dry summer or growing under cover suits it best.

Radishes

Raphanus sativus

Summer radishes

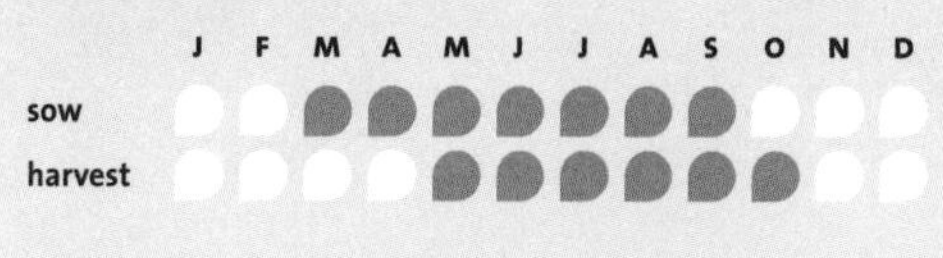

Winter radishes (including Chinese/Japanese/Mooli)

A favourite summer snack or salad standby, radishes are the quickest of all crops to grow but are not quite as foolproof as people often imagine. Besides the familiar summer salad type, there are also some unusual oriental and winter radishes which have large roots ready to pull in autumn and winter – they are worth trying if you are an adventurous gardener–cook.

HOW TO GROW

Degree of difficulty: Need good growing conditions, but little work subsequently as they grow so fast.
Sow: Summer radishes from March to early September; winter radishes from early to late July, in shallow drills where you want the plants to crop.
Spacing: Thin seedlings of summer radishes out to 2.5cm (1in) apart; allow 15cm (6in) between rows. Thin winter radishes to 5–7.5cm (2–3in) apart and allow 30cm (12in) between rows.
Routine care: Good, rich, well-drained, fertile soil is essential, but don't use ground that's recently had organic matter dug in or roots may split or fork. Keep plants watered and well weeded so that the roots grow steadily without a check. Thin out seedlings early – being so fast-growing it's easy to forget them.

EDIBLE-PODDED RADISHES

It's usually a disaster when radishes bolt, but one variety is *meant to* because it's grown especially for its edible seedpods. 'Munchen Bier' is so-called because Germans enjoy eating the pods with a glass of beer. Its alternative name – of rat's-tail radish – is less complimentary but probably more descriptive, since the dangling mass of long, slender, tapering pods do indeed look like pale green 'tails'. Consider it to be one of those oddities that is fun to grow once, just to see what it's like. Sow seeds March to late June and pick approximately 50 days after germination, while the pods are 15cm (6in) long, green and still growing – don't wait until they are old and tough. Besides being eaten raw as a cocktail snack, they can be stir fried with other veg.

VARIETIES

1 **'Black Spanish'** – a winter radish with large, round, black roots, that have a crisp white interior; used in same ways as 'Mantanghong' but with a hotter flavour. Also good for oriental stir fries.

2 **'French Breakfast'** – a traditional cylindrical summer radish with a long red top that has a small white tip at the base; very reliable.

3 **'Mantanghong'** – a large, round, Chinese winter radish growing to the size of a tennis ball, also known as 'Beauty Heart', with green rind over brilliant red flesh. It is crisp and sweet with hardly a hint of the usual hot radish taste; good in salads, but best for slicing thinly as fresh 'vegetable crisps' or cutting into batons to use as crudités.

4 **Mooli** – there are several different named varieties of the long, slim, tapering-cylindrical white roots of Japanese winter radish, known collectively as mooli. All are very similar, with crisp, white, hot-tasting roots used in oriental cookery. They can grow massive – roots 30cm (12in) long aren't unusual – one thing's for sure, you won't need many.

5 **'Scarlet Globe'** – a very popular, traditional, round, cherry-red summer radish, good outdoors but can also be sown early or late under cover for out-of-season crops.

6 **'Sparkler'** – a radish with crunchy spherical roots that are red on top and white underneath and which last particularly well.

HARVEST

Start pulling the biggest summer radishes from a row as soon as they reach useable size; check rows daily as they grow very fast – don't wait until they grow too big or they may split, or become tough and woody, or else run to seed.

Winter radishes grow very much bigger; start digging up the biggest (they are too big to pull up like summer radish) from late August and continue until mid-November. Any left in the ground by then should be dug up and stored in a dry, frost-free shed. Mooli keeps quite well until after Christmas, but use other winter radishes as soon as possible.

PROBLEMS

Poor soil, hot or dry conditions can all cause bolting. Overcrowding or leaving it too late before thinning seedlings out may mean roots can't develop – with such a fast growing plant, early thinning is essential.

Lots of small round holes in the leaves are invariably due to flea beetle; a severe attack can kill small seedlings but is unlikely to harm bigger plants. Attacks are worst when soil is dry, so keep the crop well watered. Stand a few chunks of rotting log around the crop; adult beetles (small brown and buff striped beetles that jump like fleas) often gather underneath by day and can be collected up.

Rocket

Eruca sativa

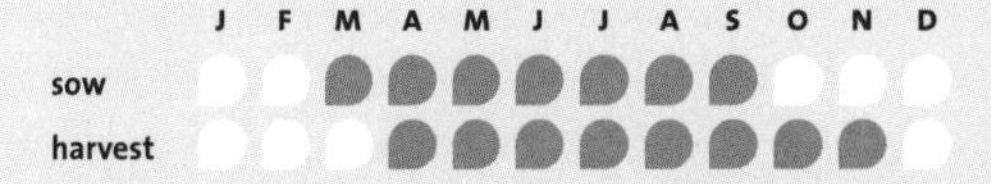

A previously almost unknown salad leaf that shot to fame when it was 'discovered' by foodies and now turns up in fashionable salad patches from Islington to Ilkley. Since it became popular, several named varieties along with the one known as 'wild rocket', have become available, but there's very little to choose between them for flavour. It's the most foolproof salad you'll ever grow, but it does live up to its name and may run to seed before you notice.

HOW TO GROW

Degree of difficulty: Easy and virtually no work.
Sow: March to September outdoors. For out-of-season leaves, sow in cold frames, under cloches or in a greenhouse border in early March, and sow early September under cover for autumn use – sow in pots later in September to grow on the windowsill.
Spacing: Thin seedlings out to 10–15cm (4–6in) apart; space rows 30cm (12in) apart.
Routine care: Water regularly and keep well weeded. Plants run to seed fairly fast in summer, so make several successional sowings through the season to replace old plants. Old plants will often re-shoot if cut back to about 5cm (2in) above ground level, but they soon run to seed again. If allowed to flower, rocket self-seeds slightly, so plants may replace themselves.

VARIETIES

1 **Rocket** (*Eruca sativa*) – deep green, oval leaves with a good flavour; they make a great salad with walnut oil and chunks of avocado and pine nuts. The cross-shaped, four-petalled, white flowers with purple veins are also edible, tasting faintly of hazelnuts and are ideal for use as edible decorations in salads and garnishes.

2 **Wild rocket** (*Dipsotaxis tenuifolia*) – a different species, with mild-flavoured, serrated-edged leaves and tall stems of yellow flowers. Self-seeds slightly; plants are longer lasting than normal rocket.

HARVEST

Start picking a few leaves as soon as they are big enough to use, leaving the rest of the plant to keep growing. Outdoors, you'll be picking rocket from April to October. Early and late sowings under cover can extend the season into March and November, or later, and you can have rocket from plants on a windowsill indoors all through the winter if there's plenty of light.

Salad leaves, mixed

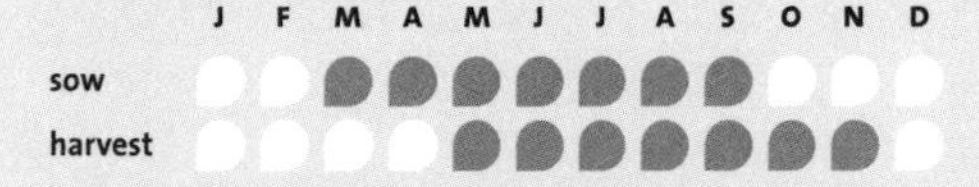

Whether it's due to people watching their weight, or the trend for healthy living and eating in general, mixed green salads and baby salad leaves have never been more popular. But instead of buying chlorine-washed, gas-packed bags of leaves from the supermarket at fancy prices, it's dead easy to set up your own production line at home – either outdoors in a small intensive salad patch with rich, fertile soil, or in a few large tubs or troughs of multi-purpose compost on the patio.

Salad leaves are also good for growing under cover early and late in the season, and you can even raise a few batches in midwinter in pots on the kitchen windowsill indoors, provided there's enough light. You can sow a pre-mixed selection (a collection which consists of several separately packed varieties inside one large outer packet) or buy your favourite individual varieties to grow and blend together in salads as you wish.

But however you grow them, these are short-lived crops that need re-sowing regularly if you want to keep yourself constantly supplied.

HOW TO GROW

Degree of difficulty: Easy and takes little time.
Sow: Late March to August outdoors, under cover in early to mid-March and September, and try some on windowsills indoors in October and February. Choose good, fertile soil with plenty of organic matter; in containers use multi-purpose compost. Sow thinly where you want the plants to crop; in containers and intensive salad beds where there's unlikely to be a problem with weed seeds, scatter the seeds evenly over the surface of the area and barely cover with more compost; elsewhere, sow in rows for easy hoeing.
Spacing: Thin seedlings out to 2.5–5cm (1–2in) apart. If growing in rows, space them 15cm (6in) apart.
Routine care: Keep well watered and weeded so they grow without a check. Protect outdoor crops with fleece in cold or windy weather early in the season.

GROWING INDOORS IN WINTER

Fill a windowsill trough, half-sized seedtray or a series of small pots or punnets with multi-purpose compost and sprinkle the seeds on top. Water sparingly and keep on a bright windowsill. Thin out seedlings slightly as they come up, but allow them to grow relatively thickly. As they start to reach a useable size, begin snipping small quantities from one end of the container, as for mustard and cress. When the container is half used, sow another one – this should come on stream just as you finish the first. During very dark days results are less good and plants far slower growing, so if you only do two batches, sow them in October and February.

VARIETIES

1 **'Mesclun'** – a varied mixture of oriental and Mediterranean salad leaves whose exact composition may vary from different suppliers, but it usually includes rocket, red kale, Chinese mustard, chervil and mizuna.

2 **Oriental mixtures** – may include mizuna (pictured), pak choi, red mustard and Chinese cabbage.

3 **Italian mixtures** – include basil (pictured), leaf chicory, 'Lollo Rossa' lettuce, rocket and radicchio.

Recipe – Home-grown green salad

Take a screw-top jam jar, add equal quantities of freshly squeezed lemon juice, walnut oil and pumpkin seed oil, season with salt and pepper and add half a teaspoonful of runny honey. Shake well until mixed.

Rinse your selection of salad leaves, choosing only perfect unblemished, young green leaves, dry gently in a salad shaker and tip into a large bowl. Sprinkle with chopped spring onions, drizzle with the dressing and mix gently with clean hands. Finish off by sprinkling pine nuts, whole green pumpkin seeds or hulled sesame seeds (or one of the mixed-seed sprinkles sold in health food shops) sparingly over the top.

HARVEST

Start snipping individual leaves or cut whole plants 6mm (¼in) above their base as soon as they are big enough to use, leaving the rest of the plant to keep growing. The same plants can usually be snipped over several times before they start running out of steam or going to seed.

INDIVIDUAL SALAD LEAVES TO GROW SEPARATELY AND MIX TO TASTE

Baby spinach leaves (for spring and autumn)
Basil
Beetroot 'Bull's Blood' (grown for red leaves for salad use, not roots)
Chinese cabbage
Claytonia (for winter salads)
Corn salad/lamb's lettuce (for winter salads)
Dandelion
Land cress
Lettuce
Pak choi
Purslane
Rainbow chard (use whole leaves of small seedling plants)
Red mustard (the new variety 'Red Frills' has delicate lacy leaves, the best for salads)
Rocket
Watercress
Winter purslane

PROBLEMS

Often one variety will take over pre-mixed salad blends and it's very common to find the strongest, fastest-growing variety comes up first and swamps everything else. Thin seedlings out while they are young and weed out some of the strongest to maintain a better balance of varieties.

Bolting is usually caused by hot or dry conditions, poor soil with insufficient organic matter or growing plants too thickly, though salad leaves don't go on for too long before you need to sow a new batch.

Shallots

Allium cepa

	J	F	M	A	M	J	J	A	S	O	N	D
plant		●	●									
harvest							●	●				

Once thought of as something only for pickling, shallots are now seen as the upmarket cousins of onions, preferred by foodies for having a finer, milder flavour. They are easy to grow, just like raising onions from sets, except that instead of only getting one larger onion, each shallot surrounds itself with a cluster of 5–6 offsets that all plump up equally. And though, in theory, most of the diseases that affect onions and leeks can also affect shallots, in practice I find them quite trouble free.

HOW TO GROW

Degree of difficulty: Easy and little work.
Plant: In February or March, with a trowel, so that they are just covered by soil. If you leave the tips showing, blackbirds will pull them out.
Spacing: 20cm (8in) apart, in rows 30cm (12in) apart.
Routine care: Keep well weeded and water in prolonged dry spells. Take care not to break the leaves when weeding.

VARIETIES

1. **'Golden Gourmet'** – large, good-quality, golden-brown bulbs. Good flavour and a good keeper.
2. **'Griselle'** – a French variety renowned for its flavour, best for planting in autumn and used straight from the garden while still growing the following summer, in much the same way as overwintering onions.
3. **'Hative de Niort'** – the very neat, small, globular, identical shallots you see on show benches, fun to grow for the village show and quite okay to eat afterwards, though this variety does not 'bulk up' like most shallots. It can be difficult to find stock for planting.
4. **'Jermor'** – produces rather tall, lean, upright shallots of the type often seen in greengrocers as 'banana shallots'. Coppery skins and pink-tinged flesh with a superb flavour.
5. **'Red Sun'** – rounded bulbs with rich, reddy-brown skin; good flavour and a long keeper.

HARVEST

When the leaves turn yellow naturally in July or August, lift the clumps carefully with a fork, shake off the worst of the soil and place them on the ground to dry off in the sun.

STORAGE

Spread out in layers in shallow trays in a frostproof shed or garage, or hang them up in a net in the shed roof to store for winter use.

PROBLEMS

Sets may get pulled up shortly after planting – but I did tell you to cover them up ever so slightly. Birds are the culprits, presumably in the hope that there'll be grubs underneath, so go round and replant them with a trowel so that the tops are just covered. The problem usually stops after a few weeks when the shallots start taking root.

Sets failing to grow large or make many offsets are usually troubled by poor growing conditions; maybe poor, infertile soil or weather that is too cold and wet, or in a hot summer lack of water may cause the same trouble. But mostly it's due to allowing the crop to be overrun by weeds – shallots don't like competition.

Sorrel

Rumex acetosa

	J	F	M	A	M	J	J	A	S	O	N	D
plant			●	●								
harvest						●	●	●	●			

Usually thought of as a herb, sorrel is 'on the cusp' since, when you grow enough of it, it can be used as a vegetable and cooked like an upmarket spinach; it has a similar texture but a rather sharper, more lemony flavour. It's good for growing in a herb garden, salad bed or in tubs on the patio.

HOW TO GROW

Degree of difficulty: Easy and hardly any work involved.
Sow: March/April in rows or in a small patch to grow into a clump; it can also be sown straight into pots to go on the patio.
Spacing: Thin seedlings out to about 10–15cm (4–6in) apart, and once well established they can be thinned out further if you want large leaves – to 15–20cm (6–8in).
Routine care: Little needed, other than watering in long dry spells. Plants are more drought tolerant than most salad leaves or spinach. Remove flower spikes when the plants attempt to run to seed; they will continue producing leaves as long as you don't let them flower. The same plants will often come up again in their second year, but it's better to sow a new row, as young plants are more productive.

VARIETIES

There is only one species, ***Rumex acetosa***, and it resembles a small-leaved dock. It is a British native plant, but don't pick it from the countryside as it'll like as not be tough and possibly contaminated by pollution or traffic fumes. Cultivated plants will grow leafier and more tender.
Buckler-leaved sorrel *(Rumex scutatus)* – this is a relative of regular sorrel but much smaller, with neat, arrowhead-shaped leaves about 2.5cm (1in) across, growing on a small, slightly spreading plant about 20cm (8in) wide and 15cm (6in) high. The leaves have a thick, succulent texture and a fresh, lemony-spinach flavour. This is an excellent, long-lived, salad-leaf plant for using fresh in mixed green salads or in sandwiches. Grow the same way as for normal sorrel; one or two plants in a pot or a tub on the patio are enough, or grow them in a salad bed. Plants often reappear the following year, but they self-seed very slightly so you'll usually need to grow your own replacements. Seedlings will transplant if you take care.

HARVEST

Start picking individual leaves when they are big enough to use; young but just about fully expanded leaves are best for culinary use. Leave the rest of the plant to keep growing.

PROBLEMS

None really.

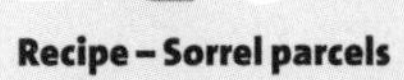

Recipe – Sorrel parcels

Large undamaged leaves of sorrel are brilliant for wrapping round food such as lamb noisettes or spoonfuls of rice mixtures when you are baking them in the oven. (Cover the tray with tin foil to stop the outer layer of leaf drying out and blackening.) Use them also for wrapping lamb chops before putting them on the barbecue; the leaf blackens and blisters but peels away leaving a lot of the flavour behind, while stopping the meat from drying out. Sorrel also makes a delicious soup.

Spinach

Spinacia oleracea

Summer spinach

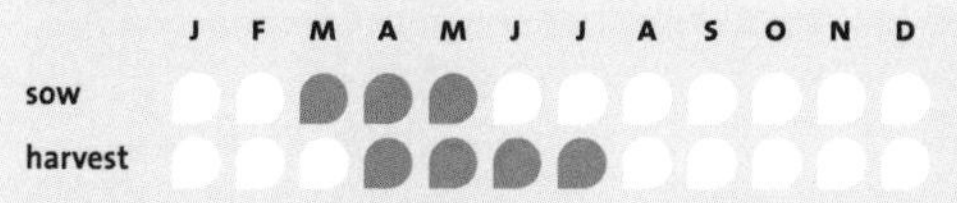

Autumn and spring/autumn spinach

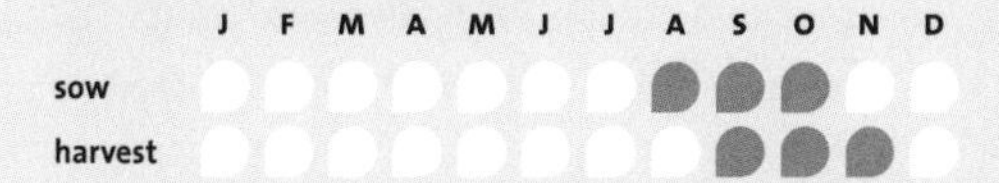

Popeye's favourite has seen a sharp surge in popularity since the arrival of neat pre-packed bags of whole young spinach leaves in the shops. They only need a quick rinse before popping them in the pan and they cook in minutes. Baby spinach leaves have also taken off as ingredients of mixed salad bags.

But spinach is quick and easy to grow and for the price of a packet of seeds you can be eating it through spring and early summer, with another flush or three in autumn. The range of varieties on sale in seed catalogues is constantly changing, so check sowing times – not all are suitable for autumn sowing. Most of the newer varieties can be eaten raw as baby leaves in salads, then as they grow bigger you can cook them.

The great thing about home-grown spinach is that you never get fed up with it, simply because it won't grow in summer when the weather is hot and dry.

HOW TO GROW

Degree of difficulty: Not difficult and takes up little time.

Sow: March to May, and (some varieties only) in late August until early October. Sow thinly in shallow drills where you want the plants to crop, or scatter in large tubs on the patio or in an intensive salad bed. If you want them for baby salad leaves, keep sowing a new row or tub every 4–6 weeks during the sowing seasons.

Spacing: Thin out seedlings to 2.5cm (1in) apart for baby salad leaves, or 7.5cm (3in) apart for use as a green leafy vegetable.

Routine care: The secret of success with spinach is rich, fertile soil containing enough nitrogen, and regular watering and weeding. It's essential for the plants to grow steadily and fast without a check. Spinach rarely succeeds in summer, simply because the weather is too hot and the soil tends to dry out; plants bolt so quickly, too, that you're better off growing more tolerant salad leaves instead during that time.

Recipe – Spinach in filo parcels

Steam and finely chop the spinach. Stir in a generous knob of butter, some crushed fresh garlic and season to taste. (As an optional variation; for a real luxury finish add several spoonfuls of double cream or crème fraîche as well to make creamed spinach.)

Take four oblong sheets of filo pastry; lay one in the centre of a greased baking sheet and brush with melted butter, lay the next on top and again brush with melted butter – keep going until you have a pile of four on top of each other, all layered with melted butter. Place a neat mound of spinach towards one end of the filo and either sprinkle with pine nuts for a vegetarian version or sit a piece of boneless salmon fillet on top of the spinach.

Finally, wrap the vacant end of the filo pastry over the top and tuck under the far end to make a parcel. Tuck the other ends under to make a good seal, then brush the top with – you've guessed it – more melted butter.

Bake in an oven preheated to 190°C/375°F/Gas Mark 5 for 20–25 minutes until the pastry is crispy and browned on top. Make one large pastry to share, or several small individual ones.

VARIETIES

1 **'Bordeaux'** – a novel variety with bright red leaf stalks and leaf veins, which makes for very colourful baby leaves when used in salads. Sow from February to May and July to September.

2 **'Galaxy'** – a good mildew-resistant variety for baby leaves that can be sown September or March outdoors, but also through the winter in a greenhouse, a cold frame or walk-in polytunnel, or in a seed tray or trough on a bright windowsill indoors for fresh, out of season crops.

3 **'Mediana'** – useful all-round variety for sowing in spring or summer to produce 'baby' spinach leaves, also in autumn to grow under cover for cutting the following spring.

Popeye was right ...

Spinach is good for you – but it's not the iron content that works the magic; spinach is virtually no different in that respect from other green leafy veg. It is, though, rich in dietary fibre, vitamins A and C, beta-carotene and lutein, which is thought to help protect against deteriorating eyesight in later life due to macular degeneration.

HARVEST

Start picking baby leaves as soon as they are big enough to use, usually within a month of sowing, and cut the crop little and often. If you are growing a regular, large-leaved variety for use as a green vegetable you can still use small plants for salads when you're thinning a row of seedlings. Start cutting large leaves when the plants reach a suitable size – don't wait too long or the plants will start running to seed and the quality of the leaves deteriorates. Expect to start picking full-sized leaves 6–8 weeks after sowing, depending on the time of year and temperature, but they will keep going for a few weeks only.

PROBLEMS

Downy mildew can affect crops, particularly those grown under cover, if conditions are less than perfect or in cold, dull, humid weather, but growing modern disease-resistant varieties prevents all but the worst incidences.

Bolting is easily triggered by high temperatures, shortage of water or poor soil with insufficient organic matter, but all spinach plants run to seed when they've reached the end of their useful life. Spinach plants are not long-lived and need sowing frequently if you want to keep yourself supplied through the season.

Spinach, New Zealand

Tetragonia expansa

	J	F	M	A	M	J	J	A	S	O	N	D
sow	○	○	○	●	●	●	○	○	○	○	○	○
harvest	○	○	○	○	○	●	●	●	●	●	○	○

Although 'proper' spinach won't grow in summer and hates hot dry weather or poorish soil, New Zealand spinach positively thrives in those conditions. It's an unusual sprawling, bushy, leafy vegetable that's a great choice for a Mediterranean-style summer, and it's very good to grow under cover in a year that's less hot and dry. The same plants keep going all summer and continue cropping right up to the first proper frost, which finishes them off. Under cover, without heat, the cropping season is slightly longer and the plants often shed seed, so next year's batch might come up all on its own. The seedlings can be transplanted to wherever you really want them if you are careful, but it's as well to have plenty of spares as they won't all take.

To eat New Zealand spinach, steam the tender young shoots, chop them and use exactly as you would normal cooked spinach – this is not a vegetable to eat raw in salads.

HOW TO GROW

Degree of difficulty: Not difficult and hardly any work involved.

Sow: Late April/May or June in rows where you want the plants to crop. Choose a hot, sunny, sheltered spot with well-drained soil for best results.

Spacing: Thin the seedlings out to 15–20cm (6–8in) apart. Allow 45cm (18in) between rows.

Routine care: Water sparingly, if needed at all; keep well weeded until plants cover the ground, when they will smother out all but the most determined upright weeds by themselves. The most important job is to pick regularly as this keeps plants bushy and leafy; if they are left unpicked they develop long, stringy stems with few usable shoots.

VARIETIES

New Zealand spinach is a botanical species and no named varieties are available. Plants are low and sprawling, about 30cm (12in) across and as high, covered in thick, fleshy, triangular leaves about 2.5cm (1in) across.

HARVEST

Start picking as soon as plants are a few centimetres/inches high; cut or snap off the tips of shoots about 2.5–5cm (1–2in) long. Pick little and often, at least two or three times a week to keep plants productive. You'll usually be picking constantly from June to October.

PROBLEMS

If the plants grow slowly or fail completely, this is a sign of a cold, wet season or overwatering.

Spinach, perpetual

Beta vulgaris subsp. *cicla* var. *cicla*

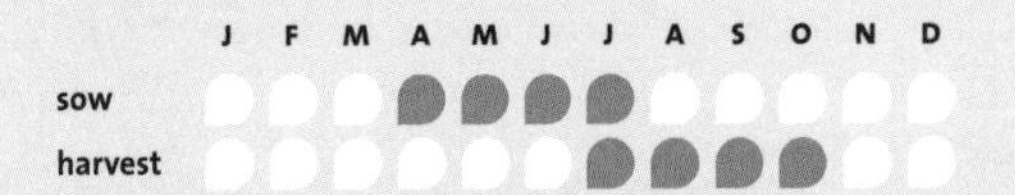

If you like spinach as a cooked vegetable, you don't have to do without it in late summer or winter when 'real' spinach isn't possible, just grow its big brother perpetual spinach instead. It's very much like Swiss chard but without the thick white midribs, and you can use it in just the same way. You don't need to keep sowing it as you do with 'real' spinach, either; the same plants will keep going for many months, including right through the winter if the weather conditions stay reasonably mild, so it requires very little labour.

HOW TO GROW

Degree of difficulty: Easy and little work.
Sow: April to mid-July; sow thinly in rows where you want the plants to crop.
Spacing: Thin seedlings out to 15cm (6in) apart and allow 30cm (12in) between rows.
Routine care: Keep plants watered in dry spells and weed regularly.

WINTER CROPS

For a good crop through the winter, sow a row or two of perpetual spinach in mid-July under a walk-in polytunnel or cold frame, or in a soil border in an unheated greenhouse. These plants will stay in perfect condition through the winter and start growing again in spring, producing enormous crops of large, tender, perfectly unblemished leaves until the plants finally run to seed in May. If you can't grow a winter crop under cover, sow in July and cover the plants with fleece over winter, which protects them from damage from wind or birds. They will grow out in the open without protection, but many leaves will be unfit to use if they are badly weather-beaten.

VARIETIES

Perpetual spinach, also known as spinach beet, is a botanical species and named varieties are rarely available. Plants form a clump of large, short-stemmed, thick-textured leaves that are shaped like very big, robust, green beetroot leaves about 30cm (12in) high.

HARVEST

Start picking individual leaves as soon as they are big enough to use; don't over-pick plants – cut little and often from all over the row and allow plants to recover between times. From an April sowing you'll be picking from mid-July onwards; a July sowing is ready to pick September/October onwards, and overwintered plants can still be picked until mid/late May.

PROBLEMS

None really.

Sprouting seeds

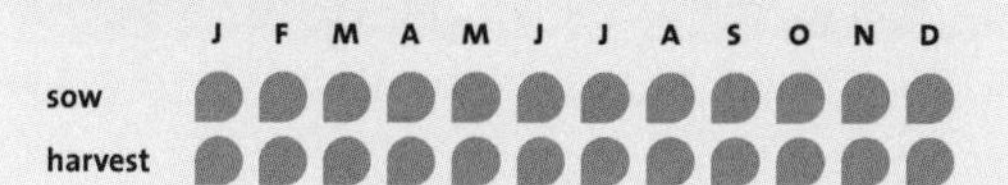

There's one edible crop you can grow without soil, pots, or even a garden – and that's sprouting seeds. Besides the well known Chinese bean shoots (mung beans), a good many other seeds can be sprouted to produce a home-grown supply of tasty, nutritious fresh salad ingredients. It only takes a few days. All you need is a jam jar (though you could splash out on a specialist sprouter from an organic supplier or a health-food shop) and some suitable seeds.

Don't try sprouting seeds that were intended for growing normally, since they may well have been treated with fungicidal seed dressing; go for those that are sold especially for sprouting. A fair selection is available in the seed racks at garden centres, but for the biggest selection look in specialist mail-order catalogues.

HOW TO GROW

Degree of difficulty: Easy, but need a little regular attention twice a day.

How to do it: Soak the seeds for an hour or so in tepid water and then rinse them well – the easiest way to do this is to put them in a fine sieve and hold them under a slowly running tap. Place them in a screw-top jar, but instead of the usual lid, cover with a piece of clean cotton or muslin fabric secured with an elastic band. Or, use a special seed sprouter, which consists of a series of perforated trays that allow you to rinse the seeds regularly without tipping them out of their container. Put the jar or sprouter in a warm place (around 18–24°C/ 65–75°F) – some people use the airing cupboard which, being dark, produces perfectly blanched salad sprouts, but you can also use a windowsill that's kept at a comfortable room temperature even in winter. This will give greener sprouts with a slightly stronger flavour.

Routine care: Sprouting seeds need rinsing two or three times every day at reasonably regular intervals – do it morning, lunchtime and evening if possible. With the jam-jar method the easiest way is to half fill the jar with water, swill it round to dislodge all the seedlings and then tip it out into a fine sieve, then hold it under a slowly running tap for a few minutes. Shake the seedlings semi-dry before returning them to their jar so they don't end up sitting in water. However, if the muslin is coarse enough, you can run the water through it into the jar and then tip it out again (holding the elastic band to make sure the contents do not fall out!) It's more reliable and quicker using a sprouter, as you simply hold the trays under the tap (some of the devices have several tiers so you can produce several different crops at once) then replace them still in position.

Recipe – Alfalfa salad

Take three handfuls of fresh alfalfa shoots, defrost one handful of frozen peas, and when just thawed but still cold, add to the shoots. Stir in a couple of finely chopped spring onions and toss with a little French dressing. You can vary this according to taste; try adding whole cherry tomatoes, chopped, de-seeded red pepper, a little watercress or corn salad, and/or thinly sliced Florence fennel or celery.

VARIETIES

1 **Mung bean *(Vigna radiata)*** – the original Chinese bean sprout, star of a million takeaways. Takes 3–5 days. Good for stir fries or using raw in salads.

2 **Aduki bean *(Vigna angularis)*** – slightly spicy-tasting beans with large seeds; they have rather thick skins that some people find a bit off-putting, unless they painstakingly separate the sprouts from the seeds and skins. Good for stir fries or salads, or add to your juicer (which gets rid of the 'bits' for you). Takes 3–5 days.

3 **Alfalfa *(Medicago sativa)*** – small, delicate shoots tasting faintly of fresh peas, and to my mind one of the very nicest for using raw in salads – brilliant with tomatoes. Takes 2–3 days.

4 **Fenugreek *(Trigonella foenum-graecum)*** – decidedly spicy, almost curry-tasting sprouts; perhaps rather an acquired taste. For stir fries. Takes 3–5 days.

5 **Radish *(Raphanus sativus)*** – small, rather hot and tangy sprouts, good for blending with a milder kind such as mung bean or alfalfa. Takes 2–5 days.

6 **Red clover *(Trifolium pratense)*** – slender sprouts similar to alfalfa, for salads. Takes 4–7 days.

HARVEST

Different types of seed vary in the length of time they take to reach a usable stage, and they develop faster at higher temperatures, but as a general rule sprouts are ready when the shoots are about 2.5cm (1in) long, though alfalfa seedlings are far smaller and ready at half that size or less. Once they are ready, give the seedlings a final rinse (getting rid of as many of the discarded seed coats as possible as they are a tad chewy), and if you don't want to use them straight away, they will keep in an open bowl in the fridge for several days without developing further or deteriorating.

PROBLEMS

If you get smelly sprouts that don't grow properly, it's usually a sign that they have not been rinsed often enough or thoroughly enough and as a result have started to ferment. Throw them away and start again after cleaning out the container.

Bent and twisty stems are nothing to worry about. The mung bean sprouts sold commercially are produced using a special technique that makes them grow very fast and very straight, which we can't reproduce at home. It's said that if you put a weight over a thin layer of mung bean seeds the shoots will grow straight as they press up against it, but I've never found a way of doing it so they aren't disturbed when they are rinsed, and they end up just as bent as usual.

Squashes and pumpkins

Cucurbita maxima, C. moschata, C.pepo

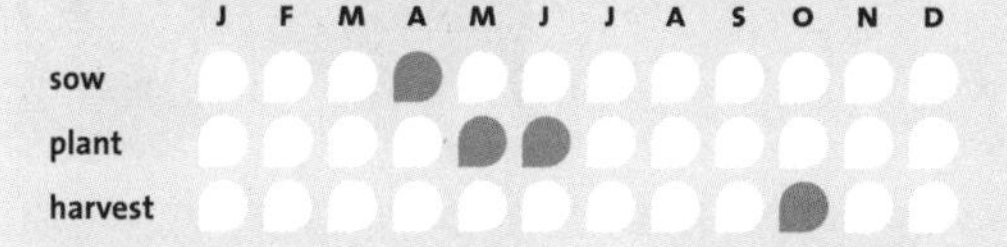

Squashes and pumpkins are some of the coolest 'new' veg around. I say new, but they are only new to us – they've been popular on the continent and in the southern USA for very many years, and in a lot of cases it's their varieties that are now reaching us. Our warmer summers are suiting them nicely.

The plants are enormous and need lots of room, but yields are not huge; expect two or three fruits from a large-growing pumpkin or squash and up to half a dozen from a small-fruited variety. But they are fast-growing, fun and quite spectacular; they don't all look like the sort of thing Cinderella turned into a coach, some are really quite curious.

Squashes generally make very good eating and are very decorative – great round the house in winter. Some pumpkins can be good eating, but they are mostly grown to use for Halloween lanterns or to enter in the village show. (For summer squashes, see also courgettes, page 107).

HOW TO GROW

Degree of difficulty: Easy and don't need much time, though extra-specially good soil preparation pays off.
Sow: Mid- to late April; sow seeds singly in pots on a windowsill indoors or a heated propagator in the greenhouse at a temperature of 18–24°C (65–75°F). Grow on in a coolish room at a temperature of 13–18°C (55–65°F), and gradually harden off young plants.
Plant: Outside when the last frosts are safely past, from mid-May to early June. For best results, plant in very rich soil; you could use a very well-manured part of the veg patch, or dig a large hole in autumn and spend winter filling it with materials you'd usually put on the compost heap, then cap it with a mound of soil in spring and plant on top of that. Where space is short, spread several inches of soil over the compost heap and plant a pumpkin or squash there. After planting, surround each plant with a low rim of soil, about 30cm (12in) out from the stem, and fill this 'basin' with water each time the plant needs watering.
Spacing: Allow a good 90cm (3ft) between plants, as they have long, trailing stems. If you have enough room for several plants, you can allow them to spread into each other's space without harm.
Routine care: Water the young plants in, and water sparingly at first. As the plants 'take off', water regularly during the summer to keep the soil moist. Regular liquid feeding is also a big benefit. Keep the area round the plants weeded until the vines cover the ground – the huge foliage will by then smother out most weeds naturally. Slowly reduce the feeding and watering towards the end of summer to encourage the fruit to start ripening, and as early autumn arrives, gently remove any foliage overshadowing the fruits to allow the sun to reach them and develop their full colour.

Culinary tip – Cooking squashes

When baking squash whole in the oven, bore a hole in it first – go right through to the seed chamber in the centre otherwise the fruit explodes as it heats up, which makes a dreadful mess of the oven, not to mention the kitchen if it blows the door open. It happens!

To roast squash, cut the fruit open and remove the seeds. Peeling is easiest done after cutting up. Cut the flesh into slices or 2.5cm (1in) chunks, then brush with olive oil. Either roast around the Sunday joint, or in a roasting pan with a little olive oil in the bottom, until tender and beginning to brown on the outside. Roast pumpkin pieces alone or with a mixture of other winter veg, such as carrots, parsnips and swede or celeriac.

To make squash soup, de-seed, peel and cube a good culinary squash and cook it in chicken stock along with a carrot, an onion or shallot and some garlic. When all the veg are tender, allow the mixture to cool then liquidize it. Add more chicken stock if it's too thick, although the consistency should be 'hearty' rather than 'thin and watery'. It will keep in the fridge for several days.

VARIETIES

1 **'Becky'** – regular, round, orange pumpkins about head-sized, ideal for Halloween pumpkins.

2 **'Butternut'** – regularly seen in greengrocers' shops and known for its bulbous-ended shape, buff colour and truly superb nutty flavour; slices are brilliant for roasting. The real thing rarely ripens even in a good summer in Britain, but now there are several butternut-type varieties of squash bred to do well in our conditions, so I'd recommend choosing one of those. Look for 'Harrier', 'Avalon' or 'Sprinter'. Their flavour is equally excellent.

3 **'Crown Prince'** – medium-sized, squat, steely blue-grey squashes with orange flesh, which are good for roasting but considered by a lot of fans to be *the best* variety for making pumpkin soup.

4 **'Sweet Dumpling'** – small, green-striped, cream squashes produced at the rate of 4–6 per plant. Delicious baked whole and then stuffed, or de-seeded, sliced and roasted.

5 **'Turk's Turban'** – a spectacular winter squash shaped a bit like a squashed cottage loaf in bright orange and patterned with cream streaks and green highlights. Very decorative but also very good eating; best roasted, when it has a distinctive roast-chestnut flavour.

6 **'Vegetable Spaghetti'** – long, cream-coloured, marrow-like squashes which, when allowed to mature on the plant until fully ripe in autumn and baked whole in the oven, contain strands of nutty-tasting vegetable spaghetti that can be removed with a fork and eaten with pasta sauces as a low-calorie alternative to real pasta. Watch out for the seeds in the centre; spoon them out carefully when you cut the cooked squash open.

GROWING GIANT PUMPKINS

If you want a whopper to wow them at the village show, the right variety is essential. 'Dill's Giant Atlantic' holds the world record of 1,021lbs, but you'd be lucky to get close to that without a bit of practice. You may also see 'Hundredweight', 'Atlantic Giant', 'Big Max' or 'Sumo Giant', depending on which seed catalogue you have.

Start plants off early indoors and prepare a large pit filled with manure or compost, capped with a mound of soil in spring. After planting, shortly after the last expected frost, cover the plant with a cloche or fleece for protection (but uncover the plant on fine days for ventilation), until it grows too big or the weather no longer requires it. Enthusiasts often grow the plants in polytunnels, but ventilation and shading are important in summer.

When the plant is in flower, wait until three fruits have set and just started to swell, then select the largest of these and remove all the rest so the plant directs all its energies into plumping up just one massive fruit. Keep the plant well watered and fed (tomato feed is ideal) and weed carefully so as not to disturb shallow surface roots.

To improve your chances, grow several plants, spaced 1.8–3m (6–10ft) apart. By having several giant pumpkins on the go at once, you can choose

whichever is biggest on the day – and meanwhile the rest will keep growing bigger, ready for the next show to come along. Really keen exhibitors help things along by barely burying the trailing stems in soil that's so well enriched it's like seed compost; the stems will then root as they run along – which all helps to swell the fruit.

HARVEST

Unless they are needed for an early autumn show or special event, try to leave pumpkins and squashes on the plants for as long as possible so they can ripen fully and gain the maximum girth and weight. If a frosty night threatens, though, play safe and cut them to bring under cover. Expect to pick them all by the end of October, when cold, damp ground could easily start the undersides of the fruit rotting.

STORAGE

Allow the fruits to dry in the sun, turning them over on their side so that the underneath can dry too, then move them into a dry, frost-free shed or garage where they should keep for several months. Check them over regularly and remove any that show signs of starting to rot. If you have space in a utility room indoors, or you can pile them up in a disused fireplace to make a winter display, they'll remain in much better condition far longer, due to higher temperatures and lower levels of humidity.

PROBLEMS

The same mosaic virus that affects marrows, courgettes and cucumbers can also affect pumpkins and squashes; if leaves develop a yellow mottling and plants become stunted, suspect a virus and pull them out.

Mice and larger rodents may nibble the skins of pumpkins, especially before they ripen and harden, so make sure compost heaps do not harbour unwanted wildlife, don't leave rubbish around and don't put food out for birds late in the day when it won't be cleared by nightfall.

Slugs may damage young fruits while the skins are very soft, and damaged areas deform as the fruits grow larger. It's not usually a problem unless you want a perfect specimen for showing, though wounds in a fruit may allow in fungal organisms, which cause the fruit to rot in a damp season.

Swedes

Brassica napus Napobrassica Group

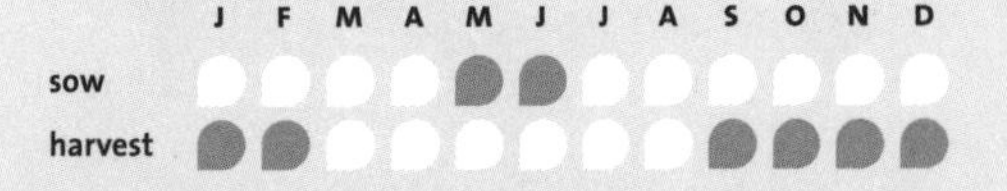

The great nineteenth-century century agriculturalist, William Cobbett, spends many pages in his epic book *Rural Rides* describing the progress of the then new-fangled 'Swedish turnip' crop on his travels through the countryside. Today, the fact that they are seen as animal fodder and are cheap in the shops makes them an unattractive proposition for gardeners, but I'm a fan of swede mixed with mashed potato and so I won't hear a word against 'em. What's more, if you want your swedes organically grown you are just as well to grow them yourself. You'll be pleased to know that they are a doddle.

HOW TO GROW

Degree of difficulty: Easy and don't need much of your time, but need the right growing conditions.
Sow: Seeds thinly in shallow drills where the plants are to crop. Sow in late May or early June – timing is critical. An open, airy situation in full sun gives best results; don't grow them crowded together or overshadowed by surrounding crops.
Spacing: Thin out seedlings in several stages until they are 23cm (9in) apart, allow 38cm (15in) between rows.
Routine care: Hoe regularly between rows to avoid competition and shading from weeds, and water in dry weather – the ground needs to stay fairly evenly moist, as wide fluctuations between wet and dry conditions can cause the roots to split and spoil.

VARIETIES

1. **'Invitation'** – a modern purple-flushed variety with similar looks to 'Marian' and resistant to clubroot and mildew; very winter hardy.
2. **'Marian'** – the old faithful with 'purple shoulders' and a creamy base, bred for flavour, but also very resistant to mildew and clubroot.

1

2

HARVEST

Start pulling swedes towards autumn as the roots become big enough to use; you don't have to wait until they are as big as the ones you see in the shops. Leave the rest of the crop in the ground as they keep better there – they'll 'stand' until the end of February. The only time you need to dig them up is if the ground is heavy clay or if a prolonged freeze is forecast.

PROBLEMS

Powdery mildew and clubroot are the chief problems, but these are easily avoided by choosing a resistant variety in the first place and by making sure the soil is well drained and not too acidic.

Recipe – Neeps and tatties

Peel and cube a swede, boil until tender then mash. Season with salt and black pepper and add a generous amount of butter. Then mix in mashed potato. (It's no good cooking the two together, since the neeps take twice as long to cook as the tatties, so the latter will have turned to soup before the neeps are tender.) To serve, sit a thick slice of haggis on a good bed of neeps and tatties, then slosh a 'wee dram' over the lot immediately before eating. Happy New Year! Well, happy any time of year...

Sweetcorn

Zea mays

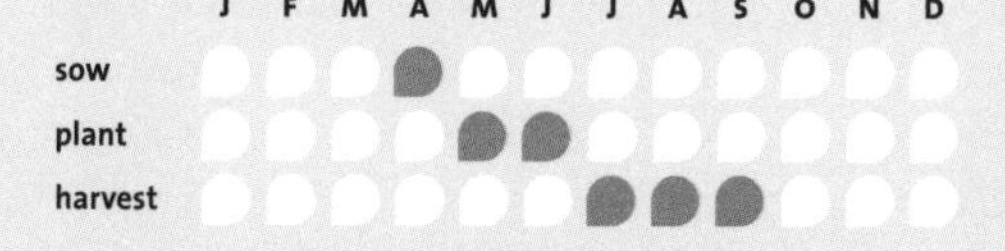

If you enjoy sweetcorn, then it's a vegetable that really is best when you grow your own because it needs cooking and eating within an hour of picking. (The sugar in the kernels starts turning to starch the minute it parts company with the parent plant, and even the journey home from a pick-your-own farm may mean it's tough and tasteless by the time it hits your plate. That's to say nothing of the whole cobs you buy from the shops which may already be several days old... This is why so many people only know sweetcorn as a frozen veg.) When you grow your own and see just how good it can really be, it's almost like discovering a totally new vegetable.

Sweetcorn isn't for everyone, though, and it does need a fair bit of space to grow; being wind-pollinated a single short row stands no chance of 'setting' properly so a decent-sized block of plants is essential. But it's an ideal crop for an allotment where there's room to do just that, and if several of your neighbours do the same, you'll be sure of a very good crop.

HOW TO GROW

Degree of difficulty: Not difficult, but needs some attention to detail and a fairly good summer.
Sow: Seeds in mid- to late April, one per small pot, on a windowsill indoors or in a heated propagator in the greenhouse, at a temperature of 16–21°C (60–70°F). Grow the young plants on at room temperature and gradually harden them off.
Plant: Outside after all risk of frost is safely past, between mid-May and early June.
Spacing: Plant in blocks, not rows, spacing plants 45cm (18in) apart in each direction.
Routine care: Water the plants in and cover with fleece on cold nights to help them establish quickly. Water in dry spells. No support is needed, even though the plants grow quite tall, as the stems hold each other up.

MINI-SWEETCORN

These are a new development; plants produce three or four very small, slim, almost pencil-like cobs 10–15cm (4–6in) long, which can be stir fried, eaten whole and raw as crudités with a creamy dip, or sliced up into rings and used raw in salads without attempting to detach the kernels – the central 'core' is fairly tender.

You'll need to choose the right varieties to produce mini-cobs; there are several and what's on offer changes frequently as new varieties come along, but 'Minipop' is widely available. Raise the plants as usual, but plant them much closer together – about 20cm (8in) apart – and pick the cobs *before* they are pollinated, while the tassels are still cream coloured and looking fresh and silky. You'll be picking earlier than with conventional varieties – test a few cobs regularly to check on progress, and don't leave them too late.

Recipe – Succotash

An old, traditional, native-American dish.

Melt a good knob of butter with some olive oil in a heavy frying pan and lightly 'sweat' a chopped red pepper and one large courgette that has been thinly sliced. Cut the kernels from two ripe sweetcorn cobs. (To do this, strip away the silks and green leafy sheaf, stand the cob up on its end on a chopping board then angle it slightly and, using a large heavy knife, press down hard to scrape each row of kernels off the cob. Rotate the cob until you've worked your way all the way round it.)

Add the kernels to the pan and continue cooking, stirring often, for another ten minutes. Sprinkle in a teaspoonful of sunflower seeds and add a little chicken stock to stop the mixture drying out. Eat hot.

You can introduce variations to the basic recipe; add chopped onions, cooked haricot or flageolet beans, garlic, a hint of fresh chilli, a handful of whole cherry tomatoes, or raw French beans cut into 2.5cm (1in) lengths at the start of cooking. Or sprinkle grated cheese over the top at the end. Enjoy!

VARIETIES

1 **'Applause'** – a very good, newish, F1 hybrid, supersweet variety.

2 **'Golden Bantam'** – an old, non-F1, rather late-ripening variety with 13cm (5in) cobs, and strong stems. Good for extending the cropping season when grown with an earlier variety, such as 'Sundance'.

3 **'Incredible'** – a sugar-enhanced variety, ripening mid-season, with a good flavour and very reliable.

4 **'Sundance'** – an F1 hybrid variety with well-flavoured kernels in 18cm (7in) cobs that mature early and crop reliably, even in the north of the country and in poor summers.

SWEET VARIETIES

These days, almost all of the sweetcorn varieties sold are either supersweet or sugar-enhanced; it's difficult to find seeds of old-fashioned, 'normal' varieties.

Supersweet sweetcorn varieties contain almost three times as much sugar as regular varieties; for some people this is just *too* sweet, so before buying seeds I suggest you buy a packet of supersweet and normal frozen sweetcorn and taste them to compare the two.

Sugar-enhanced varieties combine the best of supersweet and normal varieties, and they are said to remain sweeter for longer after picking than others.

If you choose to grow either of these types it's important not to grow them too close to regular sweetcorn varieties; allow a gap of at least 7.5m (25ft) and ideally more, since the pollen from normal varieties 'cancels out' the extra sweetness if the two have a chance to cross-pollinate.

HARVEST

The first cobs should start ripening from late July or early August, depending on the variety and the weather. Check plants regularly. Cobs are ripe when the silky tassels turn from creamy coloured to brown, but test each cob before picking it. Peel back enough of the green leafy sheath to expose a few kernels and press a thumbnail into one or two – if clear liquid spurts out the cob is not quite ripe; when it's ready to pick the juice is milky. To pick, hold the stem of the plant in one hand, close to the cob, and twist the cob cleanly off with the other – don't break or bend the main plant stems, as there will be a second cob that needs to be left to ripen later.

PROBLEMS

Poo pollination will cause gappy cobs to develop normally but when you peel back the sheaf they will have areas of bare cob with only a few kernels, or reasonably full cobs with empty spaces dotted around.

The larvae of frit-fly distort the growing tips of young plants so that they grow stunted and twisted and produce unusable, underdeveloped cobs. Protecting young plants with fine insect-proof mesh might help, but pull out and destroy affected plants.

Swiss chard

Beta vulgaris var. *flavescens*

	J	F	M	A	M	J	J	A	S	O	N	D
sow				●	●	●	●					
harvest							●	●	●	●		

Swiss chard has been an up-and-coming 'new' vegetable ever since calabrese started being rather over-used, but it's not something you can buy in the shops. It doesn't travel – the leaves wither and look very tired within a day of picking – so you have to grow it for yourself and use it fresh. It looks quite attractive in the garden, and as well as the original Swiss chard there are several versions with coloured leaf stalks which look stunning in a decorative potager. The great thing about chard is that it's a relative of beet, so if you can't grow members of the brassica family because your soil is affected by clubroot, you can still have your greens – just grow chard instead.

HOW TO GROW

Degree of difficulty: Easy; little work involved.
Sow: April to mid-July in rows outside.
Spacing: Thin the seedlings out to 15cm (6in) apart; allow 30cm (12in) between rows.
Routine care: Keep plants well watered in dry spells and weed regularly.

VARIETIES

1 **Swiss chard** – has long, thick, white, slightly flattened and ribbed stems topped by deep green, shiny, leaves with a lot of substance and texture. The plants reach 45cm (18in) high and 30cm (12in) across.
2 **Ruby chard** – has slender bright red stems topped by purplish-tinged dark green leaves; it's sometimes mistaken for rhubarb, but rhubarb that's only this big is in deep trouble. Ruby chard is not quite such a strong grower as Swiss chard and is prized more for its decorative qualities than its culinary virtues.
3 **Rainbow chard** – contains a mixture of slender red, white and yellow-stemmed varieties. None are as strong growing or robust as the original Swiss chard and, like ruby chard, the decorative qualities are its main asset. When you're thinning out rows of seedlings, save small plants to use as baby leaves in mixed green salads, or sow a few especially for this use.

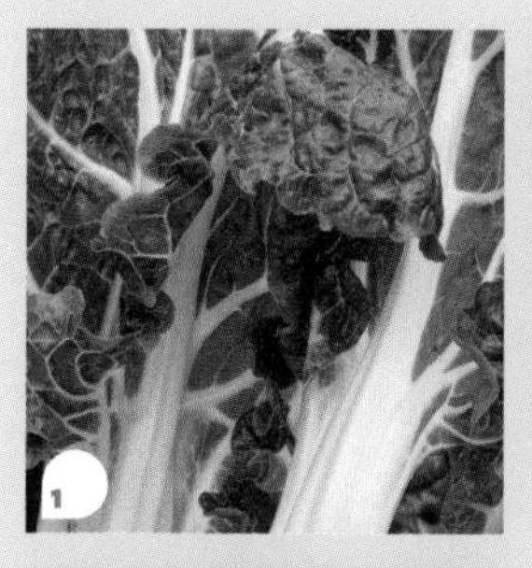
1

2

3

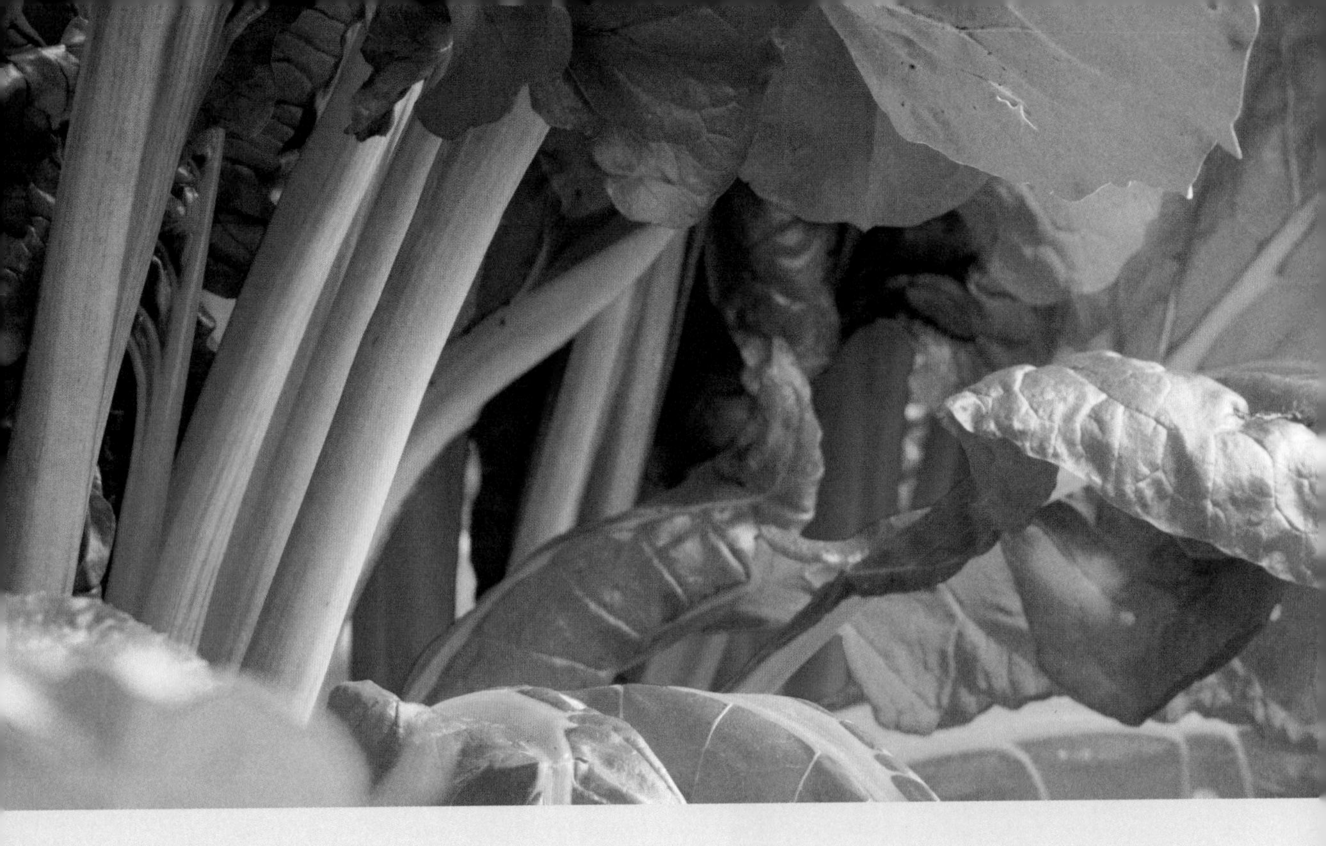

HARVEST

Start cutting a few individual leaves as soon as they are big enough to use. As plants grow older, choose only sound, healthy leaves and avoid old, yellowing, tough or tattered ones. Use a sharp knife to cut through the stem at the base of the plant, but take care not to slice into neighbouring stems. You can be cutting chard from July to October. Early-sown chard plants are usually just about exhausted by the autumn but, given a mild winter, July-sown crops often stand the winter, even though they stop growing. If there's a bad cold spell they may die down temporarily, but they grow up again in spring and produce another crop of tender young leaves until the plants run to seed in May. Either way, you get a good run for your money.

PROBLEMS

None really, it's a very healthy and trouble-free crop. Just watch out for snails, which sometimes stow away in the deep convolutions of large Swiss chard leaves in winter and create quite a stir when they turn up on someone's plate at Sunday lunch – and it's always a non-gardener with no sense of humour.

Culinary tip – Preparing and cooking chard

Once you've picked your chard and you're back in the kitchen, slice the green part of the leaf away from the thick stalks, as the latter take a lot longer to cook.

Steam the leaves for 15 minutes until tender and use just as you would spinach. (Chard does not 'cook down' as much as spinach, so you won't need to allow quite so much per person.)

Some cookery books suggest cooking the stems as a separate vegetable, claiming that they taste like asparagus when eaten with lots of melted butter. They don't; they have a mild cabbagey flavour – it's not unpleasant, but asparagus they ain't. Try them if you like, otherwise eat the leaves and send the stalks off to the compost heap. Very young and tender chard stems can be good in a stir fry if you are short of ingredients, and you can use plenty of garlic to give it flavour.

Tomatoes

Lycopersicon esculentum

Tomatoes grown in cold or frost-free greenhouse/tunnel

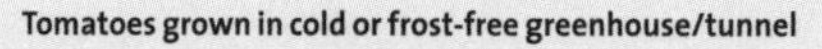
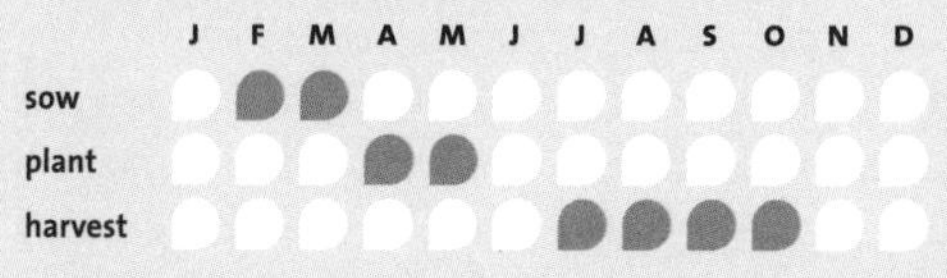

	J	F	M	A	M	J	J	A	S	O	N	D
sow		●	●									
plant				●	●							
harvest							●	●	●	●		

Outdoor tomatoes

	J	F	M	A	M	J	J	A	S	O	N	D
sow			●									
plant					●	●						
harvest							●	●	●	●		

If there's one edible crop that's attained almost cult status over the past ten years or so, it's the tomato. Where once we were satisfied with the regulation medium, round red fruit, now we want novel variations – and the more the merrier. We can have bite-sized cherry tomatoes, elongated plum tomatoes and giant beefsteak toms. Enthusiasts dabble with heirloom varieties, stripy tomatoes and unusual colours including yellow, green, white and even black (well, they're more of a rich maroony brown, really).

Growing several different kinds paves the way to some fascinating salads, but it also opens the door to some exotic tomato recipes, as you discover new ways of using up your summer tomato flush. Naturally, a lot of unusual varieties need a greenhouse to produce a proper crop, but you can still grow some very tasty toms in large pots or growing bags on a sunny patio.

HOW TO GROW

Degree of difficulty: Need attention to detail and regular commitment.

Sow: Seeds on a windowsill indoors or in a heated propagator at 18–21°C (65–70°F) from mid-February to early March (to give you plants for growing in a frost-free or unheated greenhouse) or mid- to late March (to give you plants for growing outside). Err on the late side if you are in a cold area; it makes no sense to have plants that are 60cm (2ft) tall at planting time – 30cm (12in) is better.

Plant: In a cold greenhouse or one that's only just kept frost-free from the third week in April to late May, avoiding cold spells. Plant outdoors after the last frost is safely past, which is usually mid-May to early June, depending on local conditions. You can plant into a soil border that's been well prepared in advance, or else grow one per large pot (38cm/15in in diameter) filled with a mixture of 50:50 John Innes No. 2 potting compost and multi-purpose compost, or put two or three plants in a growing bag.

Spacing: In soil borders under glass, plant tomatoes 60cm (2ft) apart; in the ground outside in a warm, sheltered spot give them slightly more room, plant 75cm (2ft 6in) apart with 90cm (3ft) between rows.

Routine care: You'll need to support the plants and feed, water, tie up and trim them regularly. Anticipate spending a bit of time on them every week on top of routine watering.

Watering: I like to sink a 10cm (4in) plastic flowerpot into the soil beside each plant so that I can pour water into it which will go straight to the roots. Immediately after planting, give each plant ½ litre (1 pint) of water, then let it go slightly short until the time the first flowers open; gradually step up watering once the first green fruits are starting to swell – by the time the plants are carrying a crop of ripening fruits they will need watering once or twice daily in containers, although those in the ground can make do with being watered every few days as they have a larger reserve to tap into.

Feeding: Start feeding with liquid tomato feed while the young plants are still in their pots, prior to planting out. After planting, feed once a week and increase this to twice-weekly as they start carrying a crop. When feeding, follow the directions on the bottle, but as a general rule about ½ litre (1 pint) of the correctly diluted product once or twice a week is about right.

Supporting plants: Push a tall cane alongside each plant of cordon (single-stemmed) varieties and tie the stem to it as it grows taller. Bush varieties do better given three shorter canes because this supports their bushier shape better, but you'll still need to tie the main stems to the canes to stop them being weighed down by the developing crop and breaking. Plants need to have their new growth tied in every week.

Training: If you are growing the normal, tall, upright (cordon) types of tomato, nip out the sideshoots that grow in the angle where each leaf joins the main stem. Again, this needs doing every week. Bush tomatoes don't need their sideshoots removed, though if they grow too big and bushy you might want to thin the growth out a bit.

COAXING LATE FRUIT TO RIPEN

A month before you expect the plants to reach the end of their cropping season, start persuading the green fruits to reach full size and ripen. Cut off the growing tip of the plant or, in bush tomatoes, take out the tip of every main stem. Cut off trusses of open flowers, as you don't want them producing marble-sized fruits that stand no chance of ripening. Reduce the amount of watering drastically. Remove a few of the lower leaves, especially those that are already yellowing, to allow more light and air to circulate. A week before you plan to pull the plants out, partly cut through the base of each stem – only about a third to half way through.

Any full-sized green tomatoes that are still on the plant when you want to take the plants out or when a frosty spell threatens should be picked and put in a dark place indoors. Don't put them on a sunny windowsill as so many people do – this just makes them shrivel up before they have time to ripen. Shutting green tomatoes up in a box with a ripe apple or banana often helps to speed up ripening, since the fruit gives off ethylene gas which hastens maturity, though they probably won't taste quite as good as ones that ripened in their own good time.

VARIETIES

1. **'Ailsa Craig'** – a traditional cordon variety for growing outside or under cover with medium-sized round, red fruits. They are known for their 'old-fashioned' flavour and are very popular and often available as young plants at garden centres at planting time.
2. **'Brandywine'** – a vintage US beefsteak variety known for its outstanding flavour. Cordon plants with large pink fruits and potato-like leaves. Grow under cover.
3. **'Gardener's Delight'** – one of the most popular varieties ever, producing huge trusses of sweet, tasty, cherry-sized tomatoes that are among the first to start ripening, due to their small size. A cordon variety for growing outside or under cover.
4. **'Green grape'** – an unusual variety with bite-sized green fruits that ripen from jade green to a light, yellowy, lime green, so you know when they are ready to pick – the flavour is exceptionally sweet and mild, almost like a real grape. A cordon variety best grown under cover or on a very warm and sheltered patio.
5. **'Marmande'** – a French variety with famously irregular, lobed, red beefsteak fruits, best grown outside. 'Stop' the plants by taking the tops out of the main stems after two or three flower trusses have set fruit on each plant, as they can't support a huge crop in our shortish season.
6. **'Roma'** – a red plum tomato with good flavour, lots of juice and few seeds, particularly good for freezing or juicing. The bushy plants 'stop' themselves naturally at about waist height. Grows outside or under cover.
7. **'Sungold'** – a cordon variety producing lots of small orange/yellow cherry tomatoes of superb flavour. Good for growing outside or under cover.
8. **'Tigerella'** – striking striped tomatoes, light and dark green at first, ripening to orange/yellow then to red with orange stripes. An unusual variety producing large crops of very tangy, tasty fruits on cordon plants, best grown under cover.
9. **'Tornado'** – an exceptionally productive outdoor bush variety that does well even in a poor summer, with very well-flavoured, medium, round, red fruits.
10. **'Tumbler'** – a very compact, trailing, bush tomato ideal for growing in hanging baskets, window boxes or tall containers; it produces a modest crop of small, round, red tomatoes. It's easy to grow and good when you don't have room for conventional tomato plants.

HARVEST

Wait until the tomatoes ripen fully on the plants before picking, so that they develop their maximum flavour. Expect to start finding your first ripe fruits on greenhouse-grown toms from early July, and on outdoor plants from early August. Try to pick tomatoes complete with their green calyx, as it helps them to keep better if you don't use them straight away.

The plants will keep producing ripe fruit steadily from then until the weather turns too cold for them to ripen; which may be late September or early October when outdoors, and not until late October when under glass.

Buying plants

When you don't have the time or facilities for raising your own tomato plants, it makes sense to buy them ready grown. Now that unusual tomatoes are so popular, many nurseries and garden centres stock quite a good selection of varieties at planting time in early summer. If you want a wider range, order young plantlets from a seed catalogue early in the season for delivery around March/April; these just need potting on arrival and growing on a bright windowsill indoors until it's time to plant them out. A firm specialising in tomatoes and unusual veg will have a huge selection to choose from. Stately gardens with large kitchen gardens containing ranges of greenhouses may often have surplus plants of interesting old or unusual varieties for sale in late spring and early summer, so check out the plant sales area if you are garden-visiting.

PROBLEMS

I fear there are rather a lot of them.

Soil-borne diseases build up over many years when tomatoes are regularly grown in the same soil, as happens in a greenhouse or polytunnel. Affected plants are gradually less vigorous and produce smaller crops year on year until it's hardly worth the bother. If you can't move tomatoes to a new place every year, either dig out and replace the soil or grow them in pots or growing bags.

Split fruit occurs as result of stresses within the skin which are caused when plants are alternately wet and dry at the roots. Avoid this by watering little and often so they never dry out or get drenched. It's a common problem outside during a dry summer when there's suddenly lots of rain. Pick affected fruits – it's mainly ripe fruits that are affected – and freeze or juice them fast so they don't go to waste.

Blossom end rot produces a sunken black, leathery, roundish patch on the underside of the tomato. This is usually found in tomatoes grown in containers – especially growing bags – and is due to irregular watering, which allows plants to dry out occasionally. To prevent it, water regularly. Plants grown in the ground are rarely affected.

Outdoor-grown tomatoes are susceptible to potato blight, the same killer disease that affects outdoor potatoes, and a wet summer often brings it along. Look for brownish patches on the upper sides of the leaves; if you spray straight away with Bordeaux mixture and repeat every fortnight you may save the crop. Once plants are badly affected, leaves start looking dead and both ripe and unripe fruit develop irregular, brown, rotten-looking patches; by then it's too late to save them. A few blight-resistant varieties are starting to appear. Tomatoes grown under cover are less likely to be affected but they aren't totally immune, since the spores can enter through open ventilators. Keep a close watch for early symptoms if there is rain in late June and July and spray outdoor tomatoes as a precaution.

Botrytis affects different parts of tomato plants in different ways; look for fluffy grey sunken patches on stems, grey mouldy flowers that drop off without setting fruit, or green tomatoes with small, round, translucent circles on the skin – known as 'ghost spot'. It's more of a problem in greenhouses and polytunnels; so try to avoid it by ventilating more and limiting watering to the base of the plants to avoid excess humidity in the air. In a particularly bad case, use a suitable fungicide.

Whitefly are little, white, Concorde-shaped flies on the underside of the leaves. They suck sap and secrete sticky honeydew on which sooty mould can grow. They are difficult to control as the young are tough little scales. Biological control (courtesy of a chalcid wasp – *Encarsia formosa*) works under glass, where the pest is more of a problem than outdoors. Try growing the fuzzy-leaved marigold, *Tagetes minuta*, alongside plants in the same soil as the tomatoes – some folk swear by its whitefly-repelling properties thanks to its pungent root exudates. Sticky traps are moderately effective, but the downside is that they will also catch 'good' insects.

Recipe – Greek tomato salad

For this salad, use large or beefsteak tomatoes that have been allowed to ripen fully on the plant in a sunny place. Slice very thinly, drizzle with your best olive oil, scatter on chopped spring onions, whole olives and cubes of feta cheese and leave to sit in the fridge for a couple of hours before serving. Eat with crusty bread as a great light lunch.

For an Italian version, do the same thing but omit the feta and olives and use slices of good-quality mozzarella and a few torn basil leaves instead. Makes a great summer starter.

Turnips

Brassica rapa Rapifera Group

Turnips sown under cover

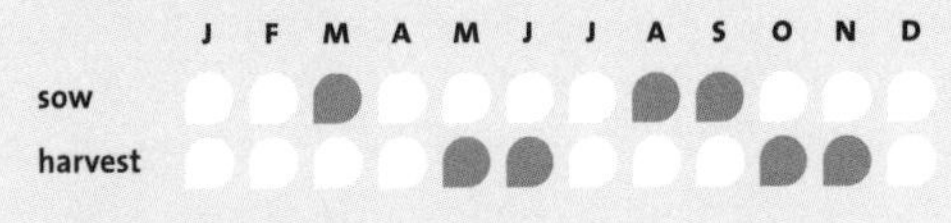

Turnips sown outside

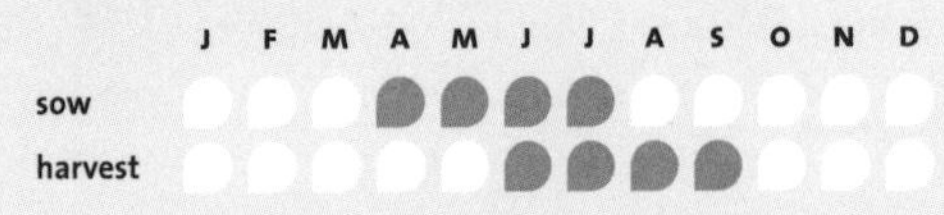

It's time we re-acquainted ourselves with the turnip. Now there's a phrase; but I stand by it. The turnip is a root veg that's rather dropped off our dinner plates of late, but it's well worth growing as a baby veg, especially under cover for an out-of-season treat, because it is both tender and delicious.

HOW TO GROW

Degree of difficulty: Easy and need little time.

Sow: March and August/early September in a cold greenhouse or polytunnel; outdoors from the beginning of April until July. Sow thinly in rows where you want the plants to crop.

Spacing: Thin out seedlings when they are big enough to handle, leaving them 2.5cm (1in) apart for baby veg, or 5cm (2in) apart for normal crops. Allow 15–20cm (6–8in) between rows.

Routine care: Water regularly and keep well weeded to avoid competition and to keep the plants growing steadily without a check. This is essential for tasty, tender turnips.

VARIETIES

1 **'Atlantic'** – a traditional-looking, flattened turnip with a purple top, but good for sowing early and late under cover as well as outdoors through the summer. Fast growing, tasty and reliable. Known to the French as 'navets'.

2 **'Golden Ball'** – an attractive, round turnip with golden skin and flesh, for growing outdoors in the summer. Superb flavour.

3 **'Snowball'** – an old, white-fleshed variety that I still rate highly.

4 **'Tokyo Cross'** – hard to find now, but one of the very best varieties for sowing under cover for an early or late crop. It's incredibly fast-growing, producing pure white, golf-ball-sized roots with excellent flavour in as little as six weeks from sowing.

HARVEST

Start pulling baby turnips when they are almost 2.5cm (1in) across; don't allow any turnip to grow much more than 4cm (1½in) across or you risk it turning woody.

PROBLEMS

Flea beetle will sprinkle leaves with tiny round holes. Although generally not that serious, severe leaf damage at the seedling stage can check a crop that's meant to grow fast and quickly so that it's young and tender when it's pulled. Prevent attacks by growing the plants under fine insect-proof mesh throughout their lives – take care the mesh is well tucked in round the edges to prevent the adult beetles crawling underneath.

Turnips belong to the brassica family, and if club root is present in the ground they'll be affected. Unlike swedes, no resistant varieties are available. Affected roots develop unsightly, bulbous, warty-looking outgrowths. Clubroot stays in the ground for 20 years, so you'll have to grow members of the brassica family elsewhere. It thrives on badly drained acid soil; alkaline or chalky ground that is well drained is rarely affected. Liming to raise the pH can help reduce its incidence.

Don't mistake clubroot (which is a solid growth) for turnip gall weevil, which is caused by a larvae living inside the hollow root and causing distortion. Cut a root open and you'll instantly see which is the cause: if it's the weevil you'll find the tunnel; if it's clubroot the flesh will stink. Gall weevil isn't a problem – just throw away affected roots.

Recipe – Turnips in cream

Peel and slice young turnips and boil for five minutes. Fry the partly cooked slices in butter until soft and lightly browned, then add chopped garlic, chives and parsley and a few tablespoonfuls of double cream so the slices are coated in herby creamy sauce. Serve on fried bread for supper or use as a creamed vegetable to go with pork or chicken. Alternatively, simply boil until tender and coat in butter and black pepper. Food of the Gods. (Well, all right then, easily pleased mortals like me!)

Watercress and land cress

Nasturtium officinale and *Barbarea vulgaris*

	J	F	M	A	M	J	J	A	S	O	N	D
sow outside			●	●	●	●	●					
harvest					●	●	●	●	●	●		

It's a commonly held view that you can't grow watercress unless you have a stretch of chalk stream running through your garden, but that's entirely wrong. Watercress grows well in pots, troughs and tubs, and you can also grow it in a damp patch of ground in a corner of the greenhouse, polytunnel or outdoors, or even in pots on the kitchen windowsill, as long as you keep it cut back so it doesn't straggle.

It's worth growing – watercress is no trouble at all and once you have a patch established you'll never be short of a sprig or three to spice up a salad, tuck into sandwiches or for a delicate soup.

HOW TO GROW

Degree of difficulty: Easy, almost no effort involved.
Sow: At any time of year in pots on a windowsill indoors, or outside in containers or well-prepared soil where you want plants to crop from late March to July. Although seeds are not easy to find at the usual outlets, they are available from specialist catalogues.
Plant: It's usually easier to root cuttings taken from the thick end of bunches of watercress bought from the greengrocer. Plant the coarse bits you normally trim off to throw away from a bunch of watercress – try to buy a bunch with a few roots on it in the first place. Plant them in a pot of multi-purpose compost and keep it good and wet on a windowsill indoors and in no time strong young shoots will appear. The initial potful can be planted out in the garden or into a tub, and cuttings from it (or self-sown seedlings, which often appear naturally around it once it's been established for a while) can be used to start new colonies. It's a good idea to do this every spring.
Routine care: The original group will keep going for a year or more as long as it's regularly watered *and well fed.* Sprinkle round a small handful of chicken manure pellets and water them in every six to eight weeks during the summer, or use a general-purpose liquid feed – the idea is to keep the plants leafy. Nip out flowers that appear at the tips of the older shoots from time to time to help the plant stay leafy and bushy.

VARIETIES

1 **Watercress** – it's very rarely that a named variety of watercress appears, so take whatever you can get.
2 **Land cress** – a large leafy cress that's something similar to baby turnip leaves in looks but with a hot spicy flavour. It is more suitable for growing on dry land than real watercress, as it survives far better in light soil and free-draining conditions.

1

2

HARVEST

Keep snipping off 2½–5cm (1–2in) lengths from the tips of the shoots to use whenever you need them. The more you cut, the more you get.

PROBLEMS

Poor growth and premature flowering and seed pod production are a sign of exhausted soil or a geriatric watercress patch. In this case, move it and start a new colony; but if it's down to insufficient feeding, give the plants a good dose of general-purpose liquid feed or sprinkle poultry manure pellets around and wash them in well. Give old plants a good trim to remove developing seedpods, flowers and weedy yellowing stems, and they should look better in a few weeks.

Fruit

Apples

Malus cultivars

	J	F	M	A	M	J	J	A	S	O	N	D
plant	●	●	●							●	●	●
harvest							●	●		●		

If you grow only one fruit tree, grow an apple – there really is no finer symbol of the English kitchen garden. It makes a good-looking tree, with superb pink and white blossom, and apples are the most multi-purpose fruit. And if you can fit in only a single tree, I'd recommend a dual-purpose variety such as 'James Grieve' that you can use for eating straight from the tree and for cooking. Otherwise, grow a good 'eater' and use that for cooking as well – the big advantage is you won't need much, if any, sugar, which is good news if you're watching your weight. 'Family trees' are also available – these have stems of several varieties grafted onto one rootstock to create a tree that will produce both cooking and eating apples. Grow one if the concept appeals to you, but be warned that the vigour of each variety can vary and you could get a lopsided tree.

Most apples sold today are grown on dwarfing rootstocks, which are prefixed with the letters M (for Malling – a nod in the direction of the Kent research station where they were first developed from the 1970s onwards). These trees followed on from the success of the earlier, semi-dwarfing MM (Merton Malling) rootstocks that had been around since the 1920s and 30s. Dwarfing rootstocks, such as M26, keep the tree small at 2.4–3.6m (8–12ft) and slow growing, with the extra advantage of bringing it into cropping early on in life. This means you should be picking your first apples within a year or two of planting. If you want a taller standard tree that you can sit under, ask for a less dwarfing rootstock, such as MM106, which should give you a tree 3.6–5.5m (12–18ft) high and the same across.

HOW TO GROW

Degree of difficulty: Free-standing trees are easy and need little work; trained trees need annual summer-pruning, which isn't difficult once you get the hang of it, but pot-grown trees will need *a lot* of regular watering.

Plant: In a fairly sheltered situation with reasonably good, deep, well-drained soil.

Spacing: A row of cordon trees can be planted as closely as 45cm (18in). Space free-standing trees, espaliers and step-overs (all grown on dwarfing rootstocks) 1.8m (6ft) or more apart, and free-standing trees 3–4.5m (10–15ft) or more apart, depending on how dwarf their rootstock.

Routine care: Water in after planting, and in subsequent years keep apple trees on dwarfing rootstocks watered in dry spells when they are carrying fruit, to prevent them shedding their crop prematurely. (This happens if they are under stress because of drought conditions.)

Well-established trees on semi-dwarfing rootstocks should not need watering, except in prolonged droughts. Each spring, mulch generously while the soil is moist and feed in mid- to late April – sprinkle general-purpose, organic fertiliser evenly around the base of each tree, covering a circle of soil 90cm–1.2m (3–4ft) wide, radiating out from the trunk.

GROWING IN CONTAINERS

Apples on dwarfing rootstocks can be grown in large tubs or pots (38–45cm/15–18in diameter) of John Innes No. 3 potting compost. Free-standing trees or upright cordons are best for growing this way but, be warned, it's far more work than growing them in the ground.

Pot-grown apple trees are very prone to drying out, so keep them very well watered (you'll need to do it daily in summer) or they'll shed their fruit. They are also top heavy and blow over easily, so keep them secured – especially in winter, when the pots also need lagging to prevent the compost freezing.

Repot or top-dress trees in containers with compost regularly in early spring (see 'Growing in containers', pages 16–17).

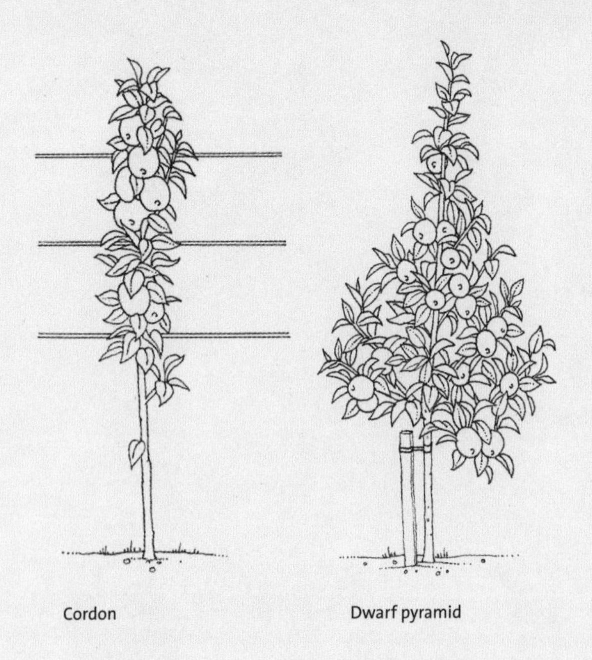

Cordon

Dwarf pyramid

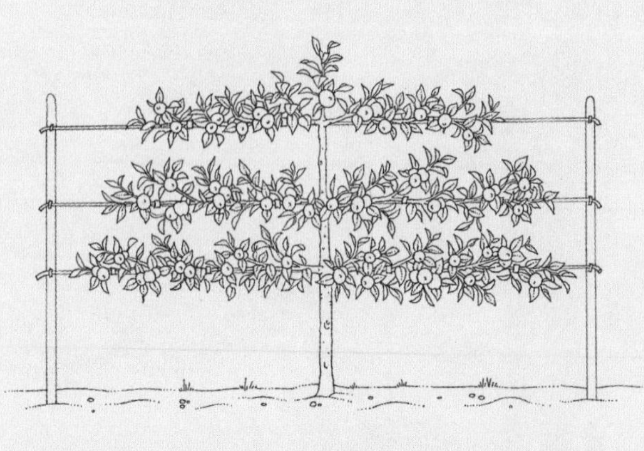

Espalier

Step-over

TRAINED TREE TYPES

Trained apple trees are a brilliant way of making the best use of limited space. Several types are available, ready-trained, from nurseries and garden centres. **Cordons** are vertical 'poles' with the fruit growing from short, branching, twiggy 'spurs' along their length. They are grown upright or inclined at an angle of 45 degrees. Plant them at the back of a perennial border to add height, or plant a row of individuals 45cm (18in) apart as a 'fruiting hedge' or plant them 75cm (2ft 6in) apart in a row at either side of a path and arch them over at the top to make a fruiting tunnel. **Dwarf pyramids** are conical trees trained to produce tiers of branches all the way up the main trunk, starting at about 60cm (2ft) above the ground. They look neat, rather like Christmas trees, and produce heavy crops. They are also easy to pick or prune since everything can be reached from the ground. **Espaliers** are good for growing flat against a wall or trellis, otherwise use a row of posts with two or three sets of horizontal wires or wooden rails for support alongside a path or at the back of a border. Each tree is an upright trunk with two or three pairs of horizontal 'arms' growing out at right angles to it. **Step-over** trees are short, single-tier horizontally trained cordons that are good for the front of a flower border, the edges of beds, or planted on the slant as 'hand rails' up a flight of steps in a sloping garden.

Pruning trained trees

Treat a cordon as follows, and treat each individual branch of an espalier, step-over or dwarf pyramid tree as a complete cordon.

In late July every year, cut out the tip of the main stem to keep it a suitable height or length for the space available (with upright cordons, make sure you can reach the top to pick the fruit).

Work your way along the main stem, stopping at each 'spur' (a cluster of short, stubby twigs). There, cut back this year's long shoots growing out of it to one leaf beyond the cluster of leaves where the new growth leaves the spur, or just beyond a developing fruit.

Check the trunk for new shoots growing directly out of it that have not yet formed a 'spur' – cut these back within 23cm (9in) of the trunk, to make them branch out and start turning into new fruiting spurs.

PRUNING

Trained trees need pruning every year in late July. You can get very technical about pruning, but freestanding apple trees grown in gardens on modern rootstocks need very little attention with the secateurs. All you need to do is cut out the three 'D's': dead, damaged or diseased wood. (Do this any time you see them, but go round checking for problems every winter, too.)

While the trees are dormant (late October to mid-March), remove a few overcrowded or badly positioned branches, if need be. Cut them back to their junction with the trunk or a larger branch, don't 'snip around' taking off lots of little bits and pieces. Avoid cutting branches hard back, close to the trunk, or simply giving the tree a 'short back and sides', since this encourages it to produce masses of strong, leafy, but useless, shoots rather than the twiggy growth that carries fruit buds. Thin, spindly growths known as 'water shoots' occasionally arise from the trunk and these should be snipped off at their point of origin.

SECRETS OF SUCCESS

Trees grown on dwarfing rootstocks (which means most of them these days) must be supported by a strong stake throughout their life as the roots are not over vigorous. They also need a circle of bare soil around them – 90cm (3ft) in diameter – that's kept clear of weeds or other plants, since they can't stand competition. With more vigorous trees, keep a 90cm (3ft) circle of bare soil around each tree for three to four years to encourage establishment. After that, the grass can be allowed to grow up to the trunk, because by then the nitrogen in the soil that is taken by the grass will prevent the trees putting on too much leaf and shoot growth.

Apple blossom needs cross-pollinating with one, or sometimes two, other varieties before it can set fruit. In practice there are nearly always enough other apples around in the area to ensure a crop (even crab apples will do), but if you live well away from other apples, consider growing two or three varieties that flower at the same time. Family trees are generally grafted with varieties which will cross-pollinate one another.

HARVEST

Pick early varieties as soon as they start ripening – from late July to September. These don't 'keep', so use them straight from the tree. Lift the fruit in your hand and gently twist; if it parts easily from the tree, it is ripe. Use windfall apples as soon as possible after they drop, because they'll soon rot if they are bruised. Leave others on the tree until mid-October, unless you are forced to pick them due to impending gales. After picking, store them until you want to use them.

STORAGE

Store picked apples in a cool, dryish place where the temperature remains constantly steady; they keep best in the salad drawer of the fridge, otherwise space a single layer in shallow 'stacking trays' so that air can circulate between them, or put them in large, loose plastic bags with air holes punched in them, and keep them in a cool, airy shed or garage. Start using stored apples as soon as you like – if the fruits are clean, dry and in perfect condition when stored they should keep until Christmas, but check them regularly and remove any that are starting to go bad before the rot spreads. Since modern houses don't have suitable facilities for storing fruit long-term, it's best to cook and freeze any surplus you can't keep this way.

Many old apple varieties were traditionally kept until long after Christmas in purpose-built fruit stores, which were once common outbuildings at country houses, or in old-fashioned pantries built onto cottage kitchens. These buildings faced north, and by adjusting small ventilators an even temperature of around 5°C (40°F) could be maintained, along with steady humidity. Perfect apples kept spaced out on wooden racks would continue to ripen slowly and to develop a more mellow flavour and softer texture.

Before the days of freezers and canned food it was essential to be able to store food in this way, but if you grow these late-keeping varieties today you'll find they still have a good flavour when eaten from October to December – it's just that they taste a tad sharper and crisper, which to modern tastes makes for better eating.

VARIETIES

Eating or 'dessert' apples:

1 **'Ashmead's Kernel'** – the connoisseur's choice; an 'antique' eating variety with sweet, crisp, aromatic, rich raisiny-nutty flavour, but low yields. Can be eaten late October onwards (and is very enjoyable) but is traditionally stored and used from December to March. Introduced around 1700.

2 **'Braeburn'** – the most popular eating apple in the shops, now available as trees for growing at home. Crisp red and yellow fruit ready mid-October until just after Christmas. Introduced 1952.

3 **'Cox's Orange Pippin'** – classic eating apple that's also a star for French open-topped apple tarts. Ready to eat October to December. It isn't really the best choice for gardens because it needs good growing conditions to do well and is very prone to pests and diseases. Bred by a retired brewer and amateur fruit breeder, Richard Cox, at his garden in Slough and introduced in 1850.

4 **'Discovery'** – early, small/medium, bright red, crisp eating apple. Ready to eat late August to late September. Introduced 1949.

5 **'Ellison's Orange'** – distinctively flavoured, medium to large, red-striped yellow fruit with pale yellow flesh. A reliably good cropper that may lapse into biennial cropping (heavy one year, light the next) if allowed to carry an over-heavy crop. Ready September and October. Raised by Rev Ellison and introduced 1911.

6 **'Egremont Russett'** – the best russet apple, with a heavy crop of suede-skinned light brown and rosy-flushed fruit with distinctively flavoured crisp flesh, superb eaten with a slice of cheddar. Ready October and at their absolute best eaten straight away, although they will keep until Christmas. Introduced 1872.

7 **'Fiesta'** – a crisp, crunchy, sharp-tasting red apple for people who like fruit with 'bite'; ready October/November. Introduced 1972.

8 **'Greensleeves'** – a reliable and very heavy cropping modern variety that's easy to grow well. Medium-sized greeny-yellow fruit is sharp and crisp early in the season, mellowing later. Ready September/November.

9 **'Jonagold'** – crisp, juicy, aromatic, flavourful, medium to large apples with cream flesh. Ready mid-October but will store until after Christmas. Raised in Switzerland in 1943 and widely grown in northern Europe.

10 **'Queen Cox'** – an improved selection from 'Cox's Orange Pippin' with more highly coloured fruit, made in 1982; modern strains claim to be self-fertile. Very popular with gardeners. Eat October to December.

11 **'Sunset'** – good, easily grown and trouble-free modern alternative to Cox, with good flavour and a heavy cropper. Ready to eat November and December. Introduced 1918.

12 **'Tydeman's Early Worcester'** – deep red, medium-sized, early-ripening apples with a rich, strawberry-like flavour. Ready to eat August/September. Raised 1929.

13 **'Worcester Pearmain'** – a sweet, juicy, aromatic, bright red apple; the slightly lopsided medium-sized fruits have lots of flavour. Ready to eat from September to October. Discovered in a market garden near Worcester and introduced in around 1876, and a major variety in commercial orchards for much of the twentieth century.

Dual-purpose eaters/cookers:

14 **'Blenheim Orange'** – large, crisp, flavoursome and heavy-cropping dual-purpose apples. Ready to eat November to January. Introduced 1740.

15 **'James Grieve'** – Medium/large pale yellow fruit streaked with red. Pick green to use for cooking from mid-August onwards; leave to ripen for eating from September to end of October. Introduced 1893.

Cooking apples:

16 **'Bramley's Seedling'** – the most popular cooking apple and the one most people grow, but it easily outgrows small gardens as it's so vigorous. It has large green fruit and often crops well in alternate years. Keeps well to use from November to March, but most people use it quite happily straight from the tree, starting with the first windfalls from August onwards (although it's a bit sharp and needs cooking with lots of sugar). Discovered growing in a cottage garden and named after the owner, a Mr Bramley. Introduced 1809.

17 **'Howgate Wonder'** – to my mind the very best cooking apple, with heavy crops of large golden fruit streaked red. They cook beautifully and taste delicious. Ready to use November to February. Introduced 1915.

18 **'Lane's Prince Albert'** – an old variety which does well in colder areas and in the north, with large, round, red-striped pea-green fruit; cooks to a pale yellow purée. Ready January to March. Introduced pre-1841 and named in honour of Queen Victoria's husband.

1
2
3
4
5
6
7
8
9
10
11
12
13
14
15
16
17
18

Crab apples:
1 **'Evereste'** – wonderful pink and white blossom, dark green leaves and red and yellow fruits ripening in autumn. Height about 4.5 (15ft), spread 2.4m (8ft).
2 **'John Downie'** – white blossom and glossy, conical fruits of red, orange and yellow carried in generous quantity. Height 4.5 (15ft), spread 2.4m (8ft).

PROBLEMS

Maggots in ripe apples are normally codling moth larvae which tunnel up through the developing core and are easily cut out when preparing fruit for cooking. (Most windfalls on examination turn out to have maggots or other damage, which has caused the tree to shed those fruits first.) In commercial orchards, fruit is routinely sprayed against maggots, but at home hang up codling moth traps from mid-May to mid-August – these are non-chemical pheromone traps which lure the males to a sticky pad, which prevents the females being fertilised – so no maggots. Allow one trap per five trees. The results are not 100 per cent effective, but pheromone traps work better if several neighbours with apple trees all use them.

Mildew can turn the tips of young shoots and bunches of flower buds powdery-grey in spring. Severe infestations can stunt trees and reduce crops; prune out badly infected twigs, and spray several times with a suitable fungicide (though it's not easy to spray even medium-sized trees thoroughly). Ignore light attacks.

Scab causes dark brown or green spots on foliage, blistered shoots and scabby, distorted fruit. Spray several times with a suitable fungicide, rake up fallen leaves and prune out badly affected shoots – burn both.

Canker causes bark to crack and shrink back in flakes, usually in characteristic concentric rings. Badly affected branches may eventually die off entirely. Some apple varieties are particularly prone to canker, especially older ones. There's no cure, but spraying with a suitable fungicide helps control it. Prune out badly affected shoots or even whole branches.

Clusters of tiny, green, wingless greenfly at the tips of shoots in spring cause severe curling and distortion of young foliage. This often only happens after the aphids have gone, as their feeding kills off pinhole-sized areas of leaf so they unfurl unevenly. You can spray if you can see the aphids, but the damage looks worse than it is and is best left to natural predators, such as blue tits.

Dense patches of grey 'fur' in nooks in the bark conceal tiny woolly aphids, which are protected by their waxy exterior coating. Colonies do little direct harm, but do create small wounds that almost always allow canker to enter. Remove as much as possible with a wire brush and spray with a suitable pesticide, then prune out badly affected areas where you can.

Flowers, leaves and shoots affected by fireblight become blackened and wilt as if they have been burnt. Oozing cankers may be found on infected stems at blossom time. This is a bacterial disease for which there is no cure, so affected branches should be cut off (making the cut 60cm/2ft back into healthy tissue) and destroyed. Pruning tools should be disinfected and any cuts sealed with wound paint. Badly affected trees should be dug up and burned. This disease also affects pears, quinces and other members of the rose family, such as hawthorn, *Photinia, Cotoneaster, Amelanchier, Chaenomeles, Pyracantha* and *Sorbus.*

Recipe – Crab apple jelly

Wash 1.8kg (4lbs) of crab apples and cut them up without peeling or coring. Put into a large saucepan. Add enough water to just cover the fruit. Simmer gently until the fruit is very soft and completely broken down. Strain through a muslin cloth or teacloth, but don't force through or the mixture will become cloudy. Measure the juice, then return it to a clean pan and warm it up. Add 450g (1lb) sugar per pint of juice and stir until dissolved. Bring to the boil, uncovered. Boil rapidly until setting point (approximately 10 minutes). Remove any scum by skimming the surface, pour into warm jars, cover and leave to set.

Apricots

Prunus armeniaca

An apricot is the ideal choice for anyone who hankers to grow exotic fruit outdoors in Britain. Apricots are likely to be far more successful than peaches and nectarines – the trees are heavy cropping and free from the leaf-curl disease that seriously affects their more glamorous relatives. Apricots thrive even in average conditions, though undoubtedly they're enjoying our newly milder climate, and the fruit ripens reliably even when the plants are grown as free-standing trees – unlike peaches and nectarines, which really need training flat against a south-facing wall.

Apricots are self-compatible when it comes to pollination, so one tree is enough – you don't need a partner for cross-pollination. If space is short, go for one of the very compact kinds sold as 'patio apricots', which are suitable for growing in pots.

The very best reason for 'growing your own' is that home-grown apricots taste far better than any you can buy in the shops. When they are picked fully ripe from the tree they are rich, juicy, unctuous and honey flavoured, with none of the unpleasant 'woolliness' and blandness of bought fruit.

In cold areas apricots can be grown as a fan-trained tree against a wall in the same way as a peach (see page 241).

Once you have a tree established you can enjoy apricots every summer for life. Take your time choosing a few perfectly ripe fruits, and eat them straight off the tree to enjoy them at their absolute best.

HOW TO GROW

Degree of difficulty: Easy and needs little attention.
Plant: Choose a sunny site that's as sheltered as possible, with fertile, well-drained soil. Stake apricot trees securely, using a stake the full height of the trunk, secured top and bottom with a proper tree tie.
Spacing: Allow 1.2m (4ft) between compact patio varieties if you choose to grow them in a border instead of pots, and 3.6m (12ft) between normal varieties grown on the usual semi-dwarfing rootstocks.
Routine care: Keep new trees watered during their first summer; then mulch generously each spring and apply a general-purpose feed such as blood, bone and fishmeal to the ground under each tree in late April. During a dry summer it's worth watering trees that are carrying a good crop of fruit to prevent premature fruit-drop, but don't water little and often, give them a thorough soaking once a week, (sink a vertical pipe into the ground alongside the tree when you plant it).

Apricot flowers open very early in the spring, so if the weather is very cold at the time, protect the flowers from frost with old net curtains or fleece (if this is practical) and hand pollinate them if there are no bees about. If a very heavy crop of fruit sets, thin them out – remove every third or fourth fruit from all over the tree so that the remainder are left well spaced out, and remove any that are obviously deformed or damaged, as they'll only fall off later anyway.

Preserving apricots

If you regularly produce huge crops of apricots that you can't use fresh, rather than freezing them (which ruins perfectly good apricots) prick whole or halved fruit and preserve them in jars of sweetened brandy to eat at Christmas, or invest in a dehydrator. This small electric device is sometimes available at specialist cookware outlets or from specialist seed or organic gardening catalogues, and comes with full instructions for drying crops. Since the temperature and timing is carefully controlled, results are far more successful than trying to dry fruit in a low oven. Though fairly expensive, the same gadget can also be used to make sun-dried tomatoes, and to dry herbs and also veg such as onions, celery and sweet peppers.

APRICOTS IN CONTAINERS

Apricots suitable for pots are genetic dwarfs; several varieties are available, though the trees are sometimes just sold as unnamed 'patio apricots', and their small size makes them ideal for growing in pots. The plants grow very slowly indeed, reaching roughly 1.2m (4ft) high and maybe 90cm (3ft) wide after ten years. Keep them in a spot where their branches are unlikely to be broken as they will take a long time to regrow.

Although the plants themselves are dwarf, they do produce full-sized fruits and, given good care, you can expect a fair return for the space. You'll start picking a few fruits within the first two or three years, and by the time a tree is five years old you could expect up to 4.5kg (10lbs) of fruit. When grown in pots the trees need frequent watering, especially in summer when they are carrying fruit – if they go short at this stage the fruit often falls off prematurely. Feed pot-grown plants every week from April to mid-August using liquid tomato feed, diluted to the usual rate recommended for tomatoes.

Patio apricots can also be planted in a border, which makes them easier to look after, though they'll still need watering in dry spells during the summer.

PRUNING

Avoid pruning apricot trees if possible – choose a well-shaped tree with evenly spaced branches in the first place. If you need to remove dead or damaged stems, or shorten long branches that spoil the shape of the tree, do so in spring when growth first starts, to minimise the risk of trees being infected by silver leaf disease (see opposite).

HARVEST

Different varieties of apricot ripen in turn from late July to early September. Leave fruit on the tree to ripen fully – you'll usually know when picking time is imminent since the fruit stops growing any larger and develops a rich, warm flush on top of its usual buff-orange colour. Ripe fruit will lift off easily in your hand without needing to be pulled or tugged.

The fruit on any one tree will ripen over a period of several weeks, not all at once, so check individual fruits before picking them.

VARIETIES

1 **Flavorcot** – a new variety, bearing plum-sized orange fruits with a reddish flush when ripe; ready in August. Ideal for growing in cooler regions, since it flowers later than other varieties. This means that frost damage is far less likely to affect the flowers and that pollination will be better, without artificial help, as there will be more insects about. Often grown on the modern 'Torinel' rootstock which produces a tree about 2.1–2.4m (7–8ft) tall and roughly 3m (10ft) across that is heavy cropping and starts producing fruit early in its life.

2 **'Moorpark'** – the traditional favourite and very reliable, with large, roundish, pale yellow fruits which develop a reddish-mahogany flush and a scattering of matching dots when ripe, towards the end of August. Often grown on the traditional semi-dwarfing rootstock St Julien A, which will give you a tree 3m (10ft) high and 4.5m (15ft) across.

3 **'Tomcot'** – a relatively new variety with heavy crops of plum-shaped orange fruits with a red flush when ripe; ready mid-July. Usually grown on 'Torinel' rootstock.

PROBLEMS

Apricot trees are perfectly hardy, but if the flowers are nipped by frost they turn brown and may not be pollinated because bees lose interest in faded flowers. Throw 'fleece' or old net curtains temporarily over a flowering tree if a cold night or windy weather is forecast. Uncover the trees on warm sunny days so that bees can visit the flowers, otherwise you'll need to hand-pollinate.

Apricot trees are self-compatible (they can set fruit from their own pollen) but the pollen needs moving from flower to flower in order to ensure fertilization. If the weather is cold at flowering time and there are few bees about, hand pollinate by dabbing a small artist's watercolour brush into all the open flowers in turn throughout the flowering season – ideally around the middle of the day – to ensure a good crop.

Tips of the stems or occasionally larger shoots die back, becoming dark coloured and brittle. Prune out affected parts, ideally in spring when growth starts, so it's easy to see exactly how much of a shoot is dead. Cut back to just above a healthy new shoot. The variety 'Moorpark' is often affected in this way.

Silver leaf is most troublesome on plums, though it can also affect apricots and cherries. The leaves take on a distinctive silvery sheen before turning brown and dying. Affected shoots, when cut, may show a brown stain running down through the central 'core', which is useful for diagnosing this particular disease. There is no cure, but if caught early it's worth cutting out affected shoots at least 15cm (6in) past the point where the core of the shoots shows the central staining. Use pruning paint on all open wounds after cutting a stem, and clean and disinfect secateurs thoroughly before using them on other trees. The reason for not pruning apricots (or other stone fruit such as plums and cherries) in winter is to try and avoid leaving any open wounds at a time when they won't heal quickly, since these are easily invaded by the fungal organism that causes silver leaf.

Blackberries

Rubus species and hybrids

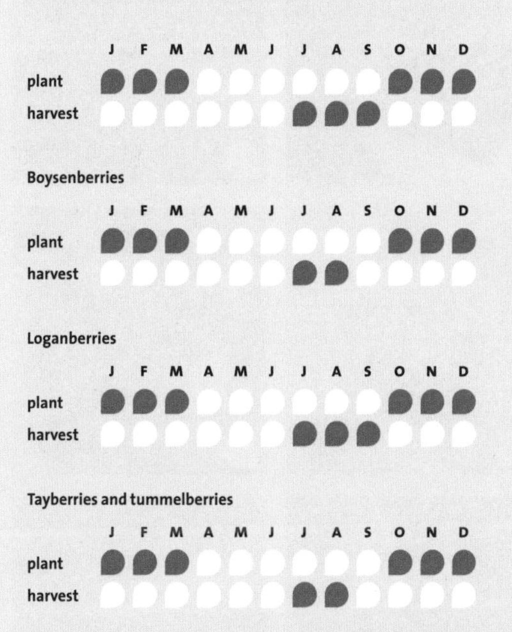

Take a drive in the countryside in September, and even in today's high-tech world you'll still find a few people out blackberrying – or 'blegging', as we used to call it in the Yorkshire Dales when I was a nipper. We'd use the handle of a walking stick to hook down out-of-reach bramble stems, which is where the best fruits are always found. But though it's free and fun, blackberrying is not very convenient for most people – traffic fumes, pollution and road safety apart. So when big, fat, juicy, cultivated blackberries started to appear in greengrocers' shops some years back, people fell on them with glee. The trouble is that they are not cheap, so it makes real sense to grow your own. Today, all sorts of compact new varieties and thornless blackberries are available (all of them ideal for small gardens), not to mention the various blackberry hybrids, such as loganberries and tayberries. All are worth growing on allotments or along fences to extend the fresh home-grown berry season. They are all easy to grow and prolific croppers.

HOW TO GROW

Degree of difficulty: Very easy and little effort needed apart from pruning, which takes a few hours each year.

Plant: One plant per family is usually enough. Unlike most fruits, blackberries are very weatherproof and do well even in rather exposed areas, including those with heavy clay soils; they also thrive in north-facing situations, though the fruit will be rather later to ripen and may not be as sweet as fruit grown in more favourable situations. Hybrid berries need a more sheltered spot and better-drained soil.

Support: If you don't have a handy shed or wooden fence to grow a blackberry against, put up a post-and-wire fence 1.2–1.5m (4–5ft) high with two or three horizontal wires spaced evenly apart, and train blackberry stems out along these for ease of management and to save space. Leave a 90cm (3ft) wide path alongside to allow easy access for weeding, picking and pruning. Where space is short, a thornless variety can be trained over an arch.

Spacing: Allow a 1.8m (6ft) run of fence for the most compact varieties, 2.4–3.6m (8–12ft) for larger modern varieties. Old-fashioned blackberries (not recommended for modern gardens) can occupy 6.1–9.1m (20–30ft) or more!

Routine care: Keep the ground well weeded all round blackberry canes; mulch heavily every spring with well-rotted organic matter, and in late April feed generously with a general-purpose feed such as blood, fish and bone. Once established, the plants only need watering in a very dry summer, but protect ripening fruit from birds by covering the plants with netting.

To keep blackberries and their hybrids under control during the growing season, knock in a 1.8m (6ft) stake alongside each plant, as well as putting up a post-and-wire fence to train the stems along. As the new stems grow it is then easy to tie them to the vertical stake and keep them entirely separate from the horizontally trained canes that grew last year, which are the ones bearing this year's flowers and fruit. When all the fruit has been picked it is easy to see which is which; you can cut out and clear the old stems that have cropped, and then lower the next batch into position and tie them in place. This saves no end of getting scratched, and confusion between old and new canes.

VARIETIES

Blackberries

1 **'Black Butte'** – a reasonably compact plant with stems 1.8–2.4m (6–8ft) long, producing heavy crops of enormous, rather tubular-shaped fruits almost 5cm (2in) long and of very good quality, which ripen from the end of July onwards.

2 **'Helen'** – a spineless 1.8–2.4m (6–8ft) blackberry with large, good-quality fruits which ripen several weeks earlier than most blackberries, from mid-July onwards.

3 **'Loch Tay'** – probably the best for a really small garden; a very compact blackberry with short thornless canes. Instead of training it along a fence, insert a 1.8m (6ft) stake alongside each plant and tie the stems to it in a bundle. Crops as well as many larger-growing varieties, with fair-sized conical fruits ripening from mid-August onwards.

4 **'Veronique'** – as ornamental a blackberry as you'll find, with pink flowers, no thorns, and a compact, semi-upright habit suitable for tying up to a post or fanning out over a 1.8m (6ft) fence panel, yet also produces good crops of large fruits in late summer.

Hybrid berries

5 **Boysenberry** – a hybrid between the blackberry and an unknown parent, with heavy crops of large blackberry-like fruits that have all the flavour of wild brambles, and which ripen in July and August. Its great strength is that it is fairly drought tolerant and often does better on light soils and in drier conditions than true blackberries. The thornless boysenberry is more often seen nowadays.

6 **Loganberry** – a hybrid of the raspberry and the dewberry (*Rubus caesius*) this is a wild, raspberry-like fruit. The loganberry has large, red, rather sharp-tasting fruit and is best used cooked in any of the ways you'd use blackberries. Allow 2.4–3m (8–10ft).

7 **LY654** – a thornless variety of loganberry with large, almost 5cm (2in) long fruits, ripening from mid-July to early September. LY59 is sometimes available, too, and though some people think the flavour of the fruit is better, the stems are very thorny.

8 **Tayberry** – a hybrid between the blackberry and the raspberry with large, rich red aromatic fruit, used raw as a fresh fruit with cream (for which it is far better suited than the much sharper-tasting loganberry) or

cooked in the same way as blackberries. Generally considered to be the very best of all the hybrid berries, so if you only have room for one I'd go for a tayberry. Usually sold simply under the name tayberry in garden centres, although named varieties are occasionally found. Ripening July and August. Allow 2.4–3m (8–10ft).

9 **'Buckingham'** – a new, thornless tayberry with large, good quality fruit, which ripens in July and August.

10 **Tummelberry** – a cross between two varieties of tayberry, with tayberry-like fruit which ripens a little later in mid-July to late August, but it is a much hardier plant and is ideal for more exposed situations.

PRUNING

After all the fruit has been picked, in late summer or early autumn, prune close to ground level all the canes that carried the crop, and tie in to the horizontal wires the young, un-fruited canes that have grown during the summer; these are the ones that will fruit next year.

HARVEST

Depending which variety you grow, you could have fruit ripening in late summer, early or mid-autumn. The fruit on any one plant ripens irregularly over a period of several weeks. Allow it to ripen fully on the canes (pick every 2 days so that none becomes over-ripe and spoils). Ripe fruits are a glossy jet black and *very* slightly soft to touch – they bruise easily and bleed juice lavishly, so handle them carefully and wear old clothes when picking. Expect to pick about 4.5kg (10lbs) of fruit per year from a good modern variety, or 3.6kg (8lbs) from a very compact form, once it's well established. Loganberries and tayberries tend to ripen earlier than most blackberry varieties; they have berries something like giant raspberries but with a tarter taste and when fully ripe they turn from red to a noticeably deeper shade of purplish-crimson, and again they feel softer to the touch when they're ready for picking.

PROBLEMS

Generally fairly free of pests and diseases.

Recipe – Rumtopf, fruit macerated in alcohol

Some years back there was a craze for giving the special pots used for making this traditional Austrian or German mixed-berry preserve, but results were often disappointing when people relied on 'bought' or 'pick-your-own' fruit. For best results you need completely fresh, top-quality fruit that's been grown without spraying, so use home-grown organic fruit. You can use any berry fruit, including raspberries and strawberries, also cherries, apricots and a few halved and stoned dessert plums or greengages (these all need to be well pricked first – use a darning needle if you still have one!), but most home fruit growers will find it's blackberries and their relatives that provide the 'bulk' of surplus crops available for preserving in this way. While the original recipe called for rum (try it if you like), brandy is more to modern tastes. (Use a good quality one, or splash out and add a blackcurrant liqueur such as crème de cassis.)

Use a large and very wide-necked ceramic or stoneware jar (ideally 5 litres capacity). Take 1kg (2.2lb) of fruit. Since strawberries are the first to become available, start off with those in June. Don't wash the fruit (this is why you don't want any that has been sprayed); but if you have to, dry it thoroughly afterwards, since you don't want to dilute the alcohol with water – too weak a brew and the fruit can go mouldy instead of being perfectly preserved. Sprinkle the hulled fruit with 500g (1lb) sugar and leave overnight, then tip it into the bottom of the jar, complete with any juice. Add a litre of brandy and put a plate inside the container on top of the fruit to keep it pushed down under the alcohol. As summer moves on, subject small quantities of surplus cultivated blackberries, raspberries, tayberries and the like to the overnight sugar treatment, then remove the plate, pour them and their juice into the pot and add more brandy. Make sure that each time the fruit is covered with spirit and kept submerged by standing the plate on the top.

When the container is full almost to the brim with fruit, top it right up with brandy and put the lid on, or use cling-film securely fixed with rubber bands, and set the container aside until Christmas or some other special occasion. To serve, strain the fruit and arrange a few spoonfuls in wine glasses with cream or posh ice-cream – honey and acacia goes very nicely – along with a little of the alcoholic juice. Strain the remaining fruit-juice-and-brandy brew from the pot to drink as a home-made liqueur, or to dilute it with sparkling wine to make your own version of champagne cocktails.

Blackcurrants

Ribes nigrum

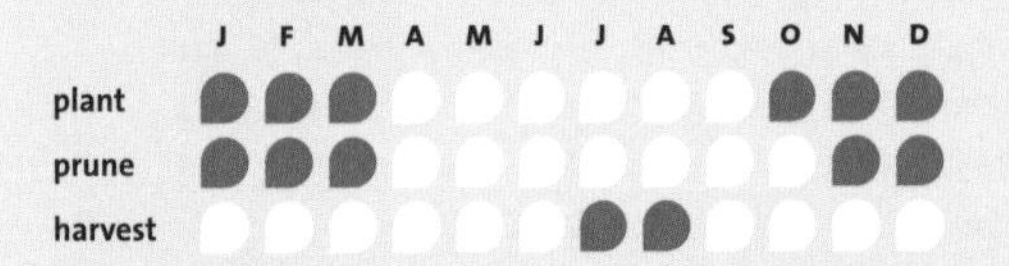

While virtually all soft fruits have soared in popularity lately, blackcurrants have been left behind, largely I think as there are so few ways of using them compared to the other, more versatile kinds. But if you have the space, do include at least one bush in a fruit garden even if you don't make jam, as the fruits are particularly high in vitamin C and they make brilliant crumbles, unusual 'foodie' sauces, and a range of tasty fruit drinks.

HOW TO GROW

Degree of difficulty: Easy and not time consuming, though correct pruning is needed, and removing the small berries from the trusses (called 'strigs') is tedious.

Plant: In a sunny spot with rich, fertile soil that stays moist – blackcurrants thrive in 'difficult' boggy spots where most fruit won't grow.

Spacing: 1.5m (5ft) apart; most blackcurrant bushes will grow about 1.2–1.5m (4–5ft) tall.

Routine care: Blackcurrants are greedy plants that really benefit from a rich diet with lots of moisture, feed and manure. Mulch generously each spring, ideally using well-rotted manure, and use a general-purpose fertiliser such as blood, fish and bone in mid- to late April. In early summer, top up nutrients by using a nitrogen-rich feed such as poultry manure pellets, giving two or three handfuls per plant, sprinkled all round. Protect ripening fruit from birds using netting in summer.

Recipe – Blackcurrant juice

This is a quick and easy way to make the most of blackcurrants without all the bother of preparing the fruit. Place the blackcurrants (still attached to their 'strigs') in a pan and add a little water. When they have softened, roughly mash them up and let the juice drip out through a mesh bag (the sort used for jelly-making). Add sugar to taste while the juice is still warm so it dissolves. (If you want to cut down on sugar, add 3–4 whole sweet cicely leaves, complete with their leaf stalks, to each 450g (1lb) of blackcurrants and remove after cooking. This should enable you to halve the quantity of sugar that's needed.) Use your fresh blackcurrant juice diluted with water or carbonated mineral water to make a vitamin-C-rich healthy cordial for children, or add a spoonful to a glass of dry white wine or sparkling wine to make your own version of Kir. The 'neat' juice can also be used as a base for all sorts of blackcurrant sauces, and it's very good with duck. Keep it in a closed jar in the fridge and use it within a week – it can also be frozen in ice cube trays to use in any of the above ways out of season.

VARIETIES

1 **'Ben Connan'** – a very compact variety that's ideal for small gardens yet has unusually large fruits, which saves a lot of time picking and preparing them for cooking. Some pest and disease resistance.

2 **'Ben Hope'** – a pest and disease-resistant variety that also withstands the gall mites that cause 'big bud' and reversion disease – once both serious problems when blackcurrants were more widely grown. It produces large crops of medium-sized fruit of good quality and is ideal for organic growers. Ripens July.

3 **'Ben Lomond'** – a popular variety with commercial growers, which is fairly compact, rather upright in shape and needs very little pruning. It bears heavy crops of large fruit which ripens at the end of July. 'Ben Nevis' is very similar but with bigger and taller bushes.

4 **'Ben Sarek'** – a heavy-cropping and very compact variety only 90cm (3ft) high, with fairly frost-resistant flowers that are rarely affected by late frosts, and very large berries that ripen from mid-July to mid-August. Can be planted at closer spacings than other blackcurrants, at 90cm (3ft) apart.

PRUNING

After allowing new bushes a season or two to settle, prune them each winter (November to March), removing one third of the bush. Choose the oldest stems (which are thickest and have the darkest bark) and any that are overcrowded or damaged, cutting them out as close to the bottom of the bush as possible. Don't take them all from one side.

HARVEST

Pick by removing whole trusses of fruit from the branches when all the fruits on each bunch have swelled up to maximum size and turned glossy jet black. Leave them on the plants as long as you dare, since the flavour becomes sweeter as they ripen more. Separate individual berries from the thin wiry stems forming each 'strig' – the traditional way was to hold the thick end of the stalk and pull the strig through the teeth of a dinner fork to strip off the berries.

PROBLEMS

Big bud is caused by gall mites living inside the buds and making them grow fat and rounded instead of long and slim – this is most obvious in spring just before new growth starts. The same mites spread reversion disease, which causes unnatural foliage and reduces fruit crops. Check plants each spring and cut off and destroy stems with affected buds. No chemical cure is available. Ideally, avoid the problem by growing resistant varieties.

Jostaberry

This is a cross between a blackcurrant and a gooseberry; the fruits look like blackcurrants but are as big as gooseberries, so they're quick to pick. The bush is fairly compact, with gooseberry-like foliage but no prickles. It has natural pest-and-disease-resistance to troubles that affect blackcurrants, and gives a heavier, sweeter crop of fruit. It's great for pies, fools and crumbles, or stewed on its own and eaten with Greek yoghurt or ice cream. I'd grow it instead of blackcurrants any time. It is best grown on a short trunk or 'leg' 15cm (6in) high and pruned by thinning out a third of the branches each winter.

Blueberries

Vaccinium corymbosum

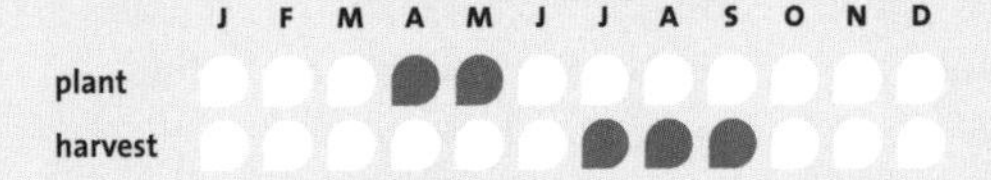

Currently the must-have soft fruit, blueberries are attractive, upright yet bushy plants with clusters of pretty white bell flowers in late spring. The flowers are followed by a fair crop of fruit and superb fiery autumn leaf colour. (This type of blueberry is known in the USA as the 'high bush blueberry'.) The only fault with blueberries for British growers is that they need acid conditions, so if you garden on chalk or limestone, grow them in large pots or tubs of ericaceous compost. They are attractive enough for the patio, but you will need two varieties flowering at the same time for effective pollination – otherwise you'll only have a smallish crop.

HOW TO GROW

Degree of difficulty: Not time consuming but needs attention to detail – the right soil type is vital.
Plant: In sun or light dappled shade, ideally in spring, though pot-grown plants can be put in during the summer. Rich, fertile, acid soil containing large amounts of organic matter is essential for success. The ground needs to be able to retain moisture but have good drainage, so work in grit as well as organic matter. Blueberries also benefit from mycorrhizal fungi, which work with the plant's roots for mutual benefit so, if possible, obtain this as a sachet or tub for tree and rose planting, and add the recommended dose to the bottom of the planting hole. Don't mix it into the soil, but sit the rootball on top so it is in direct contact with the powder. This helps plants to establish. As newly planted blueberries often suffer a high failure rate, or fade away without flourishing, every bit of encouragement helps. No supports are needed – the plants will grow naturally bushy.
Spacing: 1.5m (5ft) apart when grown in open ground.
Routine care: Water new plants regularly while they are establishing, and keep them well watered in a dry summer. They should not be drowned, but never allow them to dry right out. Avoid tapwater if possible as it contains lime; save rainwater especially, since this is naturally slightly acidic. Each spring feed the plants with a high-nitrogen feed such as chicken manure pellets, and mulch very heavily with well-rotted organic matter – use an acidic form such as pine needles or bracken if available, though bark chippings are also good. If plants don't seem to be making much new growth, feed with 25g (1oz) of sulphate of ammonia per plant in late spring or early summer, sprinkled all round the plant and well watered in.

GROWING BLUEBERRIES IN CONTAINERS

If you want to grow blueberries in pots, it's best to use compact varieties such as 'Top Hat' and 'Sunshine Blue', since large varieties can be too top heavy, and quickly become pot-bound. This means they are difficult to keep watered adequately, so fruit crops suffer and the plants can die suddenly due to water stress. Plant or re-pot blueberries in spring into 38cm (15in) pots or large tubs filled with a 50:50 mixture of John Innes ericaceous compost and peat-free ericaceous compost with a little potting grit added. This will ensure a suitable free-draining but moisture-retentive acid potting mixture. If you want to be thorough, use mycorrhizal fungi as suggested for planting in the open ground, to help plants thrive.

Saving rainwater is especially worthwhile for blueberries grown in pots, since the lime in tapwater builds up in containers and soon renders conditions unfavourable for these truly acid-loving plants. Liquid-feed regularly every week or two during the growing season with a good general-purpose feed or, if possible, one formulated specially for acid-loving plants (available from specialist suppliers by mail order if you can't find one at the garden centre). It's also worth treating the plants to two or three doses of seaweed extract during the growing season to provide a natural source of trace elements and promote strong, healthy growth. In my experience you can't afford to take your eye off the ball with blueberries. Sorry!

VARIETIES

1. **'Bluecrop'** – 1.2m (4ft) high and roughly 75cm (2ft 6in) across, with large fruits ripening in late July to mid-August. Good for containers.
2. **'Chandler'** – very large fruits on a 1.5m (5ft) plant, ripening August to mid-September.
3. **'Earliblue'** – one of the earliest of the popular varieties, ready early to late July. Grows 1.8m (6ft) high and a little over 90cm (3ft) wide.
4. **'Herbert'** – very large and exceptionally well-flavoured fruits, ripening late August.
5. **'Sunshine Blue'** – 90cm (3ft) plants with unusual pink flowers and heavy crops of superb fruits in August. Good for containers.
6. **'Top Hat'** – compact bushes 60cm (2ft) high and as wide, ideal for containers. Medium-sized fruits but good flavour.

PRUNING

No regular pruning is needed, but once plants are well established you can remove some of the branches that have carried fruit in order to thin out the plant slightly. Cut back to just above a strong, new, non-fruited shoot.

HARVEST

Pick individual berries when they have turned from green through light blue to a rich, deep blue-black under their pewtery 'bloom' – they should also feel slightly softer to the touch – just like the blueberries you buy in the shops. Depending on the variety you grow, expect to be picking from mid-July to September. Grow several varieties if you want to keep yourself supplied all summer.

STORAGE

Fruit picked without bruises or blemishes can keep for up to two weeks on the lower shelves of a fridge or in the salad drawer, but spread them out thinly – the weight of deep piles of fruit causes bruising and the fruit turns mouldy.

PROBLEMS

No real pest or disease problems in this country, but be careful about soil, water and cultivation. I think I've made my point…

Recipe – Mixed berries

Blueberries are best eaten with as little 'messing about' as possible – enjoy them the way Americans do, mixed with whole raspberries or hulled strawberries for breakfast, or as a quick, healthy dessert. Or sprinkle blueberries over the top of 'quick lemon syllabub', made by mixing 5fl oz (¼ pint) of double cream (whisked until slightly thickened) with a similar-sized tub of crème fraîche, then stirring in the juice and finely grated zest of a lemon. Spoon this into individual glass bowls and allow to set in the fridge for a few hours. Sprinkle the blueberries on top before serving.

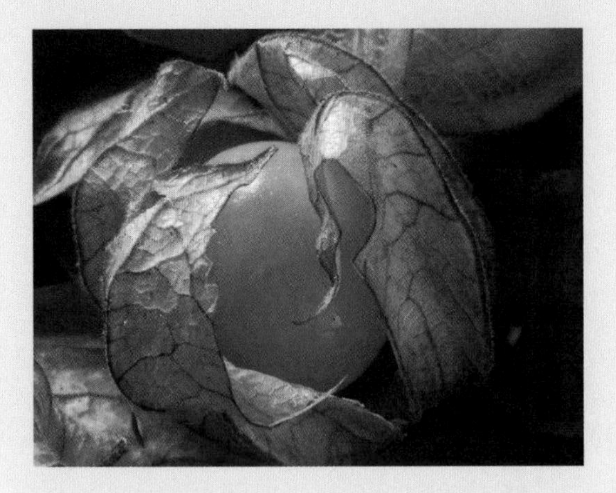

Cape gooseberries

Physalis peruviana, syn *P. edulis*

	J	F	M	A	M	J	J	A	S	O	N	D
plant		●	●									
harvest								●	●	●		
sow		●	●									

Not to be confused with the Chinese gooseberry (which is an alternative common name for kiwi fruit), the Cape gooseberry is the fruit that you'll sometimes see at the side of your plate if you order dessert in a posh restaurant. It has a buff husk where the lantern-like case has been peeled back to reveal the round orange berry inside, and though they are edible, most people leave them as decoration. There's a good reason for this – they are rather an acquired taste, and a lot of people find them a little too sharp.

But for those who enjoy them, or want to impress their foodie friends by using them decoratively, give them a try. (Their only other use is for making jam.) The plants are also very decorative, but since they are annuals you'll need to raise new plants each year if you develop a taste for them. You might also fancy trying their close cousin, tomatillo, which is currently enjoying a minor surge in popularity as an ingredient of authentic Mexican salsa.

HOW TO GROW

Degree of difficulty: Not difficult or very time consuming, but you'll have to grow your own plants from seed and you'll need a greenhouse for reliable results.

Sow: Seeds on a windowsill indoors or in a heated propagator at 16–21°C (60–70°F) in February or March. Prick out seedlings into 7.5cm (3in) pots when large enough to handle. Pot the young plants on into larger pots filled with multi-purpose compost once they fill their original pots with roots.

Plant: In the soil border of a cold greenhouse in early May or transfer to 30–38cm (12–15in) pots to grow in a conservatory. Alternatively, if you have a warm, sunny, sheltered, sun-trap patio it's worth standing pot-grown plants outside towards the end of May, or you can plant them a warm, sheltered bed of rich, well-drained soil in a sunny corner.

Spacing: 45cm (18in) apart.

Routine care: Water and liquid-feed regularly (use tomato feed); start sparingly while plants are small but increase the amounts as summer progresses and plants start producing 'lanterns'. Use canes and strings for support if needed. Treat the plants as annuals; being tender they are killed by cold during the winter, so raise new plants each spring.

Recipe – Mexican-style salsa

Take 450g (1lb) of ripe, skinned tomatoes, one red sweet pepper, a medium-sized mild onion or shallot, a small piece of ripe red chilli (enough to give some bite without overpowering the flavour) a peeled clove of garlic and two or three ripe tomatillo berries.

Put the lot in the blender with a cup of olive oil, salt and pepper to taste, and whizz it up so everything is chopped but the mixture still has some texture – it doesn't want to be a total purée. You can adjust the ingredients and proportions to suit your own taste.

Use as a dip with tortilla chips; it's just the job while you're waiting for the barbecue to fire up on a warm summer evening, and if there's any left, you can serve it as a sauce with barbecued sausages or chicken wings.

VARIETIES

1 **Cape gooseberry (*Physalis peruviana*, syn. *Physalis edulis*)** – a multi-stemmed plant with straight-ish upright stems reaching upto 90cm (3ft) high with evenly spaced leaves along them. Insignificant greenish flowers form in the leaf axils near the tips of the stems in summer, followed by green lanterns up to 5cm (2in) long which ripen to papery textured husks which each enclose a single orange berry. A botanical species, so no named varieties are available.

2 **Tomatillo (*Physalis ixocarpa*)** – almost identical plants to the above, but growing to 1.2m (4ft) under glass, and with large, yellow, sharp-tasting berries sometimes described as faintly pineapple-flavoured.

1

2

HARVEST

Watch the progress of the developing lanterns, and once they start to turn beige open one periodically to check the progress of the berry inside – Cape gooseberries turn bright orange when fully ripe, while tomatillo is more of a yellowy shade. A taste of one will soon tell you, though even when fully ripe both are decidedly sharp. They will mature from late August to October.

PROBLEMS

None.

Did you know?

The Cape gooseberry is a close cousin of the herbaceous border plant known as the Chinese lantern flower (*Physalis alkekengi*), which produces stems hung with bright orange lanterns in autumn. These are often dried to use in winter flower arrangements, but don't eat the fruits inside them.

Cherries

Prunus avium, Prunus cevasus

In my youth nobody thought of growing fruiting cherries at home – they were whopping great orchard trees that would have taken over most of the garden, especially since you needed two or more trees to cross-pollinate each other, and even then you'd still have had your crop pinched by birds. The cherries of my childhood were the pink-blossomed 'Kanzan' varieties which were planted as street trees, and whose shuttlecock-shaped head of branches would be wreathed in those double pink flowers in spring.

But fruiting cherries started to become a much more practical proposition when self-fertile new varieties grown on semi-dwarfing rootstocks became available. These were trees that pollinated themselves and grew roughly 4.5–5.5m (15–18ft) tall. However, things are even better now – the very latest ultra-dwarfing rootstocks keep cherry trees at around 1.8m (6ft) high, so they can even be grown as space-saving upright cordons, which makes them very worthwhile for virtually anyone with a garden. To my mind, though, fan-trained cherries are the most practical of all, since it is much easier to net these plants to protect the ripening fruit from birds.

HOW TO GROW

Degree of difficulty: Not difficult or time consuming, but protecting the crop from birds is the top priority.
Plant: Choose a sunny, sheltered situation with well-drained soil; if planting a fan-trained cherry, position it against a south- or south-west-facing wall. Slightly alkaline soils suit cherries best.
Spacing: Space standard tree-shaped or fan-trained cherries on ultra-dwarfing rootstocks 1.8–2.4m (6–8ft) apart; allow 3.6–4.5m (12–15ft) for a cherry tree on a semi-dwarfing rootstock. Upright cordons can be spaced 75cm (2ft 6in) apart.
Routine care: Water newly planted trees in dry spells during their first summer. Each spring, mulch with well-rotted organic matter and in late April sprinkle general-purpose fertiliser all over the soil underneath the canopy of the tree.

Cherries grown on modern dwarfing rootstocks will start carrying fruit within a couple of years of planting; net the trees to protect the fruit from birds. With fan-trained trees this is most easily done by fixing a wooden batten along the top of the fence or wall just above the top of the plant and tacking netting to it, so it can be draped down over the front of the tree and easily rolled up to give access for picking.

Useful gadget – Cherry stoner

The real joy of growing your own cherries is eating them fresh off the tree and spitting the pips out, but if you want to use the fruit in dessert recipes – they are great for filling home-made meringues – or if you grow the sour morello cherries for cooking, then taking the stones out by hand is a terrible to-do. Look in specialist kitchen equipment shops for a cherry stoner; it's very quick and easy to use and makes all the difference to your culinary creativity.

VARIETIES

1 **'Morello'** – self-fertile cherry with huge crops of large red fruits that are too sour to eat raw, but are brilliant for cooking. It has the big advantage of cropping well when grown against a north-facing wall, as long as there's reasonable light – one of the best uses for this traditionally 'difficult' garden situation. Ripens August–September.
2 **'Stella'** – tried and tested self-fertile cherry with dark red fruit which ripens mid-July to August.
3 **'Sunburst'** – self-fertile modern variety with large, sweet, black cherries ripening in early July.
4 **'Sweetheart'** – a relatively new self-fertile cherry, ripening to a dark red in late August, long after most varieties. Ideal for home gardeners, since the fruits ripen over several weeks instead of virtually all at once (which happens with many older varieties), which staggers your harvest. Just make sure you time your summer holidays so you are at home to eat them.

PRUNING

Try to avoid pruning cherries, as doing so will increase the risk of silver leaf disease. Buy well-shaped free-standing trees or ready-trained fans or cordons in the first place. Tie fan-trained trees up to trellis, or horizontal wires secured to the wall, and tie in new shoots. Each spring when new growth begins, shorten the long shoots to prevent the fan outgrowing its allocated space, and rub out emerging shoots growing directly into the wall or out from it, to maintain a neat, flat shape. In late September, shorten new sideshoots growing from the main fan of stems back to three leaves, to encourage fruiting spurs which will carry next year's fruit.

HARVEST

Leave cherries on the tree to ripen fully; this way they develop their full colour and flavour and they'll taste so much better than any you can buy in the shops.

PROBLEMS

Blackfly is a frequent pest which congregates round the tips of young shoots, particularly in late spring, and, unlike greenfly, are rarely cleared by natural predators. Wipe them off with damp kitchen roll or a J-cloth where possible, or use an organic pesticide.

Bacterial canker is a serious problem for cherries, but also affects plums, peaches and their relatives. The first symptoms are small, round, brown spots in leaves which die off to leave holes. The following spring buds fail to develop or stunted leaves appear and quickly turn yellow, then deform and wither. Branches die back and oozing cankers form on the bark. Prune out dead and cankered branches, cutting back a short distance beyond the affected areas, and spray the entire tree thoroughly with Bordeaux mixture in August and again in September. Badly affected trees, or those affected while young and poorly established, will often die completely, so remove and burn them.

Silver leaf is a major problem for plum trees, which also affects cherries and other related tree fruit to a lesser extent. Leaves take on a silvery sheen, then they turn brown and die. To make a precise diagnosis, cut an affected stem (the thicker the better) and look at the cut end – a brown stain running down through the central 'core' will confirm the cause. There is no cure, but if it is caught in the early stages, cut out affected shoots 15cm (6in) or more past the point where the central core of the shoots shows the staining, then treat all the open wounds with pruning paint.

Cranberries

Vaccinium macrocarpon

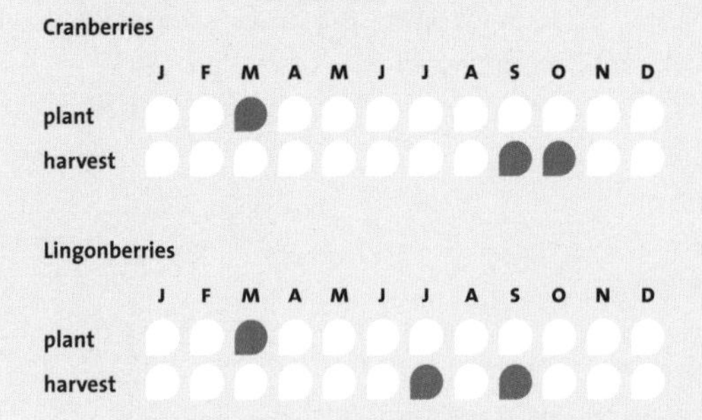

Cranberries are attractive, low-growing, sprawling, evergreen heathland plants with pink bell-flowers; they need extremely acidic semi-bog conditions, so the most practical way to grow them at home is in pots. I'm the first to admit that cranberries aren't something that most people use a lot of – for most of us it's just an annual dose of cranberry sauce with the turkey at Christmas – but if you grow your own you'll find lots of other uses.

Plants are not all that readily available but specialist fruit nurseries will sell one or two varieties.

HOW TO GROW

Degree of difficulty: Easy, given the right conditions, and no trouble.
Plant: At any time of year, though spring is best, in pots of peaty ericaceous compost with about 10 per cent added lime-free grit (look for bags of it in the sand and gravel section of a large garden centre, and if it's not described as such, read the small print).
Spacing: Allow one plant per 38–45cm (15–18in) pot or tub.
Routine care: Maintain boggy conditions by standing the pot in a wide saucer that's kept topped up with water – it's worth saving rainwater, since tap water is too limy. Feed every week or two during the growing season (April to August) using a liquid or soluble product especially formulated for acid-loving plants – this is essential since cranberries require even more acidic conditions than blueberries.

VARIETIES

'Early Black' – plants grow roughly 60cm (2ft) high in a spreading hummock shape. The medium-sized fruits are very dark red-black when fully ripe.

LINGONBERRIES *(Vaccinium vitis-idaea* var. *minus)*

Also known as cowberry, this is a more compact and reliable relative of the cranberry, which currently figures extensively in fruit catalogues for growing in pots on patios. It usually produces two crops of ripe berries per year in Britain – one in late July and the second in September. Grow and use lingonberries in exactly the same way as cranberries.

PRUNING

No pruning is needed, but cut back any long, lanky shoots in spring just to tidy the shape of the plants.

HARVEST

The fruits ripen in late September and October, turning a bright red, though they 'hold' on the plants for months once they are ripe. You can easily freeze them – just pick them into a plastic bag, tie the neck and pop them straight in the freezer, no preparation needed – they come out just as well as they went in.

PROBLEMS

None.

Recipe – Easy cranberry (or cowberry) sauce.

Put enough fresh or frozen cranberries in a small saucepan to cover the bottom evenly with fruit one or two layers deep, then add enough orange juice (from a carton) to barely cover the berries. No extra sugar is needed unless you have a very sweet tooth. Simmer until all the fruits have burst and become tender and the liquid has reduced to about half its previous volume. Add a small slug of port, stir and allow the mix to cool. Whizz in a blender. Store in a glass screw-top jar in the fridge. It keeps for up to a week.

Figs

Ficus carica

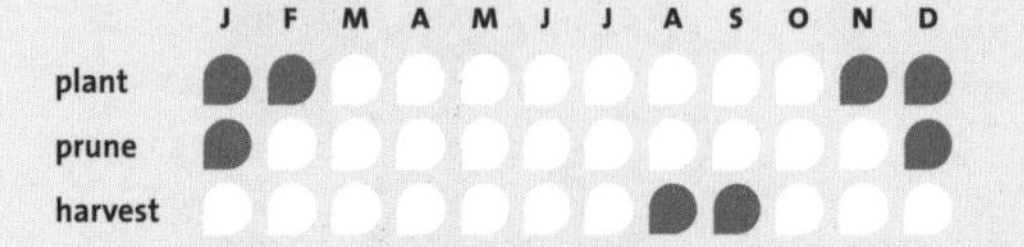

A home-grown fig fresh from the garden has to be the most sensuous fruit you'll ever eat. It is luscious and succulent – a world away from tinned figs, dried figs or the often tasteless and past-their-best 'fresh' figs that greengrocers sometimes stock in summer. If you've enjoyed really fresh, sun-ripened figs on Mediterranean holidays, you'll find growing your own well worthwhile. Currently you'll still get best results from indoor varieties grown in a greenhouse or conservatory, but now that our climate is becoming more Mediterranean, growing them outdoors is a more practical proposition, provided you have the right spot.

HOW TO GROW

Degree of difficulty: Not difficult or time consuming, but needs some attention to detail.

Plant: Plant from November to late February, outdoors or under glass. A pot-grown plant can be planted during the summer, but avoid breaking up the rootball and be prepared to keep it well watered for the rest of that growing season.

If you are planting outdoors, it's best to plant a fan-trained fig against a south-facing wall with well-drained soil – give it a well-prepared bed containing plenty of well-rotted organic matter. What a lot of *wells!*

If you are planting a free-standing tree, restrict the root-run by planting it inside a large sunken container (my old Kew landlady suggested a Gladstone bag, but a trio of paving slabs formed into a sunken 'box' against the house wall is probably more effective and more practical. Alternatively, line the planting hole with bricks and rubble. This 'restriction' stops the tree from growing large and leafy instead of fruiting well.

In a small garden it's best to grow a fig in a 38–45cm (15–18in) pot or large tub filled with John Innes No.3 potting compost on a suntrap patio. The same is true under glass; unless you have a huge greenhouse or conservatory and can spare an entire wall, grow a fig in a pot to keep it compact and save space.

Spacing: Figs are potentially large bushes or trees that need severe annual pruning to keep them to a reasonable size. A fan-trained fig can be trained to occupy whatever space is available, but allow an area at least 1.8m x 1.8m (6ft x 6ft). A fig grown as a bush in the open garden may occupy a space 3m (10ft) or more across and grow every bit as high.

In pots it's easy to keep a trained fig restricted to 1.2–1.5m (4–5ft) high and 90cm–1.2m (3–4ft) wide, though you can allow it to grow larger if space permits.

Routine care: A fig grown in the open ground can usually be left to its own devices as far as feeding and watering are concerned. A fan-trained fig in a bed in front of a wall should have its roots severely restricted by the adjacent foundations and any path or paving alongside, so it needs a spring feed (use general-purpose organic fertiliser) and a generous mulch, and to be kept well watered throughout the growing season.

GROWING A STANDARD FIG IN A CONTAINER

Train a young plant with a single stem 30–90cm (1–3ft) tall, and allow it to branch out naturally at the top. Nip out the growing tips of the shoots to make it form a 'head' in much the same way as if you were training a standard fuchsia. Figs will form near the tips of the shoots so the bushier the 'head' of the plant, the more fruit it will bear.

An outdoor variety of fig can be grown in a pot on a sheltered, sunny patio all year round and will still crop better than a bush in the open garden, but it will do better still if you are able to move it under cover for the winter – a car port is fine if you don't have a conservatory or cold greenhouse. If you have room to keep a pot-grown fig in a cold greenhouse or conservatory *all year round*, grow an indoor variety such as the delicious purple-fruited 'Rouge de Bordeaux' (available from specialist nurseries); this may bear two or even three crops of figs over the summer in separate flushes, from July until October.

Any pot-grown fig needs regular watering and liquid feeding (use tomato feed) throughout the spring to autumn period, but especially in late summer when carrying fruit – if it dries out it will shed its crop.

Prune in midwinter when the plant is deeply dormant, shortening the shoots that form the head to make the plant branch more when new growth starts in spring, and keep the plant nearly totally dry for the winter. Top-dress pot-grown figs each spring, scraping away 2.5cm (1in) of the old compost and replacing it with fresh John Innes No.3 into which you've mixed some slow-release feed granules. They will need re-potting into a slightly larger tub every four to five years, but delay it as long as possible since figs fruit best when they are potbound. Start a new tree from a cutting when your old one outgrows the available pots.

ROOTING FIG CUTTINGS

Take fig cuttings in the autumn, just as the plants shed their leaves. Take the ends of the shoots, 23–30cm (9–12in) long, making your basal cut below a leaf joint, and plant half a dozen round the side of a large, deep pot filled with multi-purpose compost, leaving the top few centimetres sticking out of the top. Stand the pot in a sheltered spot outdoors or in a greenhouse, car-port or tall cold frame. Keep the compost barely moist through the winter. By the following summer, most of the cuttings will have taken root – remove any that fail to produce leaves by July as they'll have died. Pot up the best ones individually in the autumn.

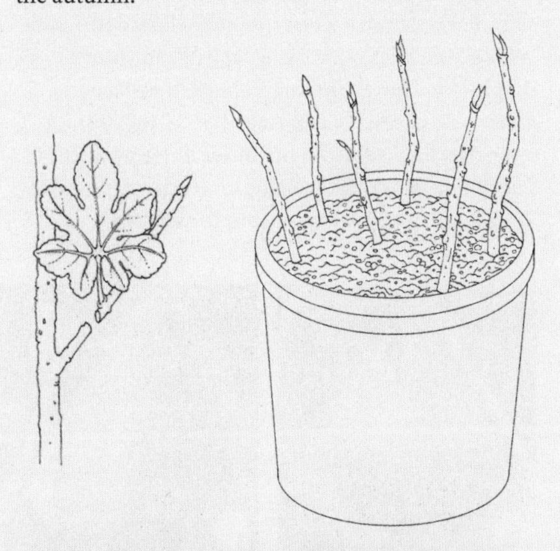

Taking fig cuttings couldn't be easier; simply cut off short fig shoots just below a leaf joint in autumn, remove any leaves and root them in pots of multi-purpose compost over winter and spring.

PRUNING

Pruning is necessary with these plants in order to keep figs within bounds, and you should do it in midwinter – in December or January. Aim to thin overcrowded growth and limit the size of the plant, improving the shape all the while. If you have a fan-trained fig, tie the fan of branches up to trellis or horizontal wires fixed securely to the wall, and rub out emerging buds that grow where you don't want them (including those growing in towards the wall or out away from it, which would ruin the flat, compact shape of the fan). Do this while they are still tiny. Shorten any main stems that have grown too long and remove any that have grown badly. Don't attempt to prune a fig (other than rubbing out unwanted buds) during the summer, as it will bleed. When winter pruning is finished, clear away dead leaves and remove any brown or mildewing figs, but don't remove any small, healthy green ones, as these will fatten and ripen next year.

VARIETIES

1. **'Brown Turkey'** – the best-known and most widely available variety for growing outdoors, producing large crops of medium-sized, purplish-tinged, brown figs with red flesh inside.
2. **'Brunswick'** – less easily found, with greeny-gold skins flushed brown when ripe and yellow flesh inside.
3. **'White Marseilles'** – a striking fig with large, very deeply lobed leaves and large, almost translucent, green-tinged, white, pear-shaped fruits. Suitable for a very warm, sheltered spot outside, for growing in pots on the patio and bringing under cover in winter, or in an unheated greenhouse or conservatory.

HARVEST

Leave figs to ripen fully on the plant; you'll know they are ready when they change colour (turning brown or purple according to variety) and feel soft; the 'eye' on the bottom of the fruit will be open and the inside should be moist, sticky and unctuous (lovely word!). If it's dry and mealy you're a few days too early. With experience you'll learn to recognise the ideal moment, which is *just* before the fruit falls off the tree naturally.

Outdoor figs will ripen at the rate of a few every day over several weeks, any time from late August onwards into autumn.

PROBLEMS

If your plants shed their foliage in autumn, don't worry – people often assume figs are evergreen, but they're not. The plants *should* be leafless from late autumn until the following spring. If plants lose their leaves in summer, though, it's a sure sign they've gone short of water. The plants rarely die and soon start making new growth once well watered again, but you've probably lost this year's crop of fruit. If you're going on holiday during the summer, set up a self-watering system or brief a kind neighbour to keep your fig watered while you're away.

Recipe – Glazed fig tart

There's really only one way to eat figs, and that's by splitting them open with your fingers and eating the flesh by scraping it away from the skin with your teeth – D. H. Lawrence (of *Lady Chatterley* fame) describes the technique very poetically in his non-rhyming poem called simply 'Figs', and they are referred to rather sensuously in Ken Russell's film of *Women in Love*. I love figs served sliced into quarters along with sliced fresh pears and Parma ham, but if you really must 'do something' with some of your crop, try this:

Cut a dozen really ripe fresh figs in half, lay them in a shallow tray, cut surface down, with a shot or two of an orange liqueur such as Grand Marnier and put them in the fridge overnight to steep. Next morning, warm several spoonfuls apricot jam in a small saucepan until runny, then mix this with the goo that's run out of the figs. Arrange the fig pieces, cut side up, in a pre-cooked cold pastry case (sweet pastry if possible, and a bought one is perfectly acceptable) and spoon the jam-and-juice mixture over the top of the fruit as a glaze, letting the rest run into the base of the pastry case. Put back in the fridge to semi-set and eat with cream or ice cream. Or both.

Gooseberries

Ribes uva-crispa

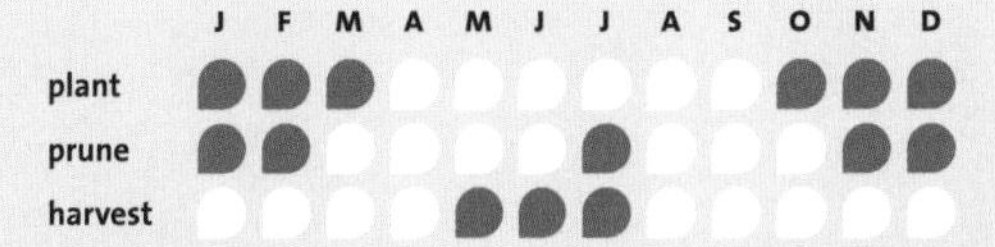

Gooseberries are one of the fast-rising stars of the soft fruit patch, and not only for making traditional fruit pies and fools – the new big thing is to grow dessert varieties, especially the red ones, and enjoy them as fresh fruit just as you'd eat grapes. If you've ever bought a punnet of dessert gooseberries at the shops you'll know they are almost worth their weight in gold.

If you are only going to grow one gooseberry, make it a dessert variety – then you can use it both for cooking and for eating as fresh fruit. And if you don't think you have room for a conventional gooseberry bush, try growing double-cordon-trained gooseberries against a warm, sunny wall, or plant a standard-trained gooseberry in a flower border; you'll even get a fair crop from one growing in a tub on the patio.

HOW TO GROW

Degree of difficulty: Just need a little time.
Plant: In a sunny, fairly sheltered situation with well-drained fertile soil.
Spacing: Allow 90cm (3ft) between conventional gooseberry bushes or standard-trained plants, or 45cm (18in) between cordons.
Routine care: Each spring, mulch the ground underneath the gooseberry bushes generously and in mid- to late April sprinkle an organic, general-purpose fertiliser all round. Water in dry spells in the summer while the plants are fruiting. Thin dessert gooseberries in stages in late May and June, cooking the small, removed fruits. Leave half the original crop evenly spaced over the bush to fill out and ripen fully.

SPACE-SAVING GOOSEBERRIES

Half-standard and double-cordon-trained gooseberries are sometimes available from specialist fruit nurseries, but you can train your own. It's easiest to start with a rooted cutting, unless you can buy a young plant that has one straight stem (for training as a standard) or two straight upright stems branching out from the base (for training as a double cordon).

Train a standard in exactly the same way as a standard fuchsia – tie the single stem up to a 90cm (3ft) cane, and pinch out the growing tip when it reaches the top. As sideshoots develop, remove any growing up the 'leg' of the plant and nip out the growing tips of those near the top once they are 5–7.5cm (2–3in) long, so that they in turn branch out. From then on, prune the bushy top of the standard in the same way as a conventional gooseberry bush.

To train a double cordon, tie both stems up to trellis so they grow upwards, remaining parallel but about 30cm or 45cm (12in or 18in) apart. Allow sideshoots to grow all along their length, then in winter take the very top off each main upright shoot and prune all the sideshoots back to 2.5cm (1in); they will start to form branching spurs which will carry the fruit in future years. From then on, summer-prune in July once the fruit has been picked, by shortening the sideshoots back to 2.5–5cm (2–3in) of the spur from which they grow. It sounds fiddly, but believe me it's very easy when you do it, and you'll get huge crops from a small space. Two double cordons would fit along a single 1.8m (6ft) fence panel, and produce *pounds* of fruit.

Recipe – Gooseberry fool

Pure nursery food – but you can't beat it. Top and tail half a saucepan full of goosegogs, then cook them in the bare minimum of water, plus 3–4 tablespoonfuls of sugar, with the lid on, then chill in the fridge. Whip 10fl oz (½ pint) of double cream until almost thick. Whizz the gooseberries and their juice to a purée in the blender – it should be roughly the consistency of apple sauce – then carefully fold into the whipped double cream. Chill in the fridge for an hour or two before serving with fingers of shortbread. And start your diet another day.

VARIETIES

1 **'Hinomaki Yellow'** – a relatively new dessert gooseberry with very large, aromatic fruit that ripens to a pale gold. It has a rich flavour with a hint of apricot and ripens in July. Mildew resistant.
2 **'Invicta'** – a heavy-cropping, traditional, green, thorny variety with well-flavoured fruit which is ready in June. Mildew resistant.
3 **'Pax'** – a thornless red dessert gooseberry which is heavy cropping, tasty and mildew resistant; it has largely taken over from the traditional old favourite,
4 **'Whinham's Industry'** – a thorny, red dessert variety.

PRUNING

Cut all the stems of young gooseberry bushes back by half after planting, but do any pruning when they are dormant in winter – from November–February. First, remove any dead or damaged stems and any overcrowded growth in the centre of the bush. Open it up almost to a wineglass shape, then shorten the main stems by half and cut back all the sideshoots to 5–7.5cm (2–3in), cutting just beyond an upward-facing bud. Remove any shoots growing out from the very base of the plants, since gooseberry bushes are traditionally grown on a short trunk or 'leg'. (This makes weeding easier and less painful when you're growing prickly varieties.) Remove any suckers that appear from below ground level. In early July shorten all the sideshoots to about five leaves to encourage the production of the buds that will carry next year's crop.

HARVEST

Start picking green gooseberries as soon as they are big enough to be worth using – you could take enough for a pie as early as May. The smaller the gooseberries, the more sour they taste, so you'll need more sugar when cooking them. Continue picking through June. Dessert gooseberries are not completely ripe until July.

STORAGE

Gooseberries are one of the few soft fruits that I reckon are worth freezing. Put whole topped and tailed fruits into bags and freeze without any more preparation.

PROBLEMS

American gooseberry mildew forms a white talcum-powder-like 'bloom' over the tips of young shoots, leaves and fruit. Wipe the 'powder' off fruits (they are still perfectly edible), and prune out the affected tips of shoots after the fruit has been picked. Prevent mildew by pruning to improve air circulation around the plants, and feed with 28g (1oz) of sulphate of potash per plant in spring instead of general-purpose fertiliser. (Excessive nitrogen is thought to encourage mildew by promoting soft, easily infected growth.) Best of all, grow resistant varieties.

Magpie moth appears as very conspicuous black, white and orange-spotted 'looper' caterpillars that can quickly strip the bushes of leaves in May and June. These need to be removed by hand on sight.

Gooseberry sawfly can be seen as dull khaki caterpillars up to 2.5cm (1in) long, usually on the undersides of the leaves from early May to late summer. A bad infestation can quickly defoliate the bushes, so remove them by hand as they appear.

Green 'cooking' gooseberries are not normally loved by birds, but if you are growing dessert varieties – particularly the red kinds – some netting may be needed to protect the crop.

Grapes – indoors

Vitis vinifera and cultivars

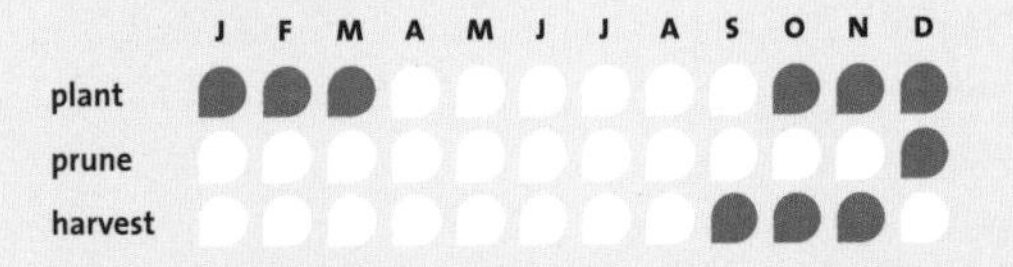

A vine makes an attractive and productive climber for an unheated greenhouse or conservatory. You can, if you wish, simply plant a vine and let it ramble over trellis on a sunny wall, but it will look a lot neater and produce much more fruit if it's properly trained. That might sound a tad tricky at first, but once you get the hang of it you'll find it becomes second nature.

HOW TO GROW

Degree of difficulty: Needs some regular care and attention to detail.

Plant: Greenhouse grape vines are traditionally planted with their roots outside in a richly prepared bed where they will be watered by natural rainfall, with the main stem led *inside* through a hole in the wall. But if this isn't practical, plant a vine in a large tub of John Innes No.3 potting compost or a rich soil bed inside the greenhouse and train it up against the glass. In a conservatory, grow it against a sunny wall.

Spacing: It will need an area the full height of the wall and at least 1.8m or 2.4m (6ft or 8ft) wide.

Routine care: Unless the roots are outside, an indoor vine will need regular watering throughout the growing season. Start gradually in March and increase the quantity throughout the summer, when the plant is in full leaf, and especially when it is carrying a crop of fruit. In a dry summer even a vine with its roots outside may need watering; in this case, thoroughly drench the area in which the roots grow once a week.

Grapes are greedy plants, so each spring sprinkle 25–50g (1–2oz) of general-purpose feed and 15–25g (½–1oz) of sulphate of potash over the soil where the roots are growing. If the vine is in a pot, scrape off the top few centimetres of soil and replace it with new John Innes No.3 into which you've mixed a dose of slow-release feed. But it's still worth using liquid tomato feed once a week during the growing season, too. Train and tie in new growth regularly during the growing season, and prune each winter.

PRUNING AND TRAINING AN INDOOR VINE

Year 1: After planting, give the vine a tall, stout cane for support and fix a series of horizontal wires across the wall 45cm (18in) apart. Select the strongest shoot, rub the rest out and train the main stem up the cane. Prevent any side shoots growing from it by rubbing them out while they are still tiny. When the plant is dormant in midwinter (December) cut the main stem down to 1.3–1.5m (4–5ft) from the ground.

Year 2: Allow the main stem (known as a 'rod') to continue growing upwards by allowing the strongest shoot at the top to grow and rubbing out the rest. Allow some sideshoots to grow (these become what's called 'laterals' in vine-growing circles), choosing ones that are positioned so that they can be trained out along the horizontal wires. Tie them carefully in place with soft string to avoid bruising the soft growth, and rub out the rest while they are small. 'Stop' these sideshoots when they are 30cm (12in) long by nipping out the growing tips with your thumb and fingernail ('pinch pruning'), and 'stop' the main upright stem in the same way when it reaches the top of the wall. You can allow a couple of bunches of grapes to develop this year, but remove the rest while they are small. In winter, prune the top of the rod back flush with the top of the wall (it will have made some more growth after you 'stopped' it in summer) and prune back the laterals to two or three buds from the rod, to leave little stubs.

Year 3: When sideshoots start to grow from the little stubs all along the rod, select the strongest one on each that is growing by a horizontal wire and rub out the rest. Tie your selected shoots in along the wires as they grow. 'Stop' these laterals by nipping out the growing tip three or four leaves beyond each developing bunch of fruit, remove tendrils, and rub out shoots growing from the top of the rod. From then on, do the same thing every year. After the first two or three years, when a vine is well established, you can allow it to carry up to two or three bunches of grapes per lateral.

Year 2
Let the main stem (rod) continue to grow upwards. Train the sideshoots along horizontal wires and stop them when they are 30cm (1ft) long.

Year 3
'Stop' lateral stems by nipping out the growing tip beyond the second or third bunch of fruit and rub out shoots growing from the top of the rod.

VARIETIES:

1 **'Black Hamburgh'**– the same variety as the Great Vine at Hampton Court, easily grown and produces large bunches of big, sweet, black grapes which ripen in mid-season. It also does well in pots.

2 **'Buckland Sweetwater'**– a reliable variety producing large amber grapes that ripen early; it also grows well in pots. Available from specialist nurseries.

3 **'Flame'** – crisp, red, seedless grapes giving you bunches just like the ones you buy in the shops. Easily grown and reliable.

4 **'Muscat of Alexandria'**– the Queen of grapes and a real aristocrat when it comes to flavour, producing big bunches of delicious, large, amber Muscat grapes. But it is also a late-ripening variety that can be rather unreliable – it really needs an Indian summer or a greenhouse that's heated in autumn to perfect the crop. Temperamental, but well worth the effort. Oh, I do love it!

5 **'Perlette'**– sweet, seedless, green grapes which ripen early (this variety can be grown indoors or outside).

SECRETS OF SUCCESS

In late January or early February, when vine buds first start developing, cut the ties that hold the rod to the wall and lay it flat on the ground for two weeks before tying it back up. This is a cunning way to encourage buds to develop right the way along the rod, so you have fruit-bearing laterals all the way up instead of only near the top of the plant.

For better bunches with bigger grapes, thin the fruit when they are the size of frozen peas. Use special vine scissors or nail scissors with long narrow blades, and carefully snip out every other grape, taking care not to damage those that are left. The remaining grapes, besides being bigger, will be more succulent, giving you more fruit and less pips per bite. Seedless grapes and grapes grown outdoors don't need thinning.

HARVEST

Leave bunches of grapes to ripen fully on the vine before cutting them, complete with their stem. If you're not sure, pick a few grapes to taste to see how they are doing.

PROBLEMS

Small 'pippy' grapes may be caused by allowing a young vine to carry more bunches of grapes than it's capable of supporting. Hard though it seems, restrict a first-year vine to two bunches, and perhaps five the following year. After that, a vigorous vine should be capable of supporting two to three bunches per lateral. Thinning the fruit also helps – you can remove up to two out of three grapes for even bigger fruit.

Powdery mildew can affect grapes indoors as well as out. Insufficient watering is nearly always the problem indoors.

Grapes – outdoors

Vitis vinifera cultivars

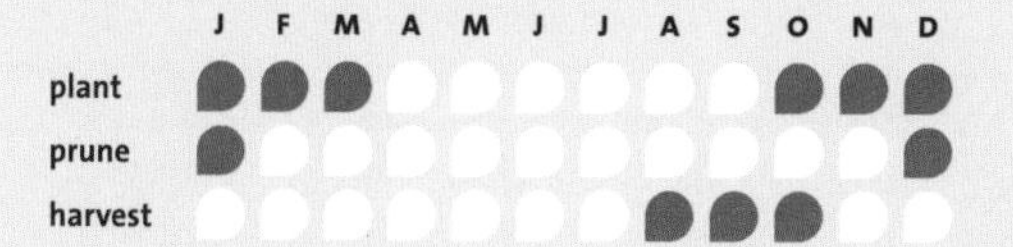

If there's one fruit that has really benefited from global warming, it's grapes. Nowadays you can expect a reasonable crop outdoors in most summers, given a reasonably mild, sunny, sheltered location and a good outdoor variety. What's more, they aren't all the sour sort that were traditionally considered only suitable for wine-making; you can grow some good 'eating' kinds, too.

Depending on how much time and effort you are prepared to put in, you can have a few 'fun' bunches on vines that are basically grown as ornamental plants and allowed to ramble over trellis or a pergola; or you can grow some serious fruit trained out properly along fences, using much the same training system as you see in commercial vineyards.

HOW TO GROW

Degree of difficulty: Can be as simple or as complicated as you like!

Plant: In early spring ideally, though pot-grown plants can be put in any time during the summer if you keep them well watered for their first growing season. Choose a warm, sunny, sheltered spot (ideally against a south-facing fence or wall) with fertile well-drained soil, and work in plenty of well-rotted organic matter first.

Spacing: 1.5m (5ft) apart. Space rows 1.8–2.4m (6–8ft) apart.

Routine care: Mulch generously each spring and sprinkle a handful of general fertiliser, such as blood, bone and fishmeal, and 15g (½oz) of sulphate of potash over the soil around the base of each vine in April. Keep vines well watered in dry spells during the summer while they are carrying fruit. This attention is most likely to be needed when growing the plants against a wall or fence, since this shelters them from some natural rainfall, especially when foundations restrict their root-run.

TRAINING AND PRUNING – THE INFORMAL METHOD

When you're growing a vine informally, allow it to scramble over trellis or secure the main stems to the uprights of a pergola and then let the rest run freely over the top.

Once the vine has had a year or two to form a woody framework it will start bearing flowers and bunches of grapes will dangle down. If it's possible to reach them easily, it's worth limiting them to four bunches per vine, at least for the first year or two of cropping, otherwise the fruits will be small and distinctly 'pippy'. Protect ripening fruit from birds with netting, or else they will eat most of them long before they are properly ripe.

In midwinter (December/January) when the vine is dormant and leafless, prune to thin out overcrowded growth – to keep it tidy and within the space you have allocated for it. Don't prune vines at any other time because they will bleed heavily, which weakens them and limits next year's crop – summer pruning can even kill a young plant.

TRAINING AND PRUNING – PROPERLY

There are several approved methods, but I recommend the 'double guyot' system, which makes the most of a vine's cropping potential and, though traditionally used for vines growing on post-and-wire fences in the open, you should use it for vines trained on trellis or wires on wall or wooden panel fence if you want dessert grapes, which need more warmth to ripen.

Year 1 – Plant vines 1.5m (5ft) apart in a row along the foot of your fence or wall in early spring. Push a stout cane in alongside each plant (this should be nearly as tall as the fence or wall), and organize three taut horizontal wires at regular intervals across the wall – the top one along the top of the structure. When the vines start to grow, select the two strongest shoots from each plant, tie them to the cane for support, and remove the rest. In the first winter, cut both shoots to 7.5cm (3in) from the point at which they grow from the base of the plant, cutting just above a bud.

Year 2 – In their second summer, select the three strongest shoots and tie them to the cane for support as before. In November, shorten the weakest of the three shoots to 7.5cm (3in), and cut the other two off at 60cm (2ft) long, cutting just above a bud. Tie these two out along the bottom wire, one in each direction.

Year 3 – New shoots will grow out along the two 'arms' and also from the main stem of the plant. Allow three strong shoots to grow from the centre, tie them to the cane as before and remove the rest. Allow each 'arm' to bear three shoots, evenly spaced along their length and rub the rest out while they are small. As the chosen three shoots grow, tie them to the middle wire with soft string, taking care not to bruise them. Nip out their growing tips when they reach 30cm (12in) beyond the top of this wire. These stems will carry bunches of flowers, so limit these to one per stem so as not to overburden the plant while it's young. As the bunches of grapes start to ripen, use the top wire to hang bird netting from, so it drapes down like a curtain. In November, cut out the horizontal 'arms' and the upright stems growing from them, and select the strongest two upright stems growing from the centre of the plant to tie down in their place. Continue in this way from then on, but as the plant is now well established you can allow each fruiting upright stem to carry two or three bunches of grapes.

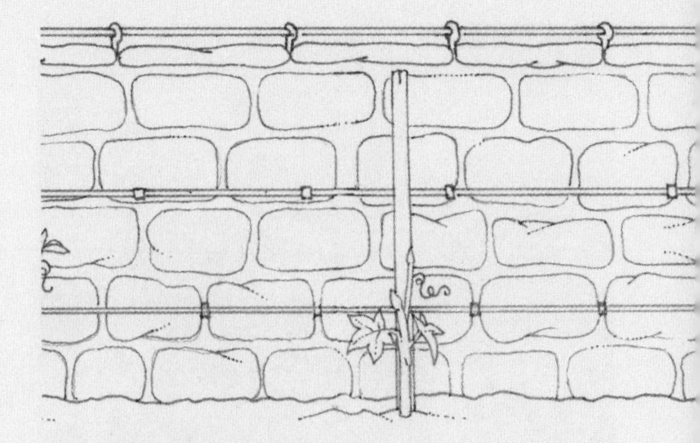

Year 1
Push in a stout cane alongside the plant and fix three horizontal wires along the wall, ready to tie in the plant as it grows.

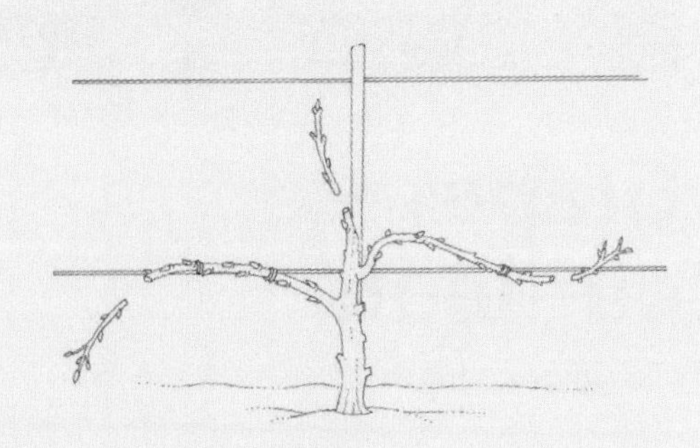

Year 2
At the end of the second year, shorten the weakest of the three remaining shoots to 7.5cm (3in). Cut back the other two to 60cm (2ft) and then tie them to the bottom wire – one in each direction.

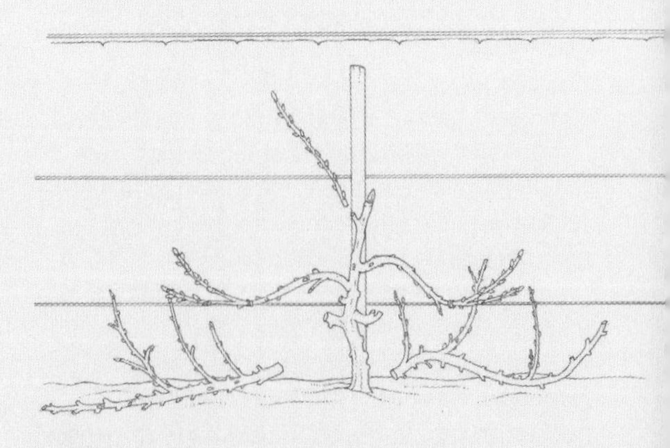

Year 3
At the end of the third year, cut out the horizontal 'arms' and select two upright stems above them as replacements, tying them in to the wires.

OUTDOOR VARIETIES

1 **'Brandt'** – a superb, dual-purpose, ornamental/dessert grape, having modest-sized bunches of small, sweet, black grapes that form reliably without any special attention, and foliage that takes on superb autumn colour. Good for a pergola or a trellis.
2 **'Dornfelder'** – a new variety with good-sized bunches of large, sweet, dark red grapes which ripen in early autumn.
3 **'Muller Thurgau'** – long known as a classic home-grown grape vine, but when grown in a warm, sunny spot or against a sunny wall in dry summers, it produces good quality, green 'eating' grapes.
4 **'Perlette'** – an easily grown and early ripening variety which produces thin-skinned, seedless green grapes, like the ones you buy in grocers and supermarkets.
5 **'Strawberry'** – an attractive vine with large leaves and bunches of coppery-pink to amber fruit which ripens late and has a most unusual flavour. Catalogues describe it as like strawberries – I'm not sure I'd agree; bubble gum springs to mind, in the nicest way.

HARVEST

Allow grapes to hang on the vine as late as possible so that they ripen completely, even after they have coloured up. Early varieties may be ready to pick in late August or September, mid-season varieties around the middle of October, and late varieties not until late October.

PROBLEMS

Powdery mildew – grey-white talcum powder-like deposits over young leaves and tips of shoots – is usually more prevalent in late summer. Often this is a symptom of insufficient water when vines are grown on a wall, so keep the plants well watered in summer and mulch heavily in spring. When it occurs in late summer it may do little harm since the leaves will fall shortly, but if there is a bad infestation earlier in the season, use a suitable fungicide. (Avoid fungicides if grapes are being grown for wine-making, since they often stop the fruit fermenting properly.)

Recipe – Stuffed vine leaves

Choose 20 young leaves and cut them off with the stalk. Rinse first then dip each leaf into boiling water until it turns khaki and becomes soft. Lift it out of the water using the stalk, then snip it off. Cool. Mix equal quantities of uncooked rice and raw lamb mince, add a tablespoon of finely chopped onion and crushed garlic, fresh chopped oregano, salt and pepper and mix. Place a tablespoon of the mixture in the centre of a cooled vine leaf, fold the sides over the meat then roll from one end to make a parcel. Stand the parcels close together in a greased baking dish. Add the juice of one lemon to 20 fl oz (1 pint) of chicken stock and cover the parcels, then add 2 tablespoons of olive oil. Put in a moderate oven for 1½ hours until the rice has absorbed all the stock and is cooked; cover with tin foil for the first hour. Eat hot or cold.

Kiwi fruits

Actinidia deliciosa

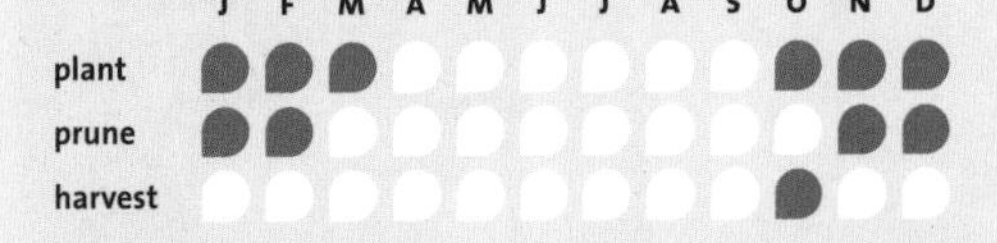

Nobody much grew the Chinese gooseberry when I was a lad; it didn't catch on until it was renamed 'kiwi fruit' and a big marketing campaign in the 1980s brought these hairy, brown, egg-shaped fruits from New Zealand to greengrocers' shops all over the country.

As plants became available, people grew them tentatively in greenhouses, where they took up a lot of room since it takes two to tango. But now self-fertile varieties are available that do well out of doors, and this is the best way to grow them – they make attractive climbers that look good and crop well on a sunny pergola, wall or fence. They are potentially very large plants that will take up a lot of space unless properly pruned; besides keeping them within reasonable bounds, doing this job also encourages them to produce more fruit.

HOW TO GROW

Degree of difficulty: Not difficult, but proper pruning takes time.

Plant: In a warm, sunny, sheltered spot with fertile, well-drained soil, after working in plenty of organic matter. One plant of a self-fertile variety is enough, unless you plan on starting a cottage industry.

Spacing: Each plant needs 3–4.5m (10–15ft) of space.

Routine care: Mulch generously in spring, and sprinkle a handful of general-purpose fertiliser such as blood, bone and fishmeal and 15g (½oz) of sulphate of potash over the soil under each plant in April. Plants need trellis or horizontal wooden rails or wires to grow on; to which you should tie in new growth regularly during the summer. Prune the plant each winter.

VARIETIES

1 **'Jenny'**– a reliable, self-fertile variety with rather small, but very sweet, fruit.

2 **Siberian kiwi or Mongolian gooseberry (*Actinidia arguta* 'Issai')** – this makes an altogether more compact and restrained plant, bearing bunches of small, gooseberry-sized fruits which ripen more reliably and slightly earlier than conventional kiwis. It's ideal for colder gardens or where there is less room. Grow the plant as a single slanting or horizontal stem tied to a supporting pole or on a wire across a fence or the shed wall, and prune the sideshoots to 20cm (8in) long each winter to form branching fruiting spurs. Or, more decoratively, train it round a large spiral shape or topiary frame in a border or a large tub on the patio.

NON-SELF-FERTILE VARIETIES

If you grow traditional varieties of kiwi fruit that have male and female flowers on separate plants, you will need one male for up to seven females to ensure the flowers are properly pollinated. Look for 'Hayward', which is a female, i.e. the fruit-bearing partner, and 'Atlas', the male pollen provider.

If you don't have room for so many kiwi fruit plants, then it's perfectly acceptable to have one of each sex and plant both in the same planting hole. Train them in the same way as described above, but have one upright stem for each plant and train alternate male and female stems out along your wires. If you're growing them informally, allow the two to scramble together over a pergola or large wall or trellis for support.

TRAINING AND PRUNING KIWI FRUIT

Year 1. Plant a kiwi fruit at the centre of the fence or wall with a cane (which needs to be as tall as the fence or wall) for support, and put up three or four horizontal wires or wooden rails, evenly spaced across the fence or wall. Train the plant up the cane and, when it reaches the top, pinch out the growing tip; this encourages it to grow sideshoots. Train a suitably positioned sideshoot out along each wire or rail, and 'stop' any other sideshoots by pinching out their growing tips after 7.5–10cm (3–4in).

Year 2. In the second summer, leave the original sideshoots (correctly known as laterals) to continue growing longer.

Year 3. In the third summer, sublaterals growing from the original laterals will start producing flowers and fruit, so 'stop' these by nipping out the growing tips roughly 15cm (6in) past the last fruit that sets.
In winter, cut back these sublaterals to about 15cm (6in) from the main stem; this keeps the plants tidier and builds up fruiting spurs. From then on, repeat the same summer 'stopping' and winter pruning every year.

HARVEST

The fruit will not be ripe until October, or later if it's been a poor summer. Pick one or two occasionally to try – they should be firm (but not rock-hard) and fairly sweet tasting, but not quite as soft as the ones you buy in the shops. If an early frost or severe weather is likely, drape fleece over the plants for protection so the fruit can continue ripening late into the season.

STORAGE

Kiwi fruit will keep for six weeks or so in a cool room; spread them one layer deep to avoid rotting. It's often thought that the flavour continues to develop over time, so don't rush to eat them straight from the vine.

PROBLEMS

Small fruit that doesn't ripen by autumn is usually due to growing in an insufficiently warm, sheltered, sunny spot. In a cold, exposed or northerly situation, grow a kiwi on the south side of a wall or train it against the sunny interior wall of a greenhouse or conservatory.

Lemons

Citrus limon

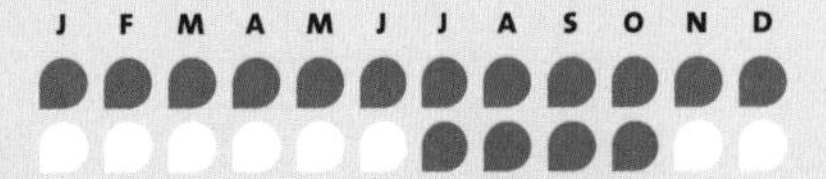

Lemons are by far the most popular citrus fruits for growing at home. They make superb plants for growing in large pots or tubs, kept in a conservatory in winter and then put outside on a patio or terrace in summer, where they create a pleasing Mediterranean feel. The plants are evergreens with scented flowers that start opening early in spring and continue sporadically all summer. Given reasonably good cultivation, it's not uncommon to see flowers and immature and ripe fruit all on the same plant at the same time. But be aware that citrus plants are spiny, so park them where you won't brush past unwisely, and where they won't be a hazard to small children or pets.

HOW TO GROW

Degree of difficulty: Not time consuming but needs attention to detail.

Plant: Buy pot-grown plants at any time of year; don't rush to repot them since they prefer to be slightly potbound.

Spacing: If you have several plants the pots can be grouped together attractively, but allow enough room for air to circulate around them.

Routine care: Correct watering is crucial; *on no account* should you do it little and often. Give citrus plants a thorough soaking when their compost is almost dry then wait until it reaches the same stage before watering again. The plants dislike lime, so avoid tap water; save up rain water or use water that's been run through a jug-type filter, or which has been boiled in the kettle and then allowed to cool to room temperature – most of the lime is then left behind as 'scale' on the element of the kettle. Feed regularly during the growing season; ideally use a special citrus feed which is available in garden centres and nurseries, otherwise use a high-nitrogen liquid feed from April to August (get this from specialist mail order suppliers) and feed monthly over the rest of the year with a general-purpose liquid feed.

All citrus plants dislike dramatic changes in their day-to-day routine, so try to avoid varying the growing conditions or cultural regime too quickly – make any necessary changes slowly. In winter the plants need a minimum temperature of 7°C (45°F) and good light, so they tend not to do well in a living room where it's usually too shady; a conservatory is far better. Citrus plants like some humidity but also need good ventilation, so open the windows whenever possible if the weather isn't too cold. If grown in a conservatory all year round, try to keep the temperature below 29°C (85°F) in summer, but it's far better to move them outside from June to late September if you can. Here, conditions stay more comfortable; they'll enjoy summer showers and they are less prone to pests outside. Stand them in a warm, sheltered spot that gets some sun, though not the strongest midday sun.

Re-pot young plants at the start of the growing season in April only if the old pot is really packed full of roots. Use a special citrus compost (available from specialist nurseries or garden centres that supply citrus plants) or use John Innes No.2 potting compost with about 25 per cent lime-free potting grit or perlite added to make a free-draining blend. Only move them up one pot size at a time – don't replant into a far bigger pot. Older plants in large pots or tubs need topdressing every year or two in spring; scrape away the top few centimetres of compost and replace it with new citrus compost or a home-made equivalent.

Did you know?

Most shop-bought lemons, unless stated otherwise, have had their skins treated with a waxy product which helps them to 'keep' well during long journeys from their country of origin. If you need lemon zest for a recipe, or whole or sliced lemons complete with their skin, remove the wax first by scrubbing the fruit gently with a nail brush in warm soapy water then rinsing it well. Better still, use your own!

VARIETIES

1 'Lemonade'– a compact plant producing good crops of medium-sized fruit with delicately flavoured juice.

2 'Meyer's Lemon' – the easiest, most reliable and cold-tolerant variety, which is also more compact than most and can flower all year round. The ideal citrus for beginners. Widely available.

3 'Variegated' – an outstandingly handsome plant with green and cream leaves and striking fruit with green and yellow stripes. Needs slightly warmer conditions than many; best kept at 10°C (50°F) in winter and should only be put outside in an exceptionally warm sheltered spot, if at all.

OTHER CITRUS FRUIT

Grow other citrus fruits in the same way as for lemons, although some kinds need more warmth in winter (10°C/50°F ideally) – and those with large fruits take a very long time to ripen.

4 Citron (*Citrus medica*) – big plants with large, ugly fruits which have very thick rind. Used for making candied peel; they are also highly aromatic and so worth drying to add to potpouri. Usually only available from specialist nurseries.

5 Kaffir lime (*Citrus hystrix*) – the variety whose foliage, known as kaffir lime leaves, are used in Thai cooking. Available from a few specialist nurseries.

6 Kumquat (*Fortunella japonica*) – not a true citrus, but a close relative, making a small, bushy, compact plant which is often grown as a pot plant. It produces small, bitter-tasting, oval orange fruits which are normally used for decoration or sliced to put in drinks instead of lemon or lime.

7 Lime (*Citrus aurantiifolia*) – the variety 'Tahiti' is fairly compact and the most reliable and cold-tolerant, producing scented flowers and small sweet fruits which are good for a G&T.

8 Orange (*Citrus sinensis*) – larger plants with scented flowers and fruits that are slower to grow and ripen than lemons. The variety 'Valencia' is widely grown for its large, almost seedless fruits which ripen in time for Christmas.

PRUNING

Hard pruning is not needed, but when growth starts in spring (around April), cut back long shoots that spoil the shape of the plant to encourage branching lower down.

HARVEST

Leave lemons on the plant for some time after they reach full size and appear to be ripe; you'll see a gradual change of colour from pale yellow to a deeper shade – and that's the time to pick them. Use secateurs to snip the fruit from the plant, complete with its stalk. Fresh lemons have a far deeper, sweeter flavour and much softer skins than ones you buy in the shops, which have been picked almost green and ripened in transit. Fresh, ripe lemons will keep in a fruit bowl for several weeks.

PROBLEMS

Plants that fail to grow well, look sickly and have few flowers or a sparse amount of fruit that soon turns yellow and drops off are usually suffering from various cultivation problems. These symptoms suggest you may be keeping the plant far too wet, re-potting before it's necessary or into a container that's too large, or that there are constant changes in growing conditions.

Sticky foliage, often with black powdery patches signify scale insects, which are the most frequent pest of citrus plants. Look for 1–2mm wide, buff-coloured 'limpets' on stems and undersides of leaves. Remove by hand using cotton buds, or spray with a suitable remedy – ideally an organic one. A small watercolour brush dipped in Scotch whisky and then dabbed on the pests is sometimes recommended (but seldom by whisky drinkers). The pests are far less of a problem on plants that are stood outside for the summer, probably due to lower temperatures and occasional showers of rain, and possibly because they are visited by natural predators. The black deposit on the leaves is sooty mould, which grows on the sticky honeydew secreted by the insects. It sponges off easily.

Recipe – Real lemonade

This is the best way to use home-grown lemons. Peel four small, unwaxed lemons thinly and carefully with a potato peeler to produce long spirals of skin. Place the rind from three lemons, including any broken bits, in a saucepan with enough cold water to cover, and warm slowly until it is hot but nowhere near boiling. Add 3–4 tablespoons of sugar and stir until dissolved, then leave the mixture to cool slowly with the peel in the water. Meanwhile, squeeze the juice from all four lemons and put in a large jug. Strain the sugar-and-peel water from the pan into this when it is cold, add ice cubes and top up with cold tap water that's been through a filter jug and set aside to cool in the fridge. For fizzy lemonade, use cold bottled soda or tonic water from the fridge. Taste, and if more sugar is needed, dissolve some in a little tepid water and add as required. Finally, decorate with the remaining uncooked twirls of lemon peel, a few extra slices of lemon and a few sprigs of fresh mint. Very Wimbledon-fortnight.

Melons

Cucumis melo

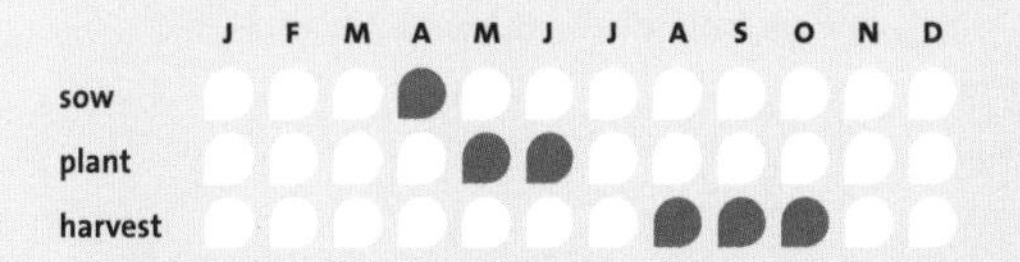

Of all the different kinds of melons you see in the shops, it's the sort known as Cantaloupe melons – the small, round ones with textured skin, or grooves or stripes that look as if they are telling you where to cut them into segments – that are the most successful for growing at home. They'll swell up and ripen quite well in an unheated greenhouse or in cold frames, and in a hot summer you can often get away with letting the plants run around on the ground outdoors, the way they grow them in countries like Greece (though over here a bit of horticultural fleece does help).

HOW TO GROW

Degree of difficulty: Needs regular care and attention.

Sow: Seeds in small individual pots on a warm windowsill indoors in early or mid-April. When the pots are full of roots, move the strongest plants on to 10cm (4in) pots (discard any weaklings at this stage), and give each plant a split cane to grow up – support the plant with plastic plant rings or tie it loosely with soft twine to avoid bruising the delicate stems. Keep the young plants on windowsills indoors in good light, or move them to a tall, electrically-heated propagator kept at 16°C (60°F) in a greenhouse or conservatory. Harden them off slowly and carefully from late April/early May onwards for at least 2–3 weeks before planting them out. Melon plants are very sensitive to cold, which stops them growing for a while, and they will be killed by the slightest hint of frost.

Plant: In a greenhouse or conservatory in mid- to late May, in a rich, well-prepared soil border, large pots or growing bags. For growing on the patio, plant in late May/early June in pots of multi-purpose compost or a well-prepared bed against a sunny wall, or plant in the veg garden under a cold frame. Alternatively, plant in early to mid-June in a warm, well-sheltered, sunny area of a veg patch or allotment in well-drained ground that has had plenty of well-rotted organic matter worked in; but be warned, results aren't so good this way.

Spacing: 45cm (18in) apart for plants to be trained vertically up canes or walls; allow 90cm (3ft) between plants running around on the ground.

Routine care: Water young melon plants very sparingly when they are first planted and increase gradually as growth 'takes off'. Too much water too soon can cause rotting at the stem base. Start liquid feeding weekly with a general-purpose feed when the first flower opens, and change to liquid tomato feed when the first fruit starts swelling. Increase this to twice weekly by the time the plants are carrying a fair crop.

Give plants suitable supports and tie the stems up regularly with soft twine as they make new growth; trellis or netting is suitable for plants growing on a fence or wall, or put an obelisk into a tub for freestanding plants growing on the patio. Plants grown in cold frames or allowed to run over the ground will do best if given some stout, plastic-coated netting

supported on bricks or logs to ramble over, to lift the stems and developing fruits off the ground to avoid rotting – have some fleece handy to cover plants if the weather takes a bad turn. Support plants grown in a greenhouse by training the stems up canes or strings attached to the greenhouse roof.

Allow sideshoots to grow, since this is where the flowers form, but pinch out their growing tips two leaves beyond a developing melon to stop plants outgrowing the space, and to encourage the plant to put its energy into swelling up the fruit.

POLLINATING MELONS

Melons *must* be pollinated or you won't get any fruit. In the garden, bees and other insects will do the job for you with no problem (remove fleece or the glass lids of cold frames during the warmest part of the day to allow insects to get at the flowers). But under cover, especially early in the season, they may not be able to get to the plants, or if the weather is bad there many be few insects around, so you will have to do the job yourself.

The best time to do this is around midday. Take a soft artist's brush and go round the plants dabbing the brush into the centre of all the wide-open flowers in turn, to transfer pollen from the male flowers to the females. (Female flowers are easily recognised since they have a small, rounded bulge just behind the ring of petals. This is the would-be melon. Male flowers simply grow on a short, thin stalk. 'Nuff said.) You'll need to hand-pollinate melons every day until bees are out in force.

Don't grow melons in the same greenhouse or polytunnel as cucumbers, since cucumbers *must not* be pollinated (if they are, the cucumbers end up 'bee stung', in other words, bulbous at one end, and inedible – bitter-tasting and full of hard seeds) and melons will cross-pollinate the cucumbers.

Recipe – Melon baskets
The great beauty of cantaloupe melons is that when they are cut in half one fruit is just enough for two people. Scoop out the seeds and fill the cavity with fresh homegrown raspberries, blueberries or strawberries, or a mixture of all three. Pile them up over the top, then splash chilled, dry sparkling wine over the lot so it fills the cavity. Tuck in, with a glass on the side.

VARIETIES

1 **'Amber Nectar',** syn. **'Castella'** – a very reliable variety for growing inside or out, producing several very reasonable-sized melons per plant if well grown; the flavour is outstanding. The fruits look very attractive, with light green 'netted' skin with green stripes marking them into segments; the flesh is orange.

2 **'Galia'**– just like the Galia melon you buy in the shops, except better when it's enjoyed fresh from the garden. Plants are best grown in a greenhouse or under a cold frame as they really need warmth. They have greeny-buff 'netted' skin and pale green flesh.

3 **'Sweetheart'** – another very reliable variety for the garden, greenhouse or cold frame. I first grew this thirty years ago and have nothing but praise for it. It produces 4–6 small but fast-ripening melons per plant. The fruits have pale green, smooth skins and the flesh is pale orange.

WATERMELONS AND MUSK MELONS

Other types of melon sometimes turn up in seed catalogues, which can be grown at home with varying degrees of success.

Musk melons – much favoured by Victorian and Edwardian head gardeners for their finer flavour and larger size, but they are difficult to grow and need well-heated greenhouses to ripen. Their heavy fruits also need supporting in individual hammock-like nets to stop their weight pulling them off the plants, so all in all they just aren't practical for today's gardeners.

Watermelons – these fruits grow absolutely huge, and while they ripen in late summer in the open fields in countries like Greece, they need to be grown under cover here. As the fruits are too heavy to grow on trellis or canes, the plants need to be allowed to ramble over the ground in the same way as marrows or pumpkins, so they take up a lot of space for a rather low and unreliable return. Occasionally a 'baby' watermelon variety appears on sale in seed catalogues; these are the ones to try if you fancy them, since the small fruit ripens more reliably. I'd still grow them under cover if possible, though. Treat them just like cantaloupe melons being grown on the ground, or grow them up strings or canes to save space, but rig up 'hammocks' to support the weight of the growing fruit.

HARVEST

The time to pick melons is August/early October. Your nose will tell you when your first melon is ripe, so follow it – ripe melons give off a delicious musky scent. A ripe melon changes colour, but don't rely on it turning yellow since a lot of varieties remain green or even deep green when ripe, but you'll always notice some colour change. Hold a suspected ripe fruit in cupped hands and with the tips of your thumbs gently press the far end of it (from which the flower dropped off) – a ripe fruit 'gives' slightly. When you've found a good one, cut it off carefully; use secateurs to snip through the short stem that connects it to the plant.

PROBLEMS

If newly planted melon plants wither and die, pull them out and you'll see the roots have vanished or have rotted off at the neck of the plant (where the root meets the shoot, at roughly soil level). This is neck rot, caused by cold, wet conditions and/or the presence of disease organisms in the soil. To avoid it, plant melons on small soil mounds with a 'moat' at the base of each and water sparingly, into the moat, until plants are established and growing well. Avoid growing melons or cucumbers in the same soil several years running.

Mulberry

Morus nigra

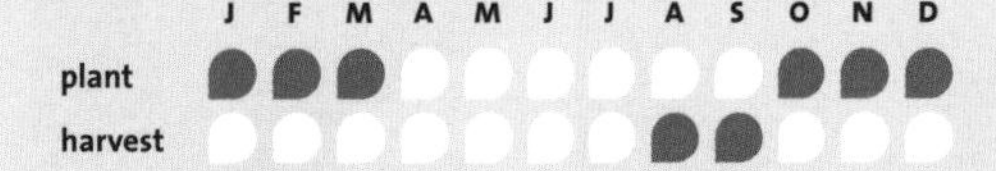

At the start of the seventeenth century, King James I flooded the country with mulberry trees in an attempt to make the nation self-sufficient in silk, having seen the scandalous prices charged by foreign producers. Alas, he had some duff advice and instead of the white mulberry (*Morus alba*) whose leaves are eaten by silkworms, he bought the black mulberry, the fruiting sort that silkworms won't touch. So if you're investing in a tree, make sure you get the right sort for your needs!

A black mulberry makes a superb medium-sized tree with bags of character; it has craggy bark, a stunning spreading shape and large heart-shaped leaves, as well as it's crowning glory – the fruits, which are splendid when ripe. Think of a dark, fragrant raspberry doused in light red wine. They are very exclusive; you'll seldom taste them if you don't grow your own, since mulberries don't travel. In fact, the distance from the branch to the mouth is too far for many – be warned, you'll get covered in purple juice that stains, so wear old clothes to pick them. The flavour is well worth the laundry inconvenience.

HOW TO GROW

Degree of difficulty: Easy, apart from picking.
Plant: In a mild, sheltered site with fertile well-drained soil; a mulberry makes a superb specimen tree to grow in a lawn. You can also start with a young one and espalier-train it to grow against a warm, sunny wall when you want it mainly for fruit production – it's far easier to pick.
Spacing: Expect a standard tree to grow slowly to 6.1m x 6.1m (20ft x 20ft); it can grow larger, but it takes a long time to do so.
Routine care: Stake a newly planted tree, but once it is well established (after 4–5 years) you can do away with support. An espalier-trained tree will need tying up to trellis or a wooden framework throughout its life. Water a new plant during the first summer if it's hot and dry, and thereafter a freestanding tree only needs watering in a dry summer when it is bearing fruit. An espalier-trained tree will need watering regularly.

VARIETIES

1 **'Chelsea'** – (considered to be the same as the variety often sold under the name 'King James'); is a modern variety which starts cropping within about 3–4 years of planting, as against the straight botanical species, *Morus nigra*, which may still be thinking about it 10–12 years later.

HARVEST

Wait until the fruit is completely ripe, in late August and early September, when it turns a deep red and becomes soft and juicy. The traditional way to harvest mulberries is to spread white sheets under the branches then shake the tree. This used to be done to keep the fruit clean so that it doesn't need washing, since the soft fruits disintegrate easily in water and half the taste runs away. (The sheets certainly need washing afterwards, though.) If you can reach the fruit, it's far better to hand pick if possible, as this prevents it from being bruised. Eat ripe mulberries immediately with cream and don't mess about with them in any way.

PROBLEMS

None, apart from the laundry issue.

Nuts

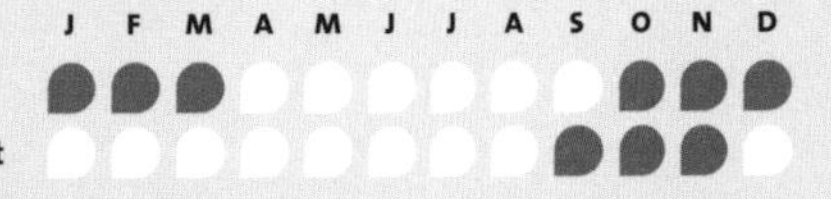

Years ago, few people would have entertained the idea of growing nuts in their gardens, but several things have happened lately to change that.

The move to natural gardening and wildlife gardening means that hazels have a useful part to play in the garden ecosystem – besides providing home-grown pea-sticks, bean poles, herbaceous-plant-supports and even fuel for wood-burning stoves. New varieties of walnuts and sweet chestnuts mean we can be picking nuts within 3–4 years of planting, instead of waiting perhaps 15 or more as we once did; and by growing compact varieties or going further and restricting the roots it's possible to keep them much smaller and slow-growing. This brings them within far more peoples' grasp. What's more, milder winters have made almonds, and indeed many other nuts, a far better bet to grow in the UK.

All sorts of nut trees and bushes quite commonly appear as plants for sale in seed catalogues and at specialist nurseries, and the new rise of fruit and veg growing means that nuts are a logical 'next step' for anyone interested in growing their own.

A nut tree or bush would fit happily into the back of a border or make a specimen plant to grow in a lawn, but if you really want to go nuts (sorry!) then an allotment would be quite a good idea.

HOW TO GROW

Degree of difficulty: Easy and not at all time consuming.

Plant: Nuts need much the same conditions as fruit – deep, fertile but well-drained soil in a sunny, sheltered spot, though when they are grown mainly as ornamentals with a bonus it's not quite so vital to provide precisely the right conditions. Even so, for best results they don't want to be overshadowed by nearby trees or buildings.

Spacing: Whether you are growing a row of nut trees or bushes on an allotment or you're dotting them around in a mixed border or in a lawn or wild garden, leave a fair distance between nuts and their nearest neighbours to allow for future growth. Give cobnuts and filberts 2.4–3m (8–10ft), dwarf sweet chestnuts or dwarf walnuts 4.5–6.1m (15–20ft), almonds (best fan-trained, unless you grow a free-standing tree in a really warm sheltered spot) 3.6–4.5m (12–15ft). Allow a hazel grown as a tree in a wild garden 3.6–4.5m (12–15ft). A row of hazels 1.2m (4ft) apart makes a good shelter belt, or you can allow several to grow 3m (10ft) apart as trees through a mixed country hedge.

Routine care: Give newly planted, single-stemmed nut trees a stake for their first year or two, until they are well established, and keep them watered in dry spells for the first summer. Bushes will usually be self-supporting. Every spring, mulch with well-rotted organic matter or bark chippings to retain moisture and suppress weeds which would otherwise compete with the tree, and in mid- to late April, sprinkle a general-purpose feed over the soil underneath the entire canopy of branches (since this coincides with the area taken up by the 'feeding' roots). Once the trees/bushes start carrying crops, be prepared to protect them from squirrels.

RESTRICTING THE ROOTS

Some specialist nurseries and mail-order organic gardening suppliers sell large-capacity, fine-mesh bags which control the root-spread of potentially large trees such as walnuts and sweet chestnuts, by allowing the fine 'feeding' roots out, but preventing the large 'nuisance' roots from escaping into the surrounding soil. In effect it's like growing the tree in a big tub, but without the bother of having to water it so much. The idea is to keep the trees more compact and slow growing than usual, and to encourage a heavy-cropping habit early on in life. I've never used them, so I can't say how well they work, but it's certainly worth a try and if you're ordering plants anyway it's very little extra expense.

PRUNING

Pruning is optional for walnuts and sweet chestnuts, and compact varieties really shouldn't need it. Buy a well-shaped plant with a good spread of branches that are evenly spaced all round the top of the trunk in the first place. A fan-trained almond will need regular attention to maintain its flat shape, pruning from May through the summer. Hazels are usually left to their own devices as hedgerow or wild garden trees or bushes (unless they are being coppiced for pea-sticks, bean poles or firewood), and cobnuts and filberts can be left alone when they are grown as dual-purpose edible-ornamental plants. However, it's worth the effort of pruning them properly in late February when you really want them for nut production.

Pruning cobnuts and filberts for nut production

Aim to create a bushy tree growing on a short, knee-high trunk with an open centre – like a short-stemmed wine glass. After planting, if the bush has a strong upright 'leader', cut the top out at about 1.2m (4ft) from the ground to encourage branching lower down. From then on, prune in late February each year; remove any stems growing up into the centre of the tree to maintain the open centre that gives the bush its wineglass shape, and shorten all the main branches by half, cutting just above an outward facing bud. Remove any suckers growing up from the base of the plant. (The reason for growing the tree on a short trunk is so that it's clear which are suckers, as all the nut-bearing branches grow out of the top of the trunk.)

VARIETIES

1 **Almond (*Prunus dulcis* var. *dulcis*)** – attractive, early-flowering trees related to peaches and cultivated in much the same way; best grown as fans on south-facing walls since they flower a fortnight earlier than peaches. Hand-pollinate if no bees are around at the time, and protect blossom from frosts using fine netting or fleece on cold nights. Manage a fan-trained almond in the same way as a fan-trained peach or nectarine (see page 241). When grown as a free-standing tree, avoid pruning because it risks the entry of disease, but if it really needs doing, undertake it in spring when new growth starts. Named varieties are rarely available, but a new one is now appearing in some catalogues:

2 **'Mandaline'** – a new almond variety bred for our climate, it is self-fertile and late-flowering, producing blossom in April after the worst of the frost is over and when there should be plenty of pollinating insects around; so hand-pollinating is not needed. Pick the nuts in early October.

Cobnuts, hazelnuts and filberts (*Corylus* species) – deciduous shrubs 2.4m x 2.4m (8ft x 8ft) with large, rounded leaves; each plant bears both male and female flowers (males are the familiar long, dangly catkins, and females are tiny red tufts that emerge from buds and are quite hard to see). Even so, you will need two different varieties to cross-pollinate each other and be sure of a crop. One cobnut and one filbert will do the trick, but if there are wild hazels growing in hedgerows nearby you may get away with just one.

These shrubs are wind pollinated, so your fruiting bushes will need to be within 'blowing' distance.

3 **Cobnut (*Corylus avellana*)** – the larger-fruited cultivated hazelnut – 'Cosford' and 'Tonda di Giffoni' have large nuts and long, yellow catkins on bare branches in spring.

4 **Filbert (*Corylus maxima*)** – a close relative of the cobnut, and similar, but with long, shaggy husks that completely cover the nut; 'Kentish Cob', despite the name, is a filbert.

5 **Red filbert (*Corylus maxima* 'Purpurea')** – probably the best dual-purpose nut bush – a most attractive shrub with long red catkins on bare branches in spring, coppery-purple foliage throughout the summer followed by a good crop of rather small nuts with frilly purple-red outer husks in autumn. Needs a sunny site for the full foliage colour to develop.

6 **Hazel (*Corylus avellana*)** – wild hazel isn't the best choice when a crop of nuts is your top priority, but when you've room for a row of them to coppice in rotation for firewood, sticks, poles and plant supports, you'll have a few nuts in the years when they aren't cut hard down.

7 **Sweet chestnut (*Castanea sativa*)** – the traditional parkland tree with the deeply indented 'spiral' bark, jagged leaves and clusters of spiny nuts; ready to harvest October. No pruning is necessary.

8 **'Regal'** – a 'compact' sweet chestnut (compared to traditional varieties!) reaching 4.5m (15ft) in ten years, though of course it will keep growing slowly and bears nuts within 2–3 years of planting.

9 **Walnut (*Juglans regia*)** – slow-growing trees known for their craggy bark and good foliage. They develop great character with age and can reach 33m (100ft) high and almost as much across in time. Named varieties are more easily accommodated, and those grown by grafting will often start cropping within 3–4 years of planting instead of the traditional 10–15. Nuts are ready October/November. No pruning is necessary.

10 **'Broadview'** – a traditional-type walnut, but one which should start producing nuts after 3–4 years. A 'compact' variety (which means compact for a walnut…) reaching 6.2m (20ft) in ten years, though it'll eventually grow bigger.

11 **'Franquette'** – the connoisseur's walnut, said to have the sweetest and best-tasting walnuts of all, but it's not a tree for a small garden; it can reach a height of 100m (30ft) and has a similar spread.

HARVEST

Almonds, sweet chestnuts and walnuts are ready to eat when they start falling from the trees naturally in the autumn. Knock the rest down then or simply wait for them all to fall and pick them up off the ground. Hazels, cobnuts and filberts are often picked slightly green in September/October and eaten fresh, and one of the great joys of growing your own is picking them and eating them 'on the hoof' while you're walking round your patch. But when you want them to store, wait until the shells are hard and brown before picking them. Pick almonds in early October, sweet chestnuts in October and walnuts in October/November.

STORAGE

Remove the fleshy outer skins of walnuts and the prickly cases from sweet chestnuts, and remove the leafy husks from cobnuts and filberts. Spread the nuts out in a shallow layer in a warm room to dry thoroughly and turn them over several times so they dry evenly.

Store whole nuts, in their shells, in trays in a cool, dry, airy, rodent-proof shed – they'll keep for six months, though they are best used earlier before they start to dry out.

PROBLEMS

Grey squirrels are the absolute bane of the nut-grower's life. Even if your garden is not very squirrel friendly due to pets roaming around, the little blighters can leap enormous distances from nearby trees to reach their favourite meals. Ideally, plant nut trees more than a squirrel's leap from surrounding trees and then devise 'rat guards' – something like you see on ships' mooring-lines – to stop them climbing the trunks. Commercial animal deterrent pepper-powders, sprays and pellets are available, but they usually need reapplying frequently, especially after rain.

Holes in shells, weevils and empty nutshells are usually caused by all sorts of boring bugs and weevils, but frankly there's absolutely nothing you can do about it, so enjoy the nuts that *are* perfect and be prepared to share a few with the resident wildlife.

Passionfruit

Passiflora edulis

Quite a few species of passionflower produce good, edible fruit, but the one that is sold as passionfruit in the shops is the purple granadilla (*Passiflora edulis*). To be honest, the plant isn't in quite the same league looks-wise as the ornamental passionflowers that have become so popular for growing on sunny walls, but as climbers go it's not bad. So if you have a really hot, sunny patio it's worth trying a purple granadilla on trellis there. But when you're more interested in growing some serious fruit, then this is a plant for the greenhouse or conservatory – but no heating is needed. Plants aren't very readily available, though they can sometimes be found in the catalogues of specialist fruit nurseries, but it's easy enough to grow your own from seed and they'll start fruiting the following year.

HOW TO GROW

Degree of difficulty: Easy and not time consuming.

Sow: Obtain seeds from specialist catalogues and sow 2–3 per small pot on a warm windowsill indoors in March/April. Keep the seedlings at room temperature and weed out the weakest seedling if all three come up, leaving two per pot. When they fill their original pot with roots, re-pot the duos of young plants into larger containers without disturbing their rootballs.

Plant: Harden off the young plants gradually and plant the best potful outside in a well-prepared border against a south-facing wall in late May or early June, after the last frost. Give plants netting or trellis to climb on – tie the existing stems in place initially, but from then on new growth hangs on using its own tendrils. Alternatively, plant in May in a rich soil border against an inner wall of a greenhouse or conservatory which gets plenty of sun, with trellis to grow on, or grow in a large 38–45cm (15–18in) tub filled with John Innes No. 3 potting compost and add a little multi-purpose compost and grit to lighten it up. When grown in a pot, give the plant a decorative obelisk or wigwam of canes to climb up, or place the container at the foot of a wall that's been covered with trellis or netting and weave the stems up into it until they start climbing naturally.

Spacing: Allow wall space of 1.8m x 1.8m (6ft x 6ft) at least. If growing on an obelisk, the structure needs to be 1.8m (6ft) high.

Routine care: During the growing season (May until September), plants need watering generously, especially when grown in containers or against a wall, since the foundations soak up a lot of available moisture. Use liquid tomato feed once a week during this time. When plants are grown in the open they usually have no problems with pollination, but under cover it doesn't always happen, so when plants start flowering use a soft artist's brush and dab it into the centre of all the open flowers in turn every day or two to pollinate them. You'll know it's worked when you start to see the round green fruits swelling.

PRUNING

In autumn, after the fruit has been picked, plants can be cut back slightly to reduce their size and tidy them up, but prune properly in February. Cut back all the long, whippy growth made during the previous summer to within two buds of the main framework of permanent stems that are trained out over the wall or over the supports.

VARIETIES

1 **Purple granadilla (*Passiflora edulis*)** – vigorous climber up to 4.5m (15ft), with large rather architectural, glossy, three-lobed leaves up to 20cm (8in) long, and white flowers, followed by 5cm (2in) diameter round purple fruits with very soft, juicy, yellow flesh filled with dark seeds.

2 **'Crackerjack'** – a cultivated variety with larger and more plentiful fruits, available from a specialist nursery; good for growing in pots on space-saving spirals or similar structures.

HARVEST

Ripe fruits may appear a few at a time from late summer into the autumn. Leave fruits on the plant after they ripen to a dark purple, and wait until they start to wrinkle slightly – just like the ones you see in the greengrocer's. Use secateurs to snip them off, complete with their own short stems. They'll keep for several days or a week in a fruit bowl if you don't need them straight away – this means you can save them up if you need lots for a recipe.

PROBLEMS

If your plants get out of control, harden your heart and prune really hard, as described earlier. During the growing season, keep wayward stems tucked in to trellis, or if grown in tubs, wind new growth regularly round the supporting obelisk in a spiral, instead of letting it shoot straight up to the top – which leaves it nowhere to go for the rest of the season.

Whitefly and greenfly are not that often a problem outside, but they are regularly seen on plants under glass – move free-standing, tub-grown plants outside for a few weeks if possible to shift pests, but when plants can't be moved out, because they are growing on trellis up the walls, ventilate the greenhouse well. This also allows in natural predators. Hose down plants regularly to flush out intruders – passionfruit plants enjoy a shower anyway.

Red spider mite is not a frequent problem, but look for dry, buff-coloured leaves or fine webs, particularly among the young tips at the top of the plants, which are often a sign that the air is too dry. Damp down paths and passionfruit foliage regularly in warm weather. Remove the worst affected leaves and introduce a biological control – *Phytosieulus persimilis* for red spider and *Encarsia formosa* for whitefly – if you can't cope by cultural means.

How to eat passionfruit

Most of us will sit down and eat a passionfruit all on its own. The technique is like eating a boiled egg. Slice the top off a nice wrinkly fruit, and dip in with a teaspoon – don't worry about the pips, eat those too – they add crunch to the rather thick, exotically flavoured juice.

But if you have enough fruits, slice the tops off and scoop the contents out into a bowl, then you can use it in more adventurous ways. Nigella Lawson uses them mixed with cream to pour over lumps of sponge cake or 'pound cake' and makes a sort of open trifle. If you only have a small quantity, perhaps from 2–3 fruits, try adding the passionfruit pips-and-goo to a fruit salad to enrich the juice, or spoon a little over posh vanilla ice cream.

Peaches and nectarines

Prunus persica, P. persica var. *nectarina*

	J	F	M	A	M	J	J	A	S	O	N	D
plant	●	●	●								●	●
harvest							●	●	●			

Although they aren't the easiest of fruits to grow perfectly, peaches and nectarines are well worth the effort. Once you've tasted your home-grown crop, left to ripen naturally in the sun and then eaten straight from the tree, you'll never look at shop-bought, imported peaches again. As a bonus the trees are very handsome; a fan-trained peach or nectarine in full bloom, with pink blossom on bare branches at the end of winter, is quite a sight.

HOW TO GROW

Degree of difficulty: Not particularly time consuming but needs attention to detail and providing the right situation is essential.

Plant: Fan-trained trees are the most reliable in our climate; plant them against a sunny, south-facing wall. If no suitable wall is available, plant a fan-trained tree against a post-and-wire fence or a wooden panel fence in a similarly warm, sunny, sheltered, south-facing situation. An enclosed area is best, since besides protecting the plant from strong winds, surrounding walls will radiate heat and light, which are essential to ripen the fruit. The soil needs to be rich, fertile and well drained with plenty of well-rotted organic matter, and ideally slightly on the chalky side. Prepare a bed especially for the trees at the base of the wall and work in plenty of well-rotted organic matter.

Spacing: Needs a wall or fence 1.5–1.8m (5–6ft) high and 1.8–2.4m (6–8ft) wide.

Routine care: Each spring, mulch generously and in mid- to late April feed with general-purpose fertiliser. Water in dry spells during the summer, especially while the tree is carrying fruit. Peaches and nectarines flower on bare branches early in spring when few insects are about, so for best results hand-pollinate by dabbing a soft artist's brush into all the open flowers every day, and if a frost or cold windy weather is predicted while the plants are in flower, cover them with fleece for protection. Keep a watch on the fruit from early July onwards, and once the first ones start colouring up, cover the plant with netting to protect from birds.

VARIETIES

Peaches

1 **'Jalousia'**– belongs to a group known as honey or China peaches; these have curiously shaped fruits which are slightly flattened with a dip in the centre – something like a doughnut, but with a melting honey flavour. Ripens early September. Several similar varieties can be found in various catalogues.

2 **'Peregrine'** – white-fleshed fruit ripening mid- to late August. Another reliable old favourite.

3 **'Rochester'** – a yellow-fleshed variety ripening early August. A traditional favourite and very reliable.

Nectarines

4 **'Lord Napier'** – probably the most reliable nectarine for growing outside, with a superb flavour and white flesh. Ripens in early August.

5 **'Pineapple'** – occasionally available from specialist nurseries, and while it's an outstanding variety with rich melting flavour, deep yellow flesh and red flushed skin, it ripens late – not until September – and needs a warm season to do well, even in a very sheltered southern location. Anywhere else it really needs growing under glass.

TRAINING AND SUPPORT

A fan-trained tree needs regular training and constant support for life. After planting, fix stout 1.8m (5ft) canes to the wall in a fan shape, with one cane for each branch of the fan, and secure the branches to their supports – tie them in regularly as they grow. When first bought, a trained fan only has a limited number of branches, so choose suitable new shoots growing out near the base of the plant to fill the gaps. After a few years the fan should cover the entire area with evenly spaced branches radiating out from near the base. Rub out any new shoots growing where you don't want them while they are small and pinch out the sideshoots if they threaten to overcrowd the framework.

During spring and summer the main fan of branches will produce sideshoots and it's these that will eventually carry the fruit, so leave them alone. But each year from May onwards, rub out any shoots that grow in towards the wall or out away from it while they are small to maintain the flat shape of the fan without the need for any more pruning. Expect a new fan-trained tree to start producing a few fruits from its second or third summer.

GROWING PEACHES IN CONTAINERS

If you haven't got room for a fan-trained peach or nectarine, you can still grow a naturally dwarf patio version as a free-standing tree in a pot in a sheltered, sunny position. Several varieties are available for this job – the range of names on offer is changing all the time, but many are simply sold as 'patio peach' or 'patio nectarine'. They are genetic dwarfs that stay exceptionally small (reaching only 1.2m/4ft after ten years) because they are slow growing. They make very attractive, early-flowering trees with pale pink blossom on bare branches, yet they produce full-sized fruit.

Buy a tree in spring and re-pot it into a 38–45cm (15–18in) pot or tub with plenty of drainage material in the bottom, and filled with John Innes No.3 potting compost – add to this 10 per cent each of multi-purpose compost and potting grit to open up the texture. Give it a stake for support. Water regularly throughout the growing season and also during very dry spells in winter. Use liquid tomato feed every week from late April to the end of August.

SECRETS OF SUCCESS

I can't emphasise it enough: the right growing conditions are absolutely essential for peaches and nectarines. Despite global warming, our climate still hasn't reached the stage where free-standing peaches and nectarines will reliably produce ripe fruit outdoors, so growing them against a warm south-facing wall is the only way of being sure of your crop. The wall stores heat by day and releases it at night, acting like a giant storage heater – and it'll even do this to some extent in winter, which helps to ripen the new wood so that flower buds are initiated. Residual warmth helps protect the flowers – which is essential as they open so early in the year. Fan-trained plants are, in any case, a lot easier to reach to pick fruit, finger-prune and protect from birds; they are also easier to spray if you need to.

If you don't have perfect peach-growing conditions, frankly you are wasting your time; the fruit either won't set, won't swell, or won't ripen, so plant an apricot instead. In a rather exposed or cold area an apricot will do better fan-trained against a sunny wall, but in most places where fruit generally does well, it will succeed as a free-standing tree (see page 241).

HARVEST

Leave the fruit on the tree until it is completely ripe; depending on the variety this is usually from late July to early September. Ripe fruit changes colour gradually, developing a warm flush, and feels slightly soft to the touch. It often has a faint scent, too.

Once the first fruits start ripening, check daily and pick any that are at just the right stage. If left too long, they become overripe and will fall off and spoil, so learn to recognise the warning signs and pick just in time for the most luscious peaches ever.

The total crop will usually ripen over a period of 3–4 weeks, with most fruit being concentrated in a two-week period, so book your summer holidays accordingly, or you'll miss the lot.

PROBLEMS

Peach leaf curl is the most serious disease of peaches and nectarines, which you can expect most plants grown in the open to get at some time. Greenhouse-grown plants are seldom affected. Large red-blistered areas appear on the foliage in summer, followed by white powdery spores and the leaves dropping prematurely. Affected trees are badly weakened, for this reason, and cropping will suffer as a result. Spray trees thoroughly with Bordeaux mixture twice, at ten-day intervals in February, and repeat in autumn just before the leaves start to drop. Rake up fallen leaves and burn or dispose of them well away from peach or nectarine trees. Spraying with foliar feed will often help the plants to recover.

Wasps are very fond of peaches and nectarines and will bore into ripe fruit – once the first few wasps discover the fruit, more arrive and a crop left untended can be ruined. Pick ripe fruit daily and don't leave damaged, fallen or overripe fruit lying around under the plants. If the insects persist, hang up old-fashioned wasp traps near the plant. (To make one, dissolve a tablespoon of jam in a pint of warm water and divide this equally between several jam jars. Put a circle of paper over the top of each jar, secured with an elastic band, and make a hole in the middle with a pencil so it's only just big enough for a wasp to get in. Empty every few days when the fluid is full of wasps and then start again.)

Recipe – Gingered peaches

Peaches are best eaten perfectly ripe, just as they are, but if you have fruit that aren't quite ripe at the end of the season, or you grow so many that you are looking for other ways of using them, try this:

Poach halved, stoned peaches gently in the sweet gingery syrup from a jar of preserved ginger, then serve chilled with cream or ice cream and some finely chopped preserved ginger sprinkled over the top. Quick and easy.

Pears

Pyrus communis

	J	F	M	A	M	J	J	A	S	O	N	D
plant	●	●	●							●	●	●
harvest									●	●		

Pears sound as though they should be as easy to grow and as reliable as apples, but only the real trouper – 'Conference' – comes into this category. Most of the upmarket 'eating' pears, with their melting flesh, can be far more tricky to please. That said, if conditions are right they can thrive, and the 'Doyenné du Comice' that we inherited in our first garden fruited reliably almost every year with very little fuss. It's the old story – give plants what they like and they will be happy.

HOW TO GROW

Degree of difficulty: Needs attention to detail, especially at harvest time.

Plant: 'Conference' will grow happily in the same conditions as apples, but if you want to grow a 'special' eating variety then a warm, sunny, sheltered spot with fertile, well-drained soil is essential. Even then results are most reliable if you train trees as espaliers or single or double cordons against a south-facing wall or fence, which acts as a 'radiator'. Pears need cross-pollinating with a compatible variety that flowers at the same time, so unless there are other pear trees within 90m (100 yards), plant a suitable pollinator. (Be sure to grow the right varieties together; not all pears 'partner' each other, and even so-called self-fertile varieties crop far better if there's another variety nearby to cross-pollinate them.)

Spacing: Free-standing trees on the moderately vigorous Quince A rootstock should be spaced 4.5m (15ft) apart, those on semi-dwarfing Quince C rootstocks 3m (10ft) apart. Those on Quince A will grow 4.5–6.2m (15–20ft) tall, and those on Quince C will grow roughly 2.4–3.6m (8–12ft) tall. (Quince C being ideal for small gardens.) Plant espalier-trained trees 2.4–3m (8–10ft) apart, single cordons 75cm (2ft 6in) apart and double cordons 1.5m (5ft) apart.

Routine care: Keep newly planted trees watered in dry spells for their first summer. Each spring mulch generously, then in mid- to late April feed with general-purpose fertiliser. In a dry summer, water trees while they are carrying a crop of fruit; this is especially important for trained trees grown against a wall or fence, since the foundations soak up much of the natural moisture in the ground. As pears ripen, cover wall-trained plants with netting to protect them from birds ('special' eating varieties are most at risk; birds aren't so interested in the more everyday kinds).

PRUNING

Prune free-standing, standard pear trees in winter in the same way as for apple trees, and prune cordons, step-over trees or each arm of an espalier in late July in the same way as a cordon apple tree (see page 190).

SPACE-SAVING PEARS

When you are short of space, it's worth planting a family pear tree. This consists of branches of several different varieties grafted on to a single trunk; so one tree produces three different varieties of pear. Only the most popular varieties are grown this way, and they are chosen to cross pollinate each other, so a family tree is a good choice when you only have room for one pear tree and there are no others growing in the immediate neighbourhood to act as pollinators.

Specialist fruit nurseries also sell pears that have been trained as espaliers (an upright trunk with two or three pairs of horizontal branches which grow out to make a flat shape suitable for growing against a wall), single cordons (which are like a short upright trunk bearing fruiting spurs along its length, and which are good for growing as 'pillars' at the back of a border) or double cordons (a pair of cordons linked by a U-shaped trunk at the base, good for growing against a wall) or step-over trees (which are basically horizontal cordons which can be planted almost as a low rail to edge a border or grown alongside a path).

VARIETIES

1 **'Beth'** – smallish, pale-yellow dessert pears with brown spots; the flesh has a sweet melting flavour. Pick early September, use late September; cross-pollinated by 'Concorde'.

2 **'Beurré Superfin'** – excellent quality, medium-sized dessert pear with rather elongated 'Conference'-shaped fruit but with sweet juicy flesh and yellow, slightly russetted skin. Pick September, ready October – the fruit does not 'keep' long. Cross-pollinated by 'Conference'.

3 **'Concorde'** – the modern equivalent of 'Conference' (it is a cross between 'Conference' and 'Doyenné du Comice') and is equally easy and reliable, with elongated light green fruit with faint brown russetting on the 'cheeks'. Pick September, use October and November. Self-fertile and a good pollinator of many other varieties.

4 **'Conference'** – the most widely grown pear, with elongated greeny-brown fruit, for eating and cooking, very reliable and heavy cropping. Pick September, use October and November. Not self-fertile, as is often suggested, but usually manages to crop well even if it is the only pear in the immediate locality (it must be particularly receptive). It is a good pollinator for many other varieties. Introduced 1894 and was our standard commercial variety for a very long time.

5 **'Doyenné du Comice'** – a large and very fat, squat dessert pear with pale green skin heavily covered with brown russetting and faint red tinged 'cheeks'. It has very rich, juicy, melting flesh; it is one of the real stars and worth making a fuss off. Pick mid-October, use November and December. Cross-pollinated by 'Conference'. A French variety, introduced to England in 1858.

6 **'Red Williams'** – a similar pear to 'Doyenné du Comice' but with red-skinned fruit when ripe. This variety is ready to pick from late August to use throughout September. Cross-pollinated by 'Conference'.

7 **'Williams Bon Chrétien'** – a very popular early dessert variety – usually simply known as 'Williams' – with classic plump, squat, soft, sweet, juicy dessert fruits. The skin is green, faintly striped with red; cross-pollinated by 'Conference'. Pick early September, use throughout September. Bred 1770.

ASIAN PEARS

These unusual fruits occasionally appear on sale in greengrocers and supermarkets; they are apple-sized and shaped, with russetty brown skin and crisp, white but faintly gritty-textured flesh inside; they are – let's say – rather an acquired taste. They are eaten fresh like a normal apple or pear, or used in salad or fruit salad. A limited range of trees is available from specialist fruit nurseries, and it is regarded more as a curiosity than a serious fruit, but if you want to try one, grow it like a normal pear tree, in a very sheltered, warm, sunny site. Some varieties are self-fertile, but the rest are pollinated by normal pears.

HARVEST

This is the tricky part, since it's not very easy to tell when pears are ready for picking. *Don't* leave pears on the trees to ripen fully as this makes them go 'sleepy' (brown and mealy-textured inside). They need to be picked at the right stage, then stored until the time is right for a particular variety to ripen. Visually, there is little warning as to when pears are ready for picking; there is only a slight colour change, the skin turning very slightly lighter, and some varieties may develop a faint warm flush. Suspect they are starting to be ready once the first few windfalls drop, and test by lifting a pear in your cupped hand; if it lifts off the tree complete with its stalk it's ready, but if it holds firm don't pull it off, leave it slightly longer.

Early varieties are ready from September onwards, but some late varieties will hang on for much longer – mid-October at least – making them difficult to harvest if bad weather knocks them down before they're ready, so you need to check constantly. If bad weather threatens after this, get the picking done, quick.

STORAGE

Store pears under cover in a cool place, such as a shed or garage, where the temperature stays fairly constant, and once they reach the peak time for that particular variety, bring a few at a time into a warm room – they'll ripen within a few days. If you simply bring pears indoors straight from the garden, late pears in particular will either stay rock hard or turn 'sleepy'. It's a common problem that's often responsible for people being disappointed with their home-grown pears. Always pick, store and ripen in three distinct stages for best results.

PROBLEMS

Compared to apples, pears are remarkably pest and disease free; they don't suffer from maggots, so if the harvest is small you can at least expect to eat all of it.

Scab is a fungal disease which causes black or brown cracks with corky edges on the fruit, which can be rather misshapen, and khaki blotches on the leaves followed by premature leaf-fall. Frequent spraying with a suitable fungicide during the growing season is the only real remedy, but most people prefer to cope by non-chemical means. To deal with the problem organically, remove affected fruit when seen, prune out cracked and scabby twigs, and rake up and destroy fallen leaves in autumn to reduce the incidence.

Small stunted fruits are invariably due to poor growing conditions or poor pollination, so feed and mulch the tree heavily each spring and try to increase its warmth and shelter it, perhaps by covering it with fleece on cold nights around flowering time. Plant a pollinator if no other pears are nearby – 'Conference' or 'Concorde' are always safe bets even when you don't know the name of your original variety. Above all, guard against poor, dry soil, which pears do not like. On chalky soil they can sometimes suffer from iron deficiency, so feed with sequestered iron in such circumstances.

Fireblight (see apples, page 194).

Recipe – Pear salad

Peel and slice a really ripe, perfect pear, then fan the slices out over a few slices of wafer-thin Parma ham on a plate and serve with vinaigrette dressing. You can easily adapt this idea to suit yourself; perhaps have alternate slices of ham and fresh fig with Parma ham and without the dressing. Or you can replace the ham with broken walnut pieces and blue cheese dressing – made by crumbling Roquefort or Stilton into the vinaigrette – and decorate with sprigs of watercress.

Plums

Prunus domestica

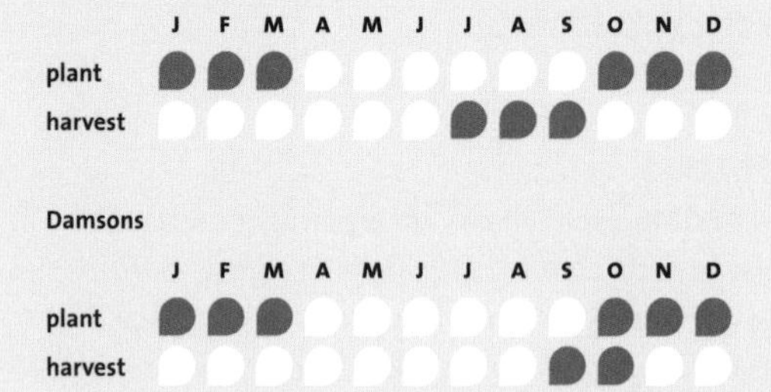

Next to apples, plums must be the most successful and widely grown garden fruit trees. But besides the familiar varieties, the plum family contains several 'forgotten fruits', including greengages, damsons, mirabelles and old-fashioned hedgerow species such as sloes and bullaces, which were once great favourites in cottage gardens and are well worth reviving.

Many plums and their relatives need another variety to cross-pollinate them, but you don't necessarily need to grow two of your own; if there are other plums growing within 90m (100 yards) in neighbouring gardens they may do the job, and wild plums, such as sloes, growing in nearby hedgerows will sometimes effect pollination, too.

HOW TO GROW

Degree of difficulty: Easy and little effort.

Plant: In a sheltered, sunny spot with fertile, well-drained soil that's ideally slightly chalky. Plums are mostly grown as free-standing trees, but fan-trained plums or cordons are occasionally available.

Spacing: A free-standing tree on a semi-dwarfing rootstock will need 4.5m (15ft) of space and grow to roughly the same height; allow 3–3.6m (10–12ft) width for a fan-trained tree.

Routine care: Keep new plants watered in dry spells for their first summer. Mulch generously each spring and in April sprinkle a double handful of general-purpose fertiliser over the soil beneath each tree.

Once the fruit has set, it's advisable to thin out very heavy crops. Some varieties, such as 'Victoria', are well known for cropping very heavily every *other* year and taking a year off in between to recover, so crop-thinning helps to even out this annoying 'biennial bearing' habit. It also helps to prevent the branches snapping under the weight of the fruit – a common occurrence with 'Victoria'. The time to thin the crop is when the plums are green and about half-size. Don't pull the surplus fruit off, instead snip them off at the stalk with secateurs leaving the biggest and best spaced fruit 7.5–10cm (3–4in) apart along the branches.

As fruits begin ripening, protect a fan-trained plum from birds by draping it with netting, though birds are less of a problem with plums than with other fruit.

PRUNING

Avoid pruning plum trees – buy a well-shaped tree in the first place with four or five strong branches spread out evenly round the top of the trunk. Remove any damaged branches during the growing season in late spring or summer and treat the wounds with pruning paint. Train and 'finger-prune' a fan-trained plum as described for fan-trained peaches and nectarines (see page 241).

Damsons should also only be pruned if absolutely necessary, otherwise you should avoid doing so. If you must prune, do so in late spring/summer, and treat as you would plums.

VARIETIES

1 Cherry plums/myrobalan plums (*Prunus cerasifera*) – very small, round plums that are more like cherries to look at, hence the name. Often simply sold as cherry plum; self-fertile, ripening late July. Some superior named varieties are currently available in specialists' catalogues, with larger fruit – these are hybrids between myrobalans and plums, with red or yellow fruit. These cross-pollinate each other, but are also cross- pollinated by 'Victoria'. The original species came from the Near East to Britain centuries ago.

2 'Czar'– reliable, heavy-cropping, self-fertile, culinary variety which ripens in early August with round, medium-sized, oval, purple fruit. Raised in 1874 and named after the Czar of Russia, who made a much-hyped state visit to England that year and some of the plums were presented to him.

3 'Old Green Gage'– small, yellowy-green fruit with superb flavour, but rather a light and irregular cropper; ripens late August/early September. Often planted in cottage garden hedgerows. Pollinated by 'Marjorie's Seedling'. Introduced pre-1724.

4 'Oullin's Golden Gage' – a reliable, heavy-cropping and well-flavoured dessert variety. Its round, light green fruit colours up on the sunny side to a red-streaked golden colour that could almost be mistaken for an apricot. Ripens early August. Pollinated by 'Victoria'. Discovered in France and introduced to Britain around 1860.

5 'Marjorie's Seedling' – sometimes simply called 'Marjorie's', this is one of the very best cooking plums with large, oval, purple fruit, yellow inside, which ripen in late September and early October. Self-fertile. An old variety, whose origins are unknown.

6 'Merryweather' – the traditional damson variety, with roundish purple fruit. A tree can be grown as specimen in a lawn or planted in a mixed country hedge in a cottage garden to save space. Self-fertile, with the biggest fruit of any damson. Good for jams and preserves such as damson cheese (which is like a very stiff jam that's turned out and sliced – ideal with a ploughman's lunch of bread and cheddar cheese). Ripening in September. Introduced 1907.

7 Mirabelles – a rare fruit with small, roundish, cherry-like plums in red or yellow, ripening in mid-August with a greengage flavour – use them for dessert or for cooking. They are very ornamental when in fruit. Mirabelles originated in France where they were once very popular, though they are little known here; trees are very occasionally offered for sale in catalogues.

8 'Victoria' – the best-known and most widely grown plum, a dessert variety that is often also used for

cooking, with large, light purplish-red and pinky-cream, oval fruit that's greeny-gold inside and very sweet and juicy. Self-fertile and very heavy cropping. Ripe around the end of August and early September most years; slightly earlier in a hot summer. The perfect dual-purpose 'lone' plum tree. Originally found in a Sussex cottage garden and introduced by a London nurseryman around 1840.

HEDGEROW PLUMS

Sloes (*Prunus spinosa*) are the fruit of the blackthorn bush, which grows widely in hedges along country lanes and round fields. The small, round, blue-black fruits it produces are ripe in October, and are often gathered for making sloe gin – enthusiasts with large gardens sometimes plant a hedge especially.

Bullaces are second cousins to damsons with fruits like small, round, blue-black damsons, ripening in October; they can be used for jams and preserves.

HARVEST

Leave plums to ripen completely on the tree. When the first few ripe windfalls drop (as against the few misshapen or diseased fruit that are sometimes shed earlier in the summer), look for large, well-coloured plums which are likely to be nearly ready. Pick plums when they are just softening but don't leave them to become over-soft – if they part company with the tree easily in your hand without needing to be tugged, they are ready.

Depending on the variety, you could be picking plums and greengages from late July to the end of September, damsons in September/October and hedgerow sloes in October.

PROBLEMS

Plums are wasps' favourite fruit of all, so pick ripe fruit daily and remove any damaged ones, including windfalls, to deter wasps, or use old fashioned wasp traps (see peaches, page 242).

Little or no fruit being produced may be due to a lack of a suitable pollinator (and even so-called self-fertile varieties will produce bigger and better crops if there's a pollinator nearby), or biennial bearing caused by allowing a traditionally heavy cropper, such as 'Victoria', to carry an enormous crop one year without thinning it to reduce the burden on the tree, which then takes a 'gap year' to recover.

Recipe – Sloe or damson gin

The same recipe can be used for sloes, damsons or bullaces, and I have to say that I prefer the damson version. Nightly.

Take 1.3kg (3lbs) of sloes or damsons, prick them all over with a darning needle or slash each fruit three or four times with a knife.

Find a large, wide-necked screw-top glass jar (an old-fashioned sweet jar is ideal – you can still sometimes get them from traditional sweet shops) or use large 2.7 litre (4 pint) plastic milk 'bottles'.

Next you need 1.3kg (3lbs) of sugar. Fill the jar with alternate layers of fruit and sugar, then top up almost to the rim with gin – a cheap supermarket own-brand gin is perfectly acceptable for this job. You'll need about a litre. Keep the empty bottle to put the finished sloe gin back into afterwards. Screw the lid on and stand the jar or bottle in a corner of the kitchen out of sunlight for 3–4 months, turning it over every day (but don't shake it) until all the sugar has eventually dissolved, then strain off the sloe gin. If you leave the fruits in for too long the mixture will turn brown rather than deep red.

Drink it on its own in small quantities like a liqueur – don't use it in gin-and-tonic. Alternatively – and very fashionable lately – is to add a slug of sloe gin to champagne or sparkling wine to make a champagne cocktail. But on its own it is the perfect nightcap. And I speak from experience...

Quinces and medlars

Cydonia oblonga; Mespilus germanica

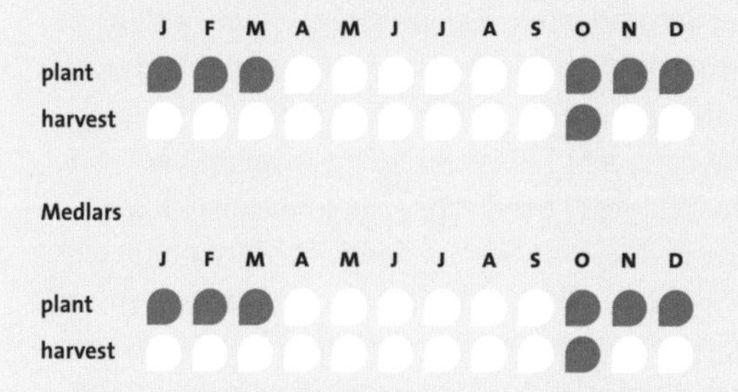

There's a certain amount of confusion about quinces, since the same common name is shared by two very different, though distantly related, plants. One is the ornamental quince or 'Japonica', *Chaenomeles japonica*, which is a medium-sized, spiny shrub grown for its spring blossom, which is followed by slightly spotted and often rather misshapen, golf-ball-sized fruit. The other is the true edible quince, which is a most attractive medium-sized tree with very large white or pale pink blossom in spring, followed by fruits that look like enormous golden pears in the autumn. It's pretty enough to grow as a decorative tree in the lawn or the back of a border, and regard the fruit as an autumn bonus.

Whilst the fruit of chaenomeles is sometimes cooked in mistake for 'real' quinces, and certainly won't do you any harm (even though it isn't all that flavoursome), real edible quinces are culinary stars. They are not a fruit you'd ever eat raw – they are far too hard and tart – but instead you can use them for making quince jelly (for glazing fruit tarts, or to eat with lamb or mutton instead of mint sauce), or stew them with apples to use in apple pies, crumbles and tarts. A hint of quince turns any dish that needs cooked apple into something altogether richer and more aromatic. Alas, unlike the owl and the pussycat, I have never eaten quince with mince.

HOW TO GROW

Degree of difficulty: Very easy and no trouble.

Plant: In deep, rich, fertile, well-drained soil. A sheltered, sunny site in a reasonably mild location is needed for the fruit to ripen reliably. Stake a new tree for its first few years, but once established it does not need support.

Spacing: Expect the tree to grow 4.5m (15ft) high and as wide. When you grow a quince tree, for both decorative and edible purposes, you can get away with growing shade-tolerant perennials underneath it and a clematis up into the canopy of branches. But wait for 2–3 years until it is well established; then it will be able to cope with the competition.

Routine care: Keep a newly planted tree watered in dry spells for its first summer, then after that it should be able to take care of itself. Mulch generously each spring, and in April sprinkle an organic general-purpose fertiliser on the soil underneath the entire canopy of branches.

PRUNING

No regular pruning is necessary, except the removal of dead, diseased or unwanted stems in winter.

VARIETIES

1 **'Meech's Prolific'** – very similar to 'Vranja'.

2 **'Vranja'**– an attractive, small- to medium-sized tree with large, pink, pear-like blossom in spring, followed by enormous crops of what look like pale green, felty-textured pears. These grow slowly over the summer into very large, fragrant, pale golden, pear-shaped fruits ripening in October. Self-fertile and very reliable.

1

2

MEDLARS

Although it's not related to quince, a medlar fulfils much the same sort of role in the garden – as a decorative, small to medium-sized, dual-purpose tree. A medlar makes a good shape, it has attractive architectural foliage that takes on wonderful fiery tints in autumn, and as a bonus it produces unusual fruits then too. The fruits are most peculiar; they look rather like russet-brown wooden Tudor roses about 4cm (1½in) across – the French name '*cul de chien*' perhaps best describes them. Delicacy forbids me from giving the English translation.

Medlars are very much an acquired taste, and one that most people would never bother to acquire. That's probably due to the fact that they are eaten after being 'bletted', which is to say that they are allowed to go half rotten in a controlled sort of way. They are probably best put in a fruit bowl and used as a natural decoration from the garden, but if you want to use them, then medlar jelly is your best bet. It is used in much the same way as quince jelly, except that besides lamb it also goes very well with pheasant or a robust free-range chicken.

The variety 'Nottingham' is most often seen for sale. Harvest the crop in mid- to late October, shortly after the leaves fall from the tree, and store clean, dry fruits in a shed for a few weeks (spread out one or two layers deep in trays) until they start to soften slightly. Traditionally they are eaten raw at this stage with cheese and a glass of port. But, as I say, it's an acquired taste.

HARVEST

Quinces don't swell up until quite late in the season, so don't worry if the fruits seem far too small at the beginning of August; they reach full size in September and don't ripen until quite late in October, turning a rich golden yellow. They are ready to pick when they lift easily off the tree in your hand, without pulling or twisting. However, if October brings windy weather and they are not properly ripe, cut the remaining fruits, complete with stalks, using secateurs and bring them indoors to finish ripening. Even when they are fully ripe they are still very hard, so they are difficult to peel and core. Remove the pips as they are poisonous.

Ripe quinces have a wonderful perfume, so even if you don't cook them, bring some indoors to put in a bowl, just for the scent. They keep for several weeks in a cool, dry place, but they don't store for anything like as long as apples or pears, so use them fast.

PROBLEMS

Ripening quinces develop brown rotting patches while on the tree; sometimes sunken, sometimes with concentric rings of white spores. These are caused by fungal diseases (mainly brown rot) which commonly affect quinces, but there's very little you can do but remove and destroy affected fruit and rake up and destroy fallen leaves in autumn. The brown rotting patches don't start to appear until quite late in the season, and unaffected fruit can be used as usual.

Fireblight is another problem (see apples, page 194).

Recipe – Quince vodka

Grate enough quinces (complete with peel but remove the core) to loosely fill a large, glass screw-top jar. Weigh out 50g (2oz) of white sugar per litre of your container, and layer sugar and grated fruit until it is filled. Top up with vodka (a cheap one is fine for this) and push the grated quince down so it's completely submerged. Leave to stand in a fairly even temperature, out of sunlight, for two months. Taste, and add more sugar if needed – tip it in the top and leave to stand for another few weeks until it's all dissolved. Strain off the fruit and pour into a screw-top bottle. Drink it 'neat'. A small glassful goes a long way.

Raspberries

Rubus idaeas

Raspberries, summer-fruiting

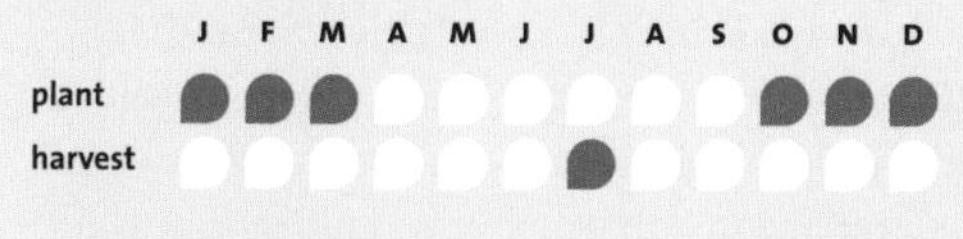

Raspberries, autumn-fruiting

Nothing goes down quite so well with afternoon tea in the garden in summer as raspberries. They are well worth growing at home, since besides being one of the most popular of soft fruits, they are always expensive to buy in the shops, and when you have your own you can indulge yourself to the hilt.

Nowadays, besides the traditional summer varieties that are picked in June and July, we can also enjoy autumn-fruiting raspberries that ripen from late August right up to the first proper frost. So grow several different varieties for continuity of crop. But if you are short of time or space, take my tip and stick to autumn-fruiting raspberries; they are a great deal less work than the summer sort, besides ripening at a time when fresh raspberry prices really peak in the shops.

HOW TO GROW

Degree of difficulty: Not difficult but need some regular care.

Plant: In good, fertile, well-drained soil containing plenty of well-rotted organic matter. Do a pH test – the soil should be neutral or very slightly acid; raspberries really don't do well on chalky ground – and, sorry, but you can't really grow them in pots. Choose a sheltered site in full sun or a spot that gets sun for at least half the day.

Spacing: Plant summer-fruiting varieties 45cm (18in) apart in the row, with 1.8m (6ft) between rows. Autumn-fruiting varieties can be planted with only 90cm (3ft) between rows if space is short, since they don't grow so tall.

Routine care: Each year in spring, mulch with well-rotted organic matter, then in April, sprinkle general-purpose fertiliser along each side of the row of canes (raspberry stems are always known as 'canes'). Keep well watered in dry spells while they are carrying a crop, otherwise the fruit will be very small and the total crop will be low. Cover rows of canes with netting as soon as the first few berries start turning pink to protect ripening fruit from birds – they will eat raspberries that are considerably more under-ripe than we would. The same set of raspberry canes will normally continue cropping well for about 10 years, after which they will have lost a lot of their vigour and should be dug up and replaced with new virus-free canes bought from a nursery. This is a good time to change to a new patch of ground, too, to reduce the risk of diseases carrying over to the new plants.

Organic growing

Ideally, grow raspberries without spraying them and pick them with clean hands, straight into a trug with a piece of kitchen paper in the bottom to keep the fruit clean. The idea behind this is that you don't need to wash the fruit before eating them, as washing raspberries turns them soggy and they lose the velvety texture that is half their appeal. Plus the fruit is so delicate that you can't help bruising it when you're washing it, so juice leaks and immediately some of the flavour and lushness is lost.

SUPPORT AND PRUNING

Summer-fruiting raspberries:

Support summer-fruiting raspberries with a post-and-wire fence made by knocking a strong 1.8m (6ft) post in at each end of the row with another one every 1.8m (6ft) along it, and stretching two strong horizontal wires between them – the first one 60cm (2ft) above the ground and the next one just below the top of the posts. As the canes grow up, tie them to both wires using soft twine, so as not to damage the stems.
If the canes grow much taller than the top wire, cut them off 15cm (6in) above it. Prune summer-fruiting raspberries shortly after you finish picking the crop, during the summer. Cut all the canes that have carried fruit down to a few centimetres above the ground – you can easily tell which these canes are as you'll see the remains of the sideshoots near the tops of fruited canes, complete with the dried-out calyces where the fruit have been picked. When you've pruned out the fruited canes, sort out the new unfruited canes that have grown during the same summer – these will carry next year's crop. Choose the strongest canes that come up in the row, roughly 45cm (18in) apart, and remove the rest, including any that have popped up in the path. If there are gaps in the row where no suitable cane has grown naturally, dig up some of the strongest unwanted canes to replant in the vacant spaces.

Autumn-fruiting raspberries:

Autumn-fruiting raspberries don't need supporting as they only grow about 90cm (3ft) tall, but if you like a neat and tidy garden you could put up short posts with a single wire about 60cm (2ft) from the ground. Don't prune them immediately after they finish fruiting in late autumn; instead wait until mid-February and then cut down *all* the canes to a few centimetres above ground level. If they don't grow back too thickly, I wouldn't bother thinning out the new growth at all, other than to remove new canes that grow up in the path instead of in the row. It's probably not 'correct', but my experience is that you can quite happily allow autumn raspberries to grow back as almost a 'hedge' in a bed 30cm (12in) or so wide instead of keeping them correctly spaced. After a few years you'll need to them thin out, but meanwhile you'll have enormous crops of fruit from a very limited space, provided you feed and mulch generously.

VARIETIES

1 **'All Gold'**– unusual yellow-fruited, autumn-fruiting raspberry, with a flavour that's often considered superior to red varieties.
2 **'Glen Ample'** – a mid-season, summer-fruiting berry, ripening from late June to early August. Spine-free.
3 **'Joan J'**– spine-free, autumn-fruiting raspberry with an exceptionally long picking season, from late July or early August until the first frost – given a mild autumn you could still be picking at Christmas. Freezes very well.
4 **'Malling Admiral'** – late season, summer-fruiting raspberry with exceptional disease resistance, ideal for organic growing. Spine-free. Ripens just after main season raspberries are over: late July to mid-August.
5 **'Malling Jewel'** – a summer-fruiting raspberry whose berries ripens in July and remain firm for a week, even when ripe. It's flavour is one of the best, though it's not the heaviest cropper. It is slow to succumb to virus.

HARVEST

Summer-fruiting raspberries ripen in late June and July or early August; autumn-fruiting raspberries from August until the first proper frost, which may not be until November. Leave raspberries on the canes to ripen fully; they'll turn from green to cream then pink before starting to redden, but give them another day or two to turn the full deep red colour that shows they are fully ripe, as the full flavour does not completely develop until then. Pick ripe raspberries between finger and thumb, pulling each one gently away from the calyx, which is left behind on the plant along with the hard 'plug' that pulls out from the centre of the fruit. (If the fruit does not come away cleanly from the plug and the calyx, it isn't ripe enough.)

PROBLEMS

Several different viruses affect raspberries, causing yellow blotches or mottled areas on the foliage; affected plants are also stunted and produce small crops. In time most raspberry canes become infected, since viruses are common, spread by greenfly, and they affect other cane fruit as well, though raspberries seem to be the main victims. There is no cure, you just have to dig up and destroy affected canes. Replace old canes after 10 years and replant in a new patch of ground to maintain strong, healthy, heavy-cropping plants.

Yellowing leaves are the classic symptom of lime-induced chlorosis. When grown in soil that is even slightly alkaline, essential trace elements (particularly manganese) are chemically 'locked up' by the chalk and raspberry plants are very sensitive to this. The resulting yellowing leads to progressively poor growth and a small or non-existent crop. If growing raspberries in soil that is not naturally neutral or slightly acid, apply sulphur powder or sulphur chips to the ground in spring (available from some garden centres but mainly from specialist suppliers; follow the directions on the pack for dosage on various soil types and pH levels), or use a chelated iron feed (sequestrene) of the sort used for rhododendrons and camellias. You'll need to repeat this treatment each year, in early spring.

Canes with black bases, or purple spots in summer that develop into cankers are signs of cane blight and cane spot – which are the two worst diseases of raspberry canes. There is no cure for either disease, although cane spot can be curbed by spraying plants regularly with fungicide – but it's far better to dig up and destroy affected plants, replant a new row, starting with 'bought' canes in a new patch of ground; or avoid it by growing a disease-resistant variety.

Recipe – Summer pudding

Slice a load of white bread (use brioche if you want the deluxe version), and remove the crusts. Make a mixture of two-thirds raspberries and one-third redcurrants (unwashed if possible, but if you have to wash them spread them out in a thin layer and allow them to dry before proceeding). If you have blackcurrants, blackberries or other soft fruit you can add any of these as well, but make sure half the mixture is raspberries. Spread the fruit out on a tray or large plate and sprinkle lightly with sugar. After an hour, strain off the juice that has run out, add a little kirsch or home-made fruit liqueur, and some leaf gelatine that has been dissolved in a little warm water. Use it quickly before it can start to set. Put a few spoonfuls aside, and mix the fruit into the rest.

Line a glass bowl with the sliced bread and moisten it with the gelatine/juice mixture; spread a layer of the fruit/juice/gelatine mixture over the bottom of the bowl, up to a couple of centimetres deep, then add another layer of bread one slice deep, cutting the pieces to fit over the fruit, and again moisten it with some of the juice/gelatine mix. Add another layer of the fruit mixture, and keep layering fruit and bread alternately until the bowl is full. Finish with a layer of bread, moistened with the last of the juice/gelatine. Put a plate on top and stand a weight on this while the pudding sets (a large tin of beans or something similar does the trick).

Chill in the fridge overnight, and if you want to turn it out to serve – which always looks good – stand the bowl in hot water for a few seconds, fit a plate over the top and hold it securely in place while you invert the whole lot sharply. With luck the pudding will stay in one piece. (If you aren't feeling brave, play safe and serve it in the original bowl.) Eat with lashings of cream.

Redcurrants

Ribes rubrum

Redcurrants and whitecurrants

	J	F	M	A	M	J	J	A	S	O	N	D
plant	●	●	●	○	○	○	○	○	○	●	●	●
harvest	○	○	○	○	○	○	●	●	○	○	○	○

If you grow raspberries you should certainly have a redcurrant bush to go with them; raspberries and redcurrants have to be one of the great fresh fruit combinations of all time. Redcurrants are also handy for making redcurrant jelly, and very good for 'stretching' small quantities of unusual fruit, such as mulberries, when you are making preserves. The plants look attractive, especially when they are trained as 'U'-shaped double cordons against a wall. But even if you just have an old neglected bush down the garden that you never get round to picking, then the fruit will be much appreciated by blackbirds – you'll get hours of amusement watching mother blackbird teaching a class of bemused babies how to pinch the fruit.

HOW TO GROW

Degree of difficulty: Not difficult or particularly time-consuming.

Plant: In a sheltered situation and fertile, well-drained soil. Redcurrants are one of the few fruits that will also do well on a north-facing wall or in a slightly shady area, though they are also happy in a sunnier spot – where the fruit ripens earlier and tastes sweeter.

Spacing: Allow 1.2–1.5m (4–5ft) between bushes; plant double cordons 60cm (2ft) apart.

Routine care: Mulch in spring and feed in April by sprinkling an organic general-purpose fertiliser around the plants. Water the bushes generously during dry spells in summer while they are carrying a crop, otherwise the fruit will be small and the harvest tiny – in a very dry year it may fail completely.

PRUNING

Prune redcurrant bushes from November to February, as with gooseberry bushes, and train and prune double cordons in late June or early July in the same way as for cordon gooseberries (see page 215). Like gooseberries, redcurrants are grown on a short trunk or 'leg'.

Recipe – Easy redcurrant jelly

You can even make this recipe in small quantities when you don't have much fruit. Pick some redcurrants, with their stalks; spread a layer a few centimetres deep over the bottom of a large, thick-bottomed saucepan and just cover them with water. Add sugar – you'll need 225g (8oz) sugar to 900g (2lbs) of fruit. Heat gently until the sugar dissolves then turn up the heat and boil quickly for 8 minutes. (Once the fruit starts to cook, taste and add more sugar if needed, since redcurrants are naturally 'tart'.) Tip the resulting thick goo into a fine sieve and hang it over a bowl until all the juice runs out – don't push it through with a spoon or the jelly tends to turn cloudy. When you've caught the lot, pour it into a clean screw-top glass jar to set and keep it in the fridge. If you want a very fancy jelly, add a few just-cooked redcurrants to it when you pour it into the jars. Alternatively, use a mixture of cranberries and redcurrants for a mixed-berry jelly, but add more sugar since cranberries are very sharp.

VARIETIES

1 **'Jonkheer van Tets'**– an oldish variety that's stood the test of time, with long trusses of large berries (which means they are less fiddly to pick), ripening early July.
2 **'Red Lake'** – a good old faithful variety, with large crops of juicy berries on long trusses, ripening early July.
3 **'Rovada'** – a heavy-cropping, late variety with large berries, ripening throughout late July and August. The flowers open late, so are not affected by late frosts.

Whitecurrants

Whitecurrants are the same shape and size as redcurrants, you grow them in exactly the same way and they ripen at the same time but the fruits are white – well, more of a warm cream-buff shade when they are completely ripe and ready to pick.

4 **'White Versailles'** (an old favourite with large, sweet fruit ripening in July) and 'Blanka' (a new very heavy-cropping variety ripening in late July and August) are reasonably readily available. They have no special uses other than to eat fresh, so they don't feature in the ranks of must-have soft fruit, but it's worth growing one bush if you have space. The fruits have a far milder, sweeter taste than redcurrants, so will 'dilute' their sharpness when added to summer puddings, fruit fools and pies. They look good in a bowl of berries to contrast with the usual reds, purples and blacks.

HARVEST

Leave redcurrants to swell up and ripen fully before picking them; the fruits increase in size quite a bit at the last minute and turn a brighter red in the day or two after first colouring up. Virtually all the berries on a truss should be red – but pick before the first fruits become so soft and over-ripe that they start to fall off the plant. When picking, take the entire truss of fruit by tearing it gently from the plant with an upward movement of your hand. Remove the individual berries from their thin green strigs in the kitchen later; the traditional way is to pull whole bunches through the prongs of a fork stabbed down onto a chopping board, but if you need to keep the fruit whole (for use in fruit salads or to go with fresh raspberries) there's nothing for it but to do it the slow tedious way, one by one, using your fingers.

PROBLEMS

Large red blisters might appear on leaves in summer, which look alarming and are caused by a species of aphid, called – with stunning simplicity – the redcurrant blister aphid. This can also affect white and blackcurrants (where, confusingly, it causes yellow blisters), though you're most likely to see it on redcurrants. The aphids are pale yellow and are found in colonies on the undersides of young leaves in spring and early summer. Check plants regularly and spray with an organic remedy when an outbreak is seen or, in late June, cut the tips of the sideshoots back almost as far as the first fruits. This removes the blistering areas and is good practice anyway, as it also helps to allow light and air to the ripening fruit, besides keeping the plants in check.

Decorating with redcurrants

Whole trusses of perfect, ripe redcurrants are very popular for using intact to garnish platefuls of 'foodie' desserts, and as decorations for cakes, puddings, cheeseboards and buffet tables – they look especially good draped over the side of a wine glass containing an individual portion of some indulgent creamy dessert.

Rhubarb

Rheum x *hybridum*

	J	F	M	A	M	J	J	A	S	O	N	D
plant	●	●	●							●	●	●
harvest			●	●	●	●	●					

Rhubarb isn't technically a fruit since it's the stems that we eat, but clearly you can't call it a vegetable as it's eaten with sugar, for pudding, so it's became a sort of honorary fruit. And very good it is too; rhubarb is enjoying a revival now that 'nursery food' such as rhubarb crumble is back in fashion. It is well known for being very easy to grow, but instead of just shoving a clump in a corner and forgetting about it until picking time, it repays a bit of effort to ensure a good supply of strong, tender, well-coloured stalks. The best rhubarb of the lot is the very tender, early, forced sort (a Yorkshire speciality) that's so expensive in the shops, but you can force your own at home given a bit of room in your airing cupboard, or simply cover a plant down the garden with a big dark pot – or even an upturned dustbin.

HOW TO GROW

Degree of difficulty: Easy and very little work.

Plant: Dormant crowns in late autumn or early spring, though pot-grown plants can also be planted during the summer if they are watered in dry spells. Although rhubarb is traditionally tucked away in any odd corner of the veg patch, you'll have much better quality stems if it's planted in a very richly manured place with reasonably heavy soil; it's good near the compost heap since it'll benefit from the rich run-off in the soil. But don't poke rhubarb away behind the shed; grow it in full sun so the stems will develop a redder colour and a sweeter and richer flavour.

Spacing: Allow 90cm (3ft) between plants and 90cm (3ft) between rows if you are growing lots.

Routine care: Remove any flowering stems that start to develop in summer; these are easily recognised as they are strong, straight, upright stems growing through the centre of the plant. Cut them out as close to the base as you can without harming the rest of the plant. Every autumn when the foliage dies down naturally, remove the dead leaves, sprinkle general-purpose fertiliser all round the plants and mulch generously. Ideally you should use well-rotted manure, but if this is not available use garden compost instead. Don't expect the same rhubarb plant to go on forever – replace oldies with new virus-free plants every ten years, and replant in a new, well-prepared site, since the old plant will have taken a lot out of the soil.

VARIETIES

1 **'Hawke's Champagne'** – also sometimes just called 'Champagne'; one of the best-flavoured varieties for pulling fresh from the garden during spring and early summer, with thick, reddish-green stems.

2 **'Stockbridge Arrow'** – a very early variety bred especially for forcing in heat (when the unfurling young leaves have a pronounced arrow shape) but which can also be grown in the garden for normal use. Very tasty, tender red stems.

3 **'Timperley Early'** – very thin red and green stems, good for forcing in the garden for cropping from February onwards; the stems remain tender late into the season so you can even pull a few in mid- and late summer – but don't over-do it or you'll weaken the plants.

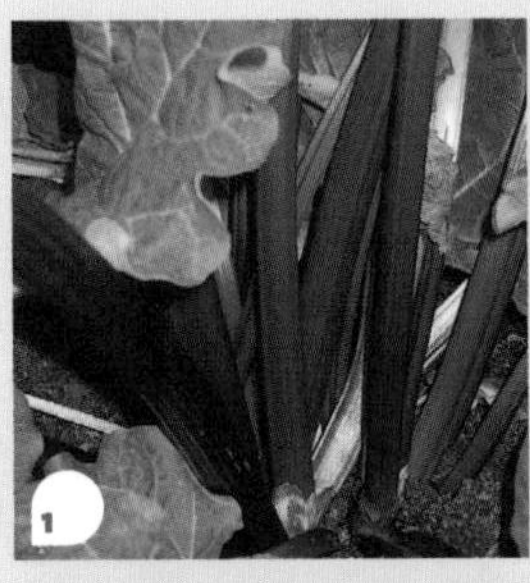
1

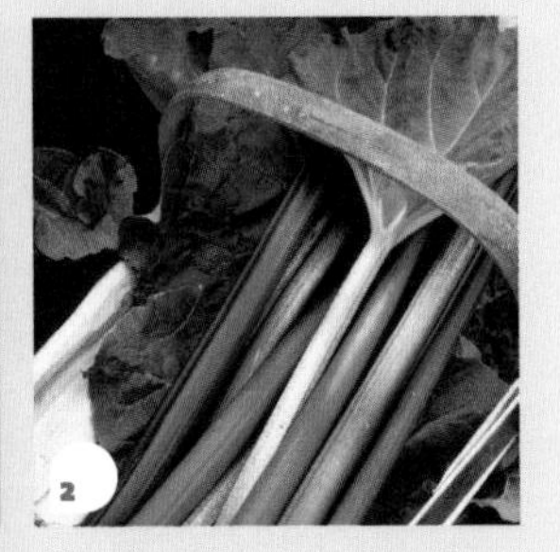
2

3

FORCING RHUBARB

Any rhubarb, even varieties not intended for forcing, can be encouraged to produce some thin, blanched stems very early in the season. All you need to do is cover a well-established crown with a special rhubarb forcing pot, or an upturned dustbin or similar large container – a bucket is not deep enough – in mid-January or early February. Cover the ground round it with straw for insulation, if possible, put the upturned bin or pot in place and a few weeks later you'll start to see long stems with pale yellowy-coloured leaves appearing. You can pull as many of these as you want until the end of March, but then uncover the plant and allow it to grow naturally, taking only a small crop of stems for the rest of that season. Rhubarb crowns are quite resilient, so as long as they are well fed and mulched each autumn, there's no reason why you shouldn't 'force' the same crown this way every year, but if you grow several plants, ring the changes to give them a rest in alternate years.

For really early rhubarb, grow a 'forcing' variety. After planting, wait one or two years so crowns are well established then dig up a crown in November and leave it on the surface of the ground for two weeks, so it's exposed to some cold and, if possible, frosty weather. Then pot it up, water it sparingly and move it to a dark, warm place indoors; the airing cupboard is ideal, but if this causes domestic difficulties, settle for the cupboard under the stairs or one in a spare bedroom or utility room. If you can't keep the pot in a cupboard, put a large container or cardboard box over the top of it to keep it in the dark. Darkness is essential. After 4–5 weeks, depending on how warm it's kept, you'll have very long, slender shoots to pull; you can keep pulling all the time the crown produces anything useable. After rhubarb plants have been forced in this way they are no use for replanting, so throw them away.

HARVEST

Rhubarb can be picked as required between March and July, but then stop harvesting it and allow the plant to grow naturally so it recovers for the rest of the season. (Many varieties become tough, stringy and tasteless in summer anyway.) For cooking, choose well-coloured stems whose leaves have *just* opened out fully, as these will be the most tasty and tender. To pick rhubarb, don't cut the stems off at the base, which leaves an open wound – instead, hold the stalk firmly near the base and tug and twist it so it comes out cleanly from the base and including the part that semi-clasps round the top of the root. This part, and the leaf, should be trimmed off when preparing rhubarb for cooking (the leaves are poisonous, but they won't hurt to go on the compost heap).

PROBLEMS

If your rhubarb fails to grow well and has dull, listless foliage, small sickly-looking stems and the growth buds die off instead of developing into stems and leaves, you've got crown rot, one of the very few problems rhubarb experiences. There is no cure for this, so you must remove and destroy affected plants and plant new ones in a different spot.

Strawberries

Fragaria x ananassa

	J	F	M	A	M	J	J	A	S	O	N	D
plant			●	●				●	●	●		
harvest						●	●	●	●	●		

If you're accustomed to the pneumatic strawberries flown in from the other side of the world out of season, your first home-grown crop will come as a very pleasurable surprise. Few people realise just how packed full of flavour, aroma and succulence strawberries can be, when they are grown naturally, allowed to ripen on the plants and then eaten within minutes – well, okay, maybe an hour or two – of picking. At home you can choose varieties for their flavour and grow them with love and care and, ideally, organically. It makes all the difference, and with strawberries it's well worth taking the trouble.

HOW TO GROW

Degree of difficulty: Not difficult but need some attention.

Plant: For best results, plant young strawberry plants in August or September so they are well established in time to start cropping next summer – if you can't obtain young plants that early, they can be planted in late autumn or spring when they're more easily available, but they'll give a rather light crop in their first year. Choose a sunny, sheltered spot with fertile, well-drained soil that's had plenty of well-rotted organic matter worked in. If the ground needs a little extra drainage, plant strawberries on top of ridges. But if your soil or situation isn't ideal or you are short of space, they'll also grow well in raised beds and large tubs or even growing bags – some people cultivate them very successfully in window boxes and hanging baskets. Each plant should normally yield roughly 225g (8oz) of fruit, so plant enough for your needs, and if there's room, grow two or three different varieties that ripen early, mid- and late season to extend the cropping season.

Spacing: On an allotment, or for a traditional strawberry bed in a big garden where there's plenty of room, space strawberry plants 45cm (18in) apart, with 90cm (3ft) between rows. In a raised bed with deep, rich soil you can grow them at closer spacing; 30–38cm (12–15in) apart with 60cm (2ft) between rows. When planting in tubs or other containers, space plants as little as 23cm (9in) apart, since the fruit and runners can hang down over the sides of the containers, but be prepared for heavy watering and feeding when the plants are grown so intensively.

Routine care: Early every spring apply 15g (½oz) of sulphate of potash to each square metre of strawberry bed, sprinkling it carefully between the plants. Being shallow-rooted, strawberries are the first fruits to suffer from dry weather and they are quickly over-run by weeds, so pay particular attention to regular weeding, and water them thoroughly in dry spells. In early summer, when the flowers are over and the first small green fruits can be seen forming, spread straw or lay synthetic strawberry mats down between the plants, tucking them under the foliage and around the 'collar' of the plants to smother weeds and protect the fruit from dirt, damp and mud splashes. (While they are carrying fruit, water carefully between the plants since splashed strawberries will often turn mouldy.)

Protect the plants from birds, too. The best way is to grow them inside a fruit cage which gives easy access for picking; alternatively, use lightweight netting-clad 'tunnel cloches' or construct a light framework of canes and drape netting over the top which can be rolled back for picking. Don't lay netting straight onto the plants, as birds can easily peck through it to reach the fruit. And if you're using netting, choose a thick heavyweight kind that makes it harder for birds to get their feet tangled – birds easily become trapped in very fine, light nylon netting and are hard to cut free, making them an easy target for predators.

After the whole crop has been picked and there's no more to come, remove the netting and store it away for next year, then clip strawberry plants over with a pair of shears to remove all the old fruit stems, leaves and runners, and finally feed the bed with an organic general-purpose fertiliser and water it well in. The plants soon make healthy, fresh, new growth which sets them up for cropping well again next year. Strawberry plants crop well for 3–4 years, after which they are best replaced with new young plants.

BUYING BARE-ROOT RUNNERS

When you order strawberry plants by mail order, they'll usually arrive as bundles of bare-root runners. Unpack them straight away and soak the roots for half an hour; separate them and plant them so that the 'crown' of the plant (where the shoots meet the roots) is level with the surface of the soil. It's a common mistake to plant too deeply and then the plants rot.

GROWING STRAWBERRIES IN CONTAINERS

Plant young plants in late summer or autumn if you can keep them under cover for protection, otherwise buy the biggest plants you can find in spring and plant them then. Use a 50:50 mixture of John Innes No.2 potting compost and multi-purpose compost, with plenty of drainage material in the bottom of the pots.

Plant four plants round the edge of a 25–30cm (10–12in) pot with one in the middle, or grow six plants in a large patio tub or a 60cm (2ft) long trough or windowbox. In the case of hanging baskets, line the basket with something that will hold moisture – cut a circle to fit from an old woollen jumper or an old bath towel – then fill it with multi-purpose compost and plant five or six strawberry plants round the edge. Put the container in a warm, sheltered spot. If you can start it off under cover in a greenhouse or car-port with good light the plants will get off to a better start. During the winter they won't need much watering, but once the plants are established keep them well watered, since you'll have a lot of plants growing very close together and they'll soon suffer if they start drying out. Hanging baskets in particular need a lot of water and regular liquid feeding to produce a good crop.

Cover containers with netting to protect them from birds while the fruit is ripening, unless they are in a very safe location close to the house where birds don't bother them. (But it's amazing how birds lose their shyness when there's a good meal on offer.)

After all the fruit has been picked for the season, cut back the plants to remove all the old foliage, give them a good drink of diluted liquid tomato feed and they will crop for another season or two before they need replacing. Alternatively, if half the reason for growing the plants is for decoration, *don't* cut them back and instead allow the runners to trail down – they'll bear some fruit the following season, but you'll need to feed and water very regularly to keep them all going.

GROWING REPLACEMENT PLANTS

Since strawberry plants need replacing every four years, it's worth raising your own from 'runners' (the young plants sent out on stems as the plant's natural means of regeneration). After all the fruit has been picked, leave a few plants that aren't cut back at the end of a row, and peg down several runners growing from them into pots of seed compost, sunk into the ground alongside. Once the runners have rooted well, snip the 'umbilical cord' connecting them to the parent plant and grow the youngsters on in a cold frame or quiet corner, watering and liquid-feeding regularly, and nipping off any runners that form until you are ready to plant them. Don't propagate sickly or diseased plants, and don't rely on getting these 'freebies' for ever; it pays to buy in new virus-free plants occasionally, especially if you've seen yields falling off, or you simply fancy trying a few new varieties.

Use bent wire to peg down strong strawberry runners into individual pots filled with multi-purpose compost, in late summer, and by autumn or early next spring you'll have young plants ready to plant into their new homes.

VARIETIES

1 **'Aromel'** – an older perpetual-fruiting variety, still quite widely available, with medium/large fruit of very good flavour which ripens in late summer and early autumn, but it is lighter cropping and more prone to disease than a lot of newer varieties.

2 **'Cambridge Favourite'** – a traditional variety that's reliable and easily grown, with good disease resistance and heavy crops of well-flavoured, medium-sized fruit which ripen from late June to late July.

3 **'Cambridge Late Pine'** – an old variety and one of the very best-flavoured ever, with sweet, dark red, aromatic fruit, ripening mid-June to mid-July. Only occasionally available.

4 **'Elsanta'** – a great favourite with pick-your-own growers and the general public, with large crops of well-flavoured berries. Ready mid-June to mid-July.

5 **'Flamenco'** – one of the 'perpetual-fruiting' strawberries; a heavy cropper with good-quality berries, fruiting in light flushes from the time the normal summer crop is over until the first autumn frosts.

6 **'Florence'** – a late variety, ripening from early July to early August.

7 **'Honeoye'** – popular and well-known variety, ripening throughout June, with firm, attractive, medium-sized fruit of good flavour and slight resistance to grey mould.

8 **'Mara des Bois'**– an old variety which is particularly delicious and aromatic, having the flavour of wild strawberries. A 'perpetual-fruiting' variety, for late summer and early autumn.

9 **'Mae'** – very early variety, ripening June and early July, but can be ripe in mid- to late May if protected with cloches in spring.

10 **'Royal Sovereign'** – the famous old variety of pre-1892, known for its outstanding flavour; but compared to modern varieties it's not easy to grow well and it's a rather light-cropper with smallish fruit. For best results, give it good growing conditions and extra care. Ripens early June to mid-July.

11 ALPINE STRAWBERRIES (*Fragaria vesca* Semperflorens Group)

These are dwarf plants producing tiny fruits about the size of your little fingernail, which are considered to be the *absolute elite* of strawberries. Plants do not produce runners, so they must be grown from seed. Sow in early spring under glass or on a not-too-warm windowsill indoors, and prick out the seedlings into individual 7.5cm (3in) pots when large enough to handle. Plant out about 23cm (9in) apart when established in summer. From a very early sowing you'll have some fruit the same summer, but plants will be most productive the following year or the one after. They are delightfully attractive when used to edge a path, potager or veg patch, forming neat clumps. Birds are far less interested in them than they are the large 'normal' strawberries so there's usually no need to net them.

11

PERPETUAL-FRUITING STRAWBERRIES

These don't – unfortunately – *really* crop perpetually, as the name suggests. They start flowering at the same time as normal summer-fruiting strawberries and they'll continue flowering and fruiting lightly, off and on, from June to October. But if you already grow summer-fruiting strawberries, it's worth nipping out the flowers of perpetual-fruiting strawberries until June, then let them flower as usual, as this makes them concentrate their efforts on a late-season crop after the summer strawberries are all over. Being later fruiting, you can get away with planting perpetual-fruiting strawberries in spring or even early summer, and you'll still have a decent crop in their first season.

HARVEST

Allow strawberries to remain on the plants until they are completely ripe. Don't pick them immediately they start to turn red; give them another day or two to finish plumping up and they'll turn a couple of degrees darker, but don't wait until they start going soft. Pick plants over daily, and don't leave small or misshapen ones behind (they are good for cooking or jam-making), since they'll go mouldy and may spread disease to the rest of your crop.

PROBLEMS

Take precautions against slugs and snails at the start of the growing season to reduce the population in and around strawberry beds – use various organic deterrents and keep the area free from weeds. In raised beds with wooden edges, one of the most practical organic methods is to tack a copper strip along the edge (this is sold in organic supplies catalogues); it gives slugs and snails a slight 'electric shock' that puts them off – there's a version with jagged teeth along one side which is doubly effective as it stops more determined molluscs climbing over the top.

Grey mould affecting fruit is a form of botrytis, which is worse in damp summers or when the fruit is splashed when you're watering the plants. Keep weeds controlled and space plants reasonably well apart to improve air circulation between them; avoid spraying with fungicide if possible, particularly if you want to use the fruit in preserves. (It's usually too late by the time fruit is going mouldy anyway.)

If you see yellow blotchy leaves and progressively low yields, suspect a virus, particularly if the plants are several years old or you propagate all your own replacement plants from runners without ever buying in new stock. Virus is spread by greenfly, so you'll usually get it eventually. Dig out and destroy affected plants any time they are seen, before it can be spread to others, and buy virus-free stock to start a new row at the first opportunity. Another good tip is to control the greenfly which spread the disease.

Herbs

Basil

Ocimum basilicum

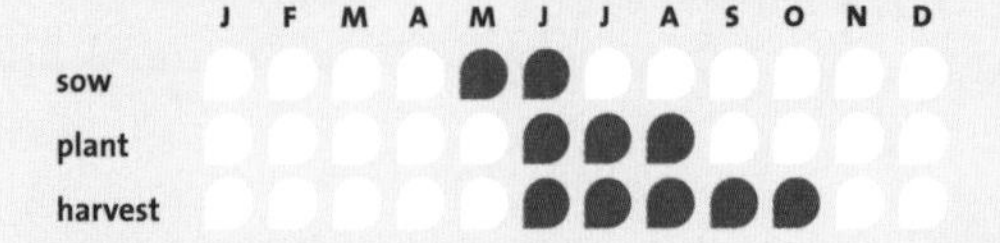

Basil has soared in popularity in recent years, and now it's easily overtaken traditional 'old faithfuls' such as parsley and chives as the herb serious cooks simply can't do without.

Basils are some of the best herbs for growing in patio containers and in pots on windowsills, since not only are they compact and attractive-looking, but they also have a delicious scent that's halfway between cloves and incense.

Plants of the basic kinds are sold by the potful in nurseries, garden centres, and even in supermarkets, and specialist seed firms stock a fair number of unusual varieties. One year I managed to grow thirteen different kinds!

Although it's renowned for use with tomato dishes, do give basil a try on home-grown, freshly cooked and buttered carrots. Delicious!

HOW TO GROW

Degree of difficulty: Not the easiest of herbs, it needs warmth and attention to detail.

Sow: In pots on windowsills indoors at room temperature all year round, or in a greenhouse or conservatory in summer. Seeds sown outside are far less successful, even in warm weather. Thin out seedlings if they are badly overcrowded, but it's a good idea to allow several young plants to grow unhindered in the same pot to make a good, bushy, 'instant' plant that's soon ready for picking.

Plant: Plant potfuls of basil outside, without breaking up their rootballs, from mid-June to early August; they'll fare best in a tub, trough or windowbox filled with good multi-purpose compost. They can be planted out in the open garden as long as the ground is well drained but contains plenty of well-rotted organic matter. A warm, sheltered, sunny spot is essential.

Spacing: Plant potfuls of basil roughly 15cm (6in) apart.

Routine care: Water sparingly, little and often; use a general-purpose liquid feed regularly on plants grown in pots or tubs, as this encourages them to remain leafy instead of trying to flower. As soon as flower buds appear at the tips of the shoots, nip them off to encourage further leafy growth.

HOW TO KEEP YOURSELF CONSTANTLY SUPPLIED

Basil is a bit slow growing at the best of times, but once it's started to flower – which it tries to do fairly early in its life – it virtually stops producing new leaves. Nipping off the flowering tips of the shoots helps for a while, but to keep up a continuous supply of leaves it is advisable to sow a new pot of basil roughly every two months. Old plants can be restored by re-potting them into a slightly larger pot with fresh compost, cutting the stems back to about 5cm (2in) high, and giving them a dose of liquid feed.

If you use a lot of basil, it's worth sowing a row in the greenhouse, which will allow you to pick enough to use as a salad leaf or for making pesto. It's ideal if you also grow tomatoes, since the two are perfect partners in the kitchen.

VARIETIES

1 **Bush basil** – makes a neat, compact, dome-shaped bush which is densely packed with very small leaves. While it can be used for cooking, this is the one that's often cultivated in large tubs outside restaurant doorways in hot countries for decoration, fragrance, and to deter flies.

2 **Lemon basil** – also known as 'Kemangie', this has much of the usual basil flavour but it is combined with a lemon-ish fragrance.

3 **'Neapolitana'** – also known as lettuce-leafed basil, is the variety with very large, slightly crinkly leaves. It's perfect for anyone who uses a lot of basil because plants are very productive and it's quick and easy to pick a decent quantity in one go. The scent is superb, too – strong and rich. Use whole leaves to add fragrance and flavour to mixed green salads, or tomato and mozzarella salads. A must-have.

4 **'Purple Ruffles'** – a glamorous variety with large, frilly-edged leaves in rich purple; great for patios, windowsills or potagers, but just as good as plain green basils for culinary use. There's also a 'Green Ruffles'.

5 **Spice basil** – an unusual variety with slightly hairy leaves, long spikes of pink flowers and a pronounced exotic, spicy scent. Can be used in Thai cookery, and it makes a good aromatic windowsill plant.

6 **'Sweet Genovese'** – a particularly fragrant variety that's traditionally used for making pesto sauce, but which is very good for general purpose use, too. Add a few leaves to a fresh fruit salad.

HARVEST

Start picking a few leaves as soon as they are big enough to use. Don't pick off whole shoots as it takes a long time for the plants to branch out and start growing again. Pick little and often and plants will be able to keep pace, but this is when it's worth having several plants or a trough of basil, so you're not over-picking the same few all the time.

Basil can be dried, but storing it this way will mean it loses much of its flavour.

PROBLEMS

Slow growth, seedlings damping off and young plants keeling over with black bases to the stems are usually symptoms of a root problem; which often originates from cold, dull or too-damp conditions, or a combination of all three. Growing plants on windowsills in pots of compost should prevent this, though even there it may occur with seedlings sown in midwinter. Try to raise plants in autumn to keep you supplied through the winter months.

Greenfly is a regular problem amongst plants grown on windowsills. Wipe them off by hand or rinse the leaves under the tap before there are too many, and repeat regularly. Moving plants outside in summer will help, because the pests are less active out there and natural predators will do their bit. Avoid spraying basil with pesticide.

Bay

Laurus nobilis

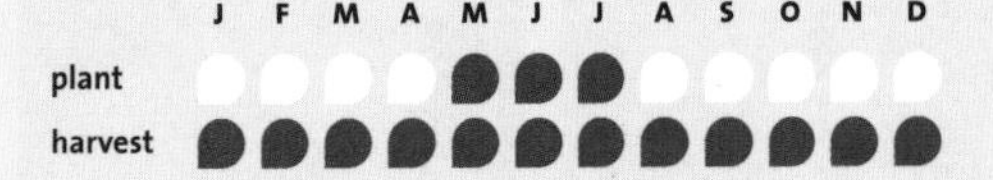

A popular evergreen for growing in tubs on smart doorsteps when it's trimmed into neat 'lollipop' or pyramid shapes, bay is also capable of growing into a large, bushy shrub when it's allowed to grow unrestricted. But if you're looking for a dual-purpose plant to form the centrepiece of a herb garden or potager and to pick to use in the kitchen, it's a winner. Being a hardy perennial, the same plant lasts for very many years, but its Mediterranean background means that it needs a mild, sheltered spot in order to thrive in Britain.

HOW TO GROW

Degree of difficulty: Not difficult in the right spot.
Plant: Plant outside after the last frosts are past, in late May or during the summer. It needs a few months of good growing weather to establish itself before winter.
Spacing: One is enough!
Routine care: In summer, pot-grown plants need regular watering and liquid feeding. In winter, stop feeding and water sparingly if the compost is dry. Trained plants in pots are best moved under cover in cold or windy spells, or if they must remain outside, move them close to a wall where they are sheltered. Plants growing in open ground are self-sufficient. In late April, prune trained specimens to retain shape when new growth starts. Don't clip with shears, as you will cut some leaves in half which will then turn brown at the edges, spoiling the look of the plant.

TRAIN A STANDARD BAY FROM SCRATCH

In a warm, sheltered spot, especially in southern gardens, you may find a few self-sown bay seedlings; the leaves are easily recognised and they are strong, upright growers. Dig one up and pot it – the strong, main stem is easy to train into a standard 'lollipop' shaped plant – or buy a small, single-stemmed plant. Push a cane in and train the main stem up it, tying in the new growth at regular intervals to keep it straight. Remove any sideshoots. When the plant reaches the top of the cane, nip out the very tip of the plant to encourage sideshoots to develop all round the top. Let them reach 7.5–10cm (3–4in) long, then nip out the tips of these to encourage them to branch out in turn. Continue doing this until the plant forms a good dense 'head' of foliage. From then on, prune to maintain shape each spring and again in late summer if it needs a slight haircut. In a year or two you'll have a plant that would have cost you pounds in the shops.

VARIETIES

There is only one species, *Laurus nobilis*, with no culinary named varieties. It makes a large, evergreen shrub with large oval leaves. My plant is dome-shaped, 3.6m (12ft) tall and 3m (10ft) wide, but specimens can be kept as small as you want by pruning.

HARVEST

Pick a few leaves when you need them for the kitchen at any time of year; choose only perfect, unblemished leaves that are full-sized but not old and coarse. Fresh leaves have far more flavour than dried ones; when you grow your own bay, fresh leaves are available all year round, so you don't need to bother drying any for storing.

PROBLEMS

Scale insects are tiny, hard-to-see, buff-coloured, limpet-like creatures which attach themselves to the stems and undersides of the leaves, and though they don't do much damage themselves, their sticky secretions allow spots of black sooty mould to develop, which can soon cover large patches. Affected leaves don't look very attractive and you wouldn't want to use them for cooking. Use an organic remedy such as soft soap to remove the pests, and when hot, dry weather causes the sooty mould to dry out, wash it off with a hose or sponge it off by hand.

Borage

Borago officinalis

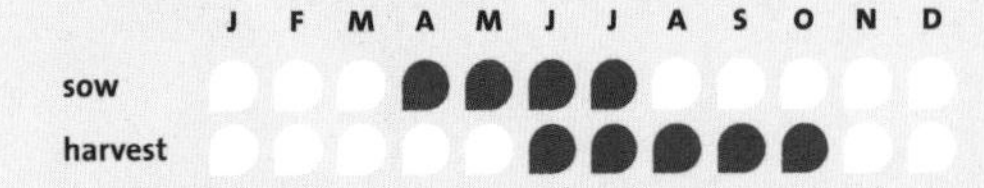

Borage is the herb that's commonly associated with Pimm's, but it also has other strings to its bow: the blue flowers are edible, good for decorating 'designer' food of all sorts – from salads and cakes to soups – and the faintly cucumber-flavoured leaves are brilliant in sandwiches or salads. (Although they are rather hairy, the leaves lose their prickliness when they're rinsed in water.)

Borage also makes a handsome plant in the garden, and it's worth growing in a potager or herb garden for its colour alone, even if you only use it occasionally in the kitchen. The unusual white-flowered form is especially stunning. Both forms are outstanding 'bee plants'.

HOW TO GROW

Degree of difficulty: Easy and little work.
Sow: In spring or summer; plants are hardy annuals and self-seed, so once you have some plants established just transplant the seedlings to wherever you want them to grow.
Plant: If you prefer to start with young, pot-grown plants from a nursery, plant them in late spring or early summer.
Spacing: Leave about 15cm (6in) between seedlings, as they won't all survive. By the time plants are in flower they'll be overcrowded as they grow tall, branching stems up to 60cm (2ft) high, but you can always thin them out a bit more later.
Routine care: Self-sown seedlings take care of themselves, but those you transplant and any plants bought in pots will need watering until they are safely established.

VARIETIES

1 ***Borago officinalis*** – the familiar form with slightly hairy, oval leaves and blue 'beaky' flowers.
2 ***Borago officinalis* 'Alba'** – the rare, white-flowered form is available as seed from specialist seed firms and sometimes as plants at herb farms.

HARVEST

Pick a few perfect leaves as you need them from seedlings or young plants before they have started to flower – after that they become a bit coarse and tough. Pick individual flowers as needed, again choosing perfect specimens.

PROBLEMS

Self-sown seed comes up best, since it's absolutely fresh and the plants 'know' precisely the right stage to shed it. If your bought seed is slow to germinate or fails to come up, be patient; and even if only a few come up, once you have a few plants established you'll be able to find your own seedlings or save fresh seed from them.

Recipe – Pimm's

A refreshing and not-very-alcoholic drink (ahem!) which is delicious for garden parties on hot summer days. It looks as good as it tastes served overflowing with ice, flowers and greenery. Pour a quarter of a bottle of Pimm's into a large jug, tip in two handfuls of ice and top with chilled fizzy lemonade or tonic water. Add in slices of oranges, cucumber, a few whole strawberries and some borage flowers. Garnish with a few borage leaves and some stems of edible flowers to complete the look.

Caraway

Carum carvi

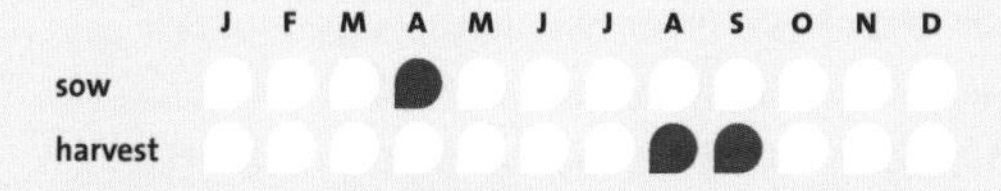

It's not often that you meet anyone who grows caraway, especially since you can easily buy the liquorice-flavoured seeds at the supermarket, and a little goes a long way in recipes. But grow it once, even out of curiosity, and you'll find that freshly grown caraway seed outshines anything out of a jar – full of vibrant aroma that sets taste-buds popping.

HOW TO GROW

Degree of difficulty: Not difficult or time-consuming, but needs a little attention to detail.
Sow: In April and caraway will usually behave like a hardy annual and flower the same summer; sow it any later and it becomes biennial – flowering in its second year. Sow in rows or where you want it to crop, since seedlings don't transplant well, or sow directly into a patio tub; a container full makes a novel display.
Spacing: Thin out seedlings to 10–15cm (4–6in) apart. Allow 20–30cm (8–12in) between rows.
Routine care: Water in dry spells and keep the plants well weeded, or you'll have flowers but no seeds.

VARIETIES

1 ***Carum carvi*** is a botanical species; there are no other varieties. These feathery annuals grow 60cm (2ft) tall; producing ripe seeds late in the summer.

HARVEST

Wait until the seeds have formed and started to dry out naturally in the seed heads, which will be towards the end of summer. Once they turn buff-coloured, and a few are turning a shade browner, cut the stems and hang them upside down in a warm, dry place out of direct sun. Tie a large, loose, paper bag over the heads to catch the seeds as they shed. After a few days, give the stems a shake to dislodge the rest. Spread the seeds out on a shallow tray and pick out any 'chaff', then leave them to finish drying off thoroughly.

STORAGE

Store the seeds in glass, screw-top jars kept in a cool, dark place. Replace your stocks every year. Fresh seeds have far more flavour.

PROBLEMS

If plants fail to produce seed it usually indicates poor growing conditions – over-wet soil, too cold or not enough sun – or that weeds were allowed to compete. It's also possible that plants may have set seed then shed them before you noticed.

If the seeds are sown late in the season the plants will behave as biennials and won't flower until next year. When this happens you'll get a stronger plant that produces more flowers and yields a bigger crop, assuming the plants don't die off over winter – they'll do best in a sheltered spot with well-drained soil.

Using caraway seed

Most famous as the 'seed' in seed cake, caraway is far better used these days – it makes a great partner for cabbage, cheese and apples. Scatter whole seeds on coleslaw, cooked cabbage and almost any vegetable in a cheese sauce, or add them to apple in pies, tarts and crumbles, as well as fancier apple desserts.

When making bread, use them in or on a loaf instead of poppy seeds to infuse it with the characteristic caraway flavour and aroma, or add them to sunflower seeds, pine nuts, bits of broken walnut or hulled sesame seeds to make a crunchy and healthy 'sprinkle' for salads.

Chamomile

Chamaemelum nobile

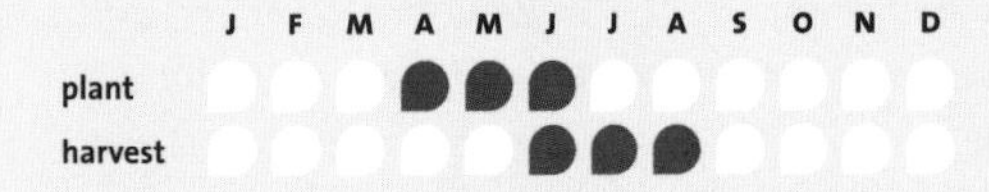

There are several types of chamomile, the best known of which is the non-flowering form, 'Treneague', which is sold for planting up as chamomile lawns, but for culinary purposes the one you want is its close cousin, Roman chamomile. It's the flowers that are used, which have cosmetic and semi-medicinal properties, as well as being delicious brewed as chamomile tea.

HOW TO GROW

Degree of difficulty: Easy and needs no time.
Plant: Roman chamomile can be grown from seed, but it's easiest to start with a pot-grown plant, as one is usually all you need. Plant it outside in a warm, sunny spot with well-drained soil, or in a pot.
Spacing: If growing several, plant 30cm (12in) apart in each direction.
Routine care: Water a new plant in initially, but once established chamomile is quite drought resistant.

How to use chamomile flowers

Make chamomile tea by pouring a cupful of boiling water over 1 dessertspoonful of dried chamomile flowers in a jug; let it brew for three to five minutes, then strain it off into a cup with a slice of lemon in the bottom. It's very pleasant on its own, but if you would prefer it sweeter, add a teaspoonful of honey or elderflower cordial (the two go very well together). It's a good last-thing-at-night drink to help you sleep and you can also use the dried flowers for making sleep-inducing herbal pillows.

Use fresh flowers on the table at a buffet or garden party; either put long-stemmed flowers in jam jars or sprinkle the flower heads – fresh or dried – over the table.

VARIETIES

1. ***Chamaemelum nobile* (Roman chamomile)** – looks very similar to the non-flowering lawn chamomile, with low-spreading, apple-scented, feathery foliage, except that unlike 'lawn' chamomile it produces pleasant white, daisy flowers. Each plant spreads to roughly 45cm (18in) wide and grows roughly 10cm (4in) high, with flowers on feathery 25–30cm (10–12in) stalks.
2. ***Chamaemelum nobile* 'Flore Pleno'** – (double Roman chamomile) is even prettier. It is a more compact plant, only 7.5cm (3in) high and 30cm (12in) across, with most attractive double white flowers on stalks about 15cm (6in) high.

HARVEST

Pick newly opened flower heads to dry, or cut flowers complete with stems to use decoratively.

STORAGE

To dry the flowers, lay them out on a clean, flat surface, ideally face down, in a cool, dry place out of direct sun. When thoroughly dry, store in thick brown paper bags in a dark, dry cupboard.

PROBLEMS

If plants fail to thrive or simply die off, it is usually because of poor growing conditions. Chamomile really must have exceptionally good drainage and full sun to do well. Grow plants in containers on the patio, in windowboxes or even hanging baskets if your garden conditions aren't suitable.

Chervil

Anthriscus cerefolium

	J	F	M	A	M	J	J	A	S	O	N	D
sow			●	●	●	●						
harvest				●	●	●	●	●	●			

This is another unsung herb that deserves to be better known, since it's easily grown and is really quite handy. You can use it as you would parsley, but it has a more delicate, semi-aniseedy flavour. Alternatively, mix chopped chervil with softened butter to make butter pats for melting over hot, cooked veg as a luxurious finishing touch. The ferny plants don't take up much space in a veg patch, or are perfect in a pot. Although they aren't the most productive of herbs when grown on the windowsill, it's still do-able if you want a fresh winter supply.

HOW TO GROW

Degree of difficulty: Easy and little trouble.
Sow: From March to late June in rows or informal patches where you want plants to crop. Chervil does best in light shade. Seedlings don't transplant well.
Plant: Pot-grown plants are sometimes available from nurseries during the summer; plant them straight away without breaking up the rootball.
Spacing: Thin seedlings or plant out plants to 15cm (6in) apart in both directions.
Routine care: Keep watered in dry spells and weed regularly. Plants will keep going all season and well into the winter if protected from frost with fleece. Expect a useful flush of new foliage early in spring before the plants run to seed. Let a few plants set seed, then self-sown seedlings will produce next year's crop.

VARIETIES

1 **Common chervil *(Anthriscus cerefolium)*** – is a biennial, cow-parsley-like plant that in its first year forms a loose rosette of slightly hairy, ferny foliage reaching no more than 20–30cm (8–12in) high and as much across. There's also a rare plain-leaved form available through specialist growers, which has foliage very much like flat-leaved parsley, but on a smaller plant.

HARVEST

Start picking entire 'fronds' as soon as they are big enough to use; if you have a row you can treat it as a cut-and-come-again crop, snipping a section at a time while the rest grows back up. This is the ideal way to treat it if you use a lot of chervil. Don't attempt to dry or freeze it, only fresh chervil keeps its unique aniseedy-parsley taste.

PROBLEMS

If a plant fails to grow well it is usually due to over-wet or over-dry soil, or because it has been grown in a hot place or in strong sun when it prefers cooler conditions in light shade.

Greenfly is a regular problem, but don't be tempted to use sprays – simply wash the pests off the plants with a jet of water when you're watering them.

Recipe – Chervil soup
Make proper chicken stock by boiling up the carcass after a roast chicken lunch with a carrot and an onion, then strain. Melt 25g (1oz) of butter in a large saucepan and gently 'sweat' a finely chopped shallot and a finely chopped stick of celery with an optional clove of garlic. When they're soft, add two handfuls of washed chervil, having removed the thicker stalks, stir briefly, then add a pint of the chicken stock. Cook for 20 minutes, then put in the blender. Serve with a swirl of double cream and a scattering of freshly chopped chervil leaves on top.

Chives

Allium schoenoprasum

	J	F	M	A	M	J	J	A	S	O	N	D
plant				●	●	●	●					
harvest					●	●	●	●	●			

Chives is one of our most useful and popular herbs; unlike many others that need to be sown afresh each spring, it is an easy-going perennial plant that comes up again each year with no fuss. It also looks good in the garden, so it is ideal for growing in a mixed herb planter on the patio or as edging around a potager. The slender stems can be chopped and used as a substitute for spring onions when you need a mild oniony flavour in salads and sauces, and the pretty, mauve-purple, pom-pom flowers can be used for decoration. In the garden they attract bees.

HOW TO GROW

Degree of difficulty: Easy and needs no time.
Plant: Plant pot-grown plants in late spring or summer.
Spacing: Space plants 15cm (6in) apart.
Routine care: Water in new plants, then keep them weeded and watered in dry spells. It's often recommended that you remove developing flower heads, but frankly I find it doesn't make much difference to the performance of the plant, so just leave them and enjoy them in the garden. Remove the faded flower heads complete with stalk to keep plants looking tidy – you'll need to do this little and often through the summer.

GROWING OUT-OF-SEASON CHIVES

Chives are so useful that it's handy to have a plant or two in pots to use in winter – at room temperature and given reasonable light the plants will continue growing on a kitchen windowsill. Buy a small potted plant from a nursery or garden centre in late summer and cut the top growth down to 2.5cm (1in) above the top of the pot, since the old foliage will be a bit 'tired' by then. Re-pot into a slightly larger container with new potting compost – it will soon produce clean, fresh, tender young growth. If you already have chives in the garden, dig up a clump in September, cut the foliage down to 2.5–5cm (1–2in) above the top of the roots, and divide it up. Pot up one or two small clumps into 10cm (4in) pots to bring indoors, and plant the rest back out in the garden. The indoor chives will have made new shoots ready to use by the time outdoor plants are dying down for the winter.

VARIETIES

1 ***Allium schoenoprasum*** – the only widely available kind, making clumps of slender, upright, spring-onion like stems with spherical, mauve-purple flowers 2.5cm (1in) across on 15cm (8in) stalks, held just above the tops of the plants.

2 **White chives (*Allium schoenoprasum* 'White Form')** – a rare white-flowered form sometimes available from specialist herb farms.

3 **Garlic chives or Chinese chives (*Allium tuberosum*)** – a perennial 30cm (12in) high and wide with clusters of white flowers at the top of the stems in summer. The leaves are flat, thin, long and narrow and have a mild garlic flavour.

HARVEST

Start snipping a few leaves as soon as they are big enough to use; there's no need to cut the whole plant at once, just select a small bunch of leaves from the clump and cut those down to 2.5cm (1in) or so above the ground. Where chives is grown for decorative purposes, cut from the back of the plant so you don't see the gap. Chives can be a little late to emerge in spring but you should be cutting from May until the end of September.

PROBLEMS

If your plants are long, lanky and tatty-looking with yellowing or broken and bent leaves, it's a sign that you don't use your chives very often. Cut the whole plant down to 2.5–5cm (1–2in) above the ground, clear away the old foliage, and they'll soon make fresh new growth. Spindly growth may also result when chives is grown in a spot that is too shady.

Using chives

It's very fashionable in restaurants right now to lay three or four whole fresh chive stems artistically at an angle over the top of a plate of 'designer' food, but traditionally the stems are chopped into tiny green 'rings' and sprinkled over creamy soups, sauces or cooked veg for both flavour and as decoration.

If you go in for altogether heartier food, try a generous spoonful of chopped chives mixed into sour cream, crème fraîche or melted butter and dolloped into the middle of an opened-up baked potato.

And for evenings in front of the television when you want something tasty that's easy to eat, mix soft cream cheese with a little crumbled Stilton and add enough double cream to make a soft dipping consistency. Flavour the mixture with a handful of chopped chives and maybe a little garlic, and sit and dip crisps or tortilla chips into it. This same recipe also makes a good dip to go with crudités at a party.

Coriander

Coriandrum sativum

Coriander, for leaves

	J	F	M	A	M	J	J	A	S	O	N	D
sow					●	●	●					
harvest						●	●	●	●			

Coriander, for seed

	J	F	M	A	M	J	J	A	S	O	N	D
sow					●							
harvest								●	●			

Coriander is two herbs in one. It's best known today for the coriander leaves that are used in 'foodie' households as a general-purpose green leaf for chopping and sprinkling, and for using in all sorts of ethnic-inspired dishes. But coriander is also a spice; the hard, round, buff, peppercorn-sized seeds are ground to make the coriander used in Indian cookery. It's a vital ingredient of garam masala – the blend of spices used for making delicately flavoured vegetable dishes as opposed to the usual blast-your-head-off hot curries. I love the whole seeds added to pasta mixed with broccoli to give mini-explosions of flavour.

Years ago you could only grow basic coriander, which you had to use for both jobs. (This is the worst of both worlds, since if you cut too many leaves the plant doesn't have the oomph to set seed, and if you want it to set seed you can't pick many leaves.) But since coriander became fashionable, we can now buy seeds of named varieties especially for leaf production, and these are the ones I strongly suggest you go for. Both this and the regular coriande – which is also fine for seed production – are frost-tender annuals, and quite easily grown. If you're growing coriander for leaves, the plants will start cropping far sooner and go on for much of the summer, while if you want it for seeds you have to look after the plants all season in order to collect a crop of dried seeds at the end of the summer.

HOW TO GROW

Degree of difficulty: Not difficult but needs some attention to detail; this is more so with seed varieties than leafy kinds.

Sow: In late May for seed coriander, and from then onwards at six-weekly intervals through the summer to have regular supplies of leaf coriander. Sow in rows where you want the plants to crop. You can also sow leaf coriander in tubs on the patio or in pots or troughs on the kitchen windowsill, and when grown indoors you can keep sowing all year round to have fresh leaves through the winter.

Spacing: Thin out the seedlings of seed-bearing varieties to 7.5–10cm (3–4in) apart, and allow 30cm (12in) between rows. Leaf coriander plants can be sown in rows or scattered over compost in patio tubs or windowsill troughs to get the maximum crop from the space – the seedlings only need thinning if they are very overcrowded, to around 25cm (1in) in each direction.

Routine care: Water sparingly as seedlings may damp off if allowed to get too wet. Keep coriander plants well weeded to avoid competition, particularly those that will be left to flower for seed production, as they are easily swamped.

VARIETIES

1 **For seed production:** Use the usual coriander species but don't cut the foliage as a herb, then it will concentrate on going to seed. Plants will grow flower stalks about 60cm (2ft) high followed by seed heads that you can grind to use as coriander spice. You'll occasionally find a named variety, 'Moroccan', in specialist seed catalogues that is especially for seed production; it's deliberately fast to flower and set seed and doesn't waste time producing unwanted leaves.

2, 3 **Leaf production: 'Cilantro'** and **'Leisure'** – both are good for later sowings. There's little to choose between them for flavour, and when grown as a cut-and-come-again crop they grow about 10cm (4in) tall, depending on how often you cut them.

HARVEST

Start snipping leaf coriander as soon as the leaves are big enough to use; they make a good cut-and-come-again crop, and the same plants should re-shoot several times if you don't cut them too closely. Cut them off 2.5–5cm (1–2in) above the base of the plant. Leave seed coriander to grow without cutting the leaves if possible. The plants will send up 60cm (2ft) flower stalks topped by light, airy, white, cow-parsley-like flowers in summer, then by the end of summer you'll see round, greenish seeds developing. Leave these until they turn buff, then cut the stems and hang them upside down to dry in a warm, airy place out of direct sun. Tie loose paper bags over the heads to catch any seeds that drop, then shake after a several days to encourage ripe seeds to shed.

STORAGE

When all the seeds have detached from the heads, spread them out to dry on a shallow tray then store in glass, screw-top jars in a cool, dark place. Grind in a spice mill or coffee grinder kept for the purpose, but do just this before use, for maximum flavour.

PROBLEMS

Plants usually fail to grow and thrive in poor growing conditions or a cold, wet summer – it's better to grow leaf coriander in pots on a windowsill indoors in a bad season. Seed coriander can be grown in the soil border of a greenhouse or polytunnel for protection as long as the air isn't too humid – otherwise the plants may be affected by fungal disease or the seed heads might go mouldy when they should be starting to dry out.

Recipe – Mild and spicy Thai-style vegetables

Thinly slice a mixture of onions, carrots, Florence fennel, kohl rabi, turnips, French beans, courgettes, and any similar veg. Take a heavy saucepan with a little oil in the bottom and cook two teaspoonfuls of ground coriander, a chopped clove of garlic and a handful of chopped fresh or dried chilli, according to your preference for taste. When soft, add the other vegetables, a teaspoonful or two or tamarind paste, and a half or whole tin of coconut milk, depending on the quantity of veg (it should just cover them) and simmer gently for 20 minutes. Season with salt and black pepper, serve with rice or crusty bread.

This is an easy-going dish, so it's fun to vary the ingredients depending on what veg you have in quantity down the garden. You can add other spices as well – cardamon and /or cumin – or omit the ground coriander and sprinkle chopped leaf coriander over before serving.

Dill

Anethum graveolens

Dill, for leaves

	J	F	M	A	M	J	J	A	S	O	N	D
sow				●	●	●						
harvest					●	●	●	●				

Dill, for seeds

	J	F	M	A	M	J	J	A	S	O	N	D
sow				●	●							
harvest							●	●	●			

Dill was once famous as the magic ingredient in gripe water, which is given to babies and small children to cure colic and hiccups, though nowadays you are more likely to encounter it in jars of pickled dill cucumbers or as the key flavouring in gravadlax – the Scandinavian answer to smoked salmon. It is used as a fresh green herb in much the same way as fennel when poaching or baking fish, especially salmon. It's a beautiful plant in the garden; tall and stately with feathery foliage topped off with airy yellow 'cartwheel' flowers that put you in mind of a very elegant and unusual-coloured cow parsley. If you have an informal rambling potager or herb garden in a warm, sunny spot, it's the perfect hardy annual herb to leave to self-seed so that the next generation can roam around at will, giving edible beds and borders an ethereal quality.

Recipe – Easy dill cucumbers
Take a deep, heatproof jar and place a few fresh-picked dill tips (including immature flowers and some leaves) in the bottom. Slice young cucumbers lengthwise and lay them on top of the dill. Add more dill, more cucumber and continue layering up to the top. When full, heat equal parts of white wine vinegar or cider vinegar with water, add a little salt and when scalding hot pour over the cucumbers to fill the jar. Leave to cool, seal and store – still in the jar of liquid – in the fridge. The pickles are ready to use after a few days, but use within a week or so, since they don't keep for long. It's a tasty way to use a glut of cucumbers.

HOW TO GROW

Degree of difficulty: Easy and needs no work.
Sow: When growing for seed production, sow in April or May outdoors where you want plants to crop. If you want dill leaves to use as a leafy green herb, it's best to sow in tubs or windowboxes from April to late June, or pots or punnets indoors. You can keep also keep sowing through the winter indoors to use as a cut-and-come-again crop.
Spacing: Thin out overcrowded seedlings to 10–15cm (4–6in) apart.
Routine care: Water sparingly in dry spells since plants dislike wet conditions; weed regularly. Keep separate batches of plants for leaf production; these are usually best grown in tubs or pots rather than out in the open garden.

VARIETIES

1 ***Anethum graveolens*** – the species looks something like fennel, with spreading, feathery, fennel-like foliage and yellow flowers, but dill is shorter at 90cm (3ft), finer and more delicate. Use for seed or leaf production, but keep separate plants for each job. Specialist seed firms sometimes supply a variety
2 such as 'Mammoth', which is especially grown for seed production. It runs to seed early and has less leaf than the usual dill. Alternatively, if you can't find this variety you can simply let your ordinary dill run to seed (by not cutting back the foliage) and harvest the seed it produces.

1

2

HARVEST

Harvest leaf dill as a cut-and-come-again crop as soon as the leaves are large enough to make it worthwhile, and though the same plants may re-grow, it's best to sow a new batch once you are about halfway through the first lot, if you want continuous supplies.

Seed crops should be sown in April and left to grow without taking more than a very few leaves for cooking; the seeds form towards the end of the summer and can be collected straight off the plants when they start to shed naturally, to avoid spoiling your garden display. Use dill seeds whole in dill vinegar, but try grinding them to use as a spice in a ground seed mixture – you can use this as seasoning when you want to cut down on salt.

STORAGE

Keep whole dry dill seeds in screw-top jars and grind freshly as needed.

DILL AS A CUT FLOWER

Strange though it sounds, dill makes a very good cut flower. You'll see it used quite a lot on the continent, where tall stems are dotted though an informal vase of mixed country-garden flowers such as delphiniums, stocks and roses to add a 'top storey'. Its angular architectural look lends itself to adventurous abstract floral art and competition work.

You might come across a variety called 'Vierling' in specialist catalogues, which has stronger stems and is sometimes grown especially for cut flowers.

PROBLEMS

Failure to grow well is usually down to poor growing conditions – probably soil that's too wet and heavy, or a semi-shady situation. To do well dill needs full sun and well-drained soil.

Fennel

Foeniculum vulgare

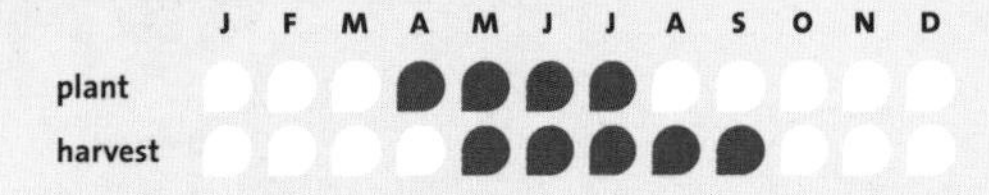

The herb fennel (not to be confused with the vegetable, Florence fennel) is a tall perennial that is decorative enough to be grown in a flower garden – especially if you grow the attractive, purplish-brown form known as bronze fennel, which is every bit as good for cooking as the plain green form. Both self-seed fairly freely, so once you've grown either of them it's just a case of pulling out seedlings that come up where you don't want them. Fennel is best known for being cooked with fish, but if you grow your own it's also well worth collecting the heavy crops of seeds that are generated in late summer to grind and use as a spice – you'll get lots.

HOW TO GROW

Degree of difficulty: Easy and no effort needed.
Sow: A few seeds in a pot in spring, thin them out and plant out the whole clump without breaking up the rootball; this produces a strong plant in the first season. 'Bought' seeds scattered around or sown in the open ground outside don't seem to do so well.
Plant: Plant out a whole clump as soon as it fills the pot, without breaking up the rootball, from late April onwards.
Spacing: Allow 60cm (2ft) between fennel and neighbouring plants in a border.
Routine care: Water in, and if need be water again in dry spells until the plant is established, by which time it will be fairly self-sufficient. At the end of the autumn, cut down dead fennel stems to 2.5cm (1in) above the ground and the plants will re-grow next spring. Self-sown seedlings left to grow where they appear naturally need no attention at all. Pull out unwanted ones while they are still a few centimetres tall; if you don't they'll develop powerful tap roots that are extremely difficult to extract and the plants will re-grow strongly if only the tops are cut off.

VARIETIES

1 ***Foeniculum vulgare*** – the green-leaved form which is usually preferred for use as a herb by serious herb fanciers. Plants have strong, feathery leaves and throughout summer and early autumn the stems are topped by yellow, cow-parsley-like flowers. Height 1.5–1.8m (5–6ft).
2 ***Foeniculum vulgare purpureum* (bronze fennel)** – the decorative form with bronze-purple foliage, good for a country flower border and very attractive growing with roses and through tall perennials. This also self-seeds, but it produces a mixture of purple-leaved and green-leaved seedlings.

1

2

HARVEST

Start snipping whole leaves as soon as they are big enough to use; take the younger ones that have only just opened out fairly close to the top of the plants, as fading, large, lower leaves won't be in the first flush of youth and will have lost much of their flavour, besides being tougher.

PROBLEMS

In many gardens excessive self-seeding can be a nuisance, so deadhead plants if you don't want to save seeds to make your own spice.

Horseradish

Armoracia rusticana

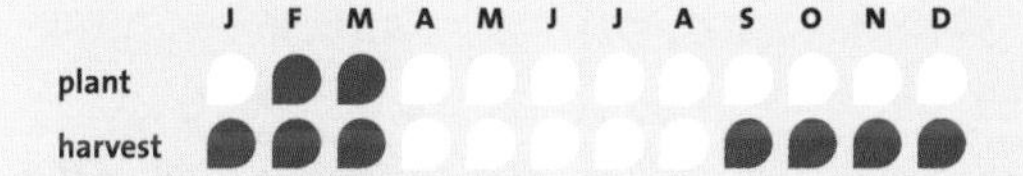

This is the plant that puts the h.h.h.h...hot into horseradish sauce, and for anyone who's ever wondered, it's the root that's used. The plant is a bit of a mixed blessing: it's great if you are a real horseradish fancier who likes to do-it-yourself, but gardeners tend to view it with a certain amount of caution as it can be an invasive perennial with deep roots that re-grow when they're broken off. This makes it notoriously difficult to shift. It's a native of Europe that's been naturalised here since before the sixteenth century, and you'll often find hefty clumps growing out in the countryside. If you want some, don't pinch a bit – grow your own.

HOW TO GROW

Degree of difficulty: All too easy.

Plant: Specialist herb nurseries stock plants which can be planted any time of year, but several seed firms also offer roots in their autumn catalogues for delivery and planting in late winter/early spring. To plant, push a cane or dibber into the ground and make a hole, then feed a single, long, thin, thong-like root into it – fattest end uppermost – leaving the tip at ground level. To prevent clumps spreading, plant them inside bottomless buckets, and sink each one to its rim in the ground leaving 2.5cm (1in) above the surface so roots can't 'climb out'. Although wild horseradish can colonise dreadful, infertile ground, it's worth choosing a deep, rich, fertile bit of ground that's been beefed up with plenty of organic matter if you want to grow good-quality roots for the kitchen.

Spacing: Plant three 'thongs' 60cm (2ft) apart in a triangular shape to make a large clump. Old roofing slates sunk vertically into the ground around the clump will contain the roots.

Routine care: Mulch well. Tidy up dead leaves when the plants die back naturally each autumn.

VARIETIES

Armoracia rusticana – a botanical species. Plants have large, dock-like leaves with slightly serrated edges and in time they form clumps 90cm (3ft) tall and the same across. It's very rare to see flowers, though they do sometimes send up stems with white flowers in spring.

HARVEST

Wait two years before collecting your first crop so that the plants have time to grow strong, thick roots. Dig up one plant in September or October, as this is when the flavour is strongest; you can dig roots at any other time in winter or early spring but they'll have less 'strength'. Use the biggest, thickest roots for the kitchen – keep back one long thin 'thong' to replant and to replace its parent. From then on, dig up one plant every year, repeating the process each time. This keeps clumps under control and means you are using tender young horseradish and not tough ancient stuff.

PROBLEMS

Overgrown horseradish can become a weed that's as difficult to shift as ground elder. To remove unwanted clumps, dig out as much as you can, and keep chopping out any re-growth, or cover with black plastic for two years or more to smother it. If you don't mind using weedkiller, spray glyphosate onto the foliage when the plant is in full growth. You may need to treat re-growth several times before it gives up.

Lavender

Lavandula species

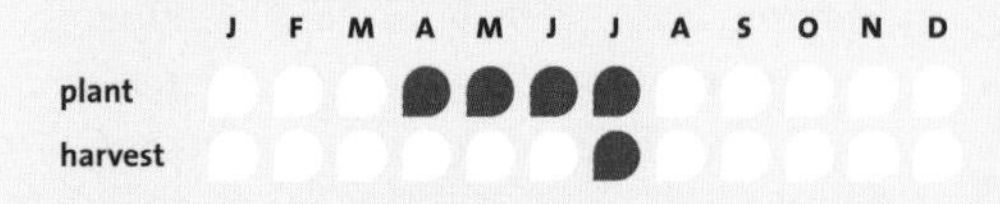

It's only since lavenders became so 'cool' as garden plants that anyone remembered to use them as herbs, but they add much-needed colour and fragrance to herb gardens, and the more compact varieties make first-class dwarf hedges for potagers. But even if you just grow them in beds and borders or along a path (and bees love them), you can still pick some flowers and leaves to use in cookery.

HOW TO GROW

Degree of difficulty: Easy, but needs some care.
Plant: Plant young, pot-grown lavenders in a sunny, sheltered spot with well-drained soil and not too much organic matter from late April to the end of July; it's advisable to get them established well before the winter.
Spacing: Plant compact varieties 20cm (12in) apart, and larger, more spreading kinds 45cm (18in) apart.
Routine care: After planting, water lavenders in and keep young plants watered in dry spells until they are well established; after that they are fairly drought tolerant. The flowers are at their best in June/July, then they dry out on the plants and continue to look good for some time afterwards, but once they start looking a bit past their best in September/October, clip the plants over to remove the dead heads and shorten the year's growth back to within a few centimetres of the thicker, older stems. Don't cut back into old brown wood, though, as this can kill the plants. Try to avoid doing any cutting or tidying up in spring, as plants should be making new growth then in order to flower.

VARIETIES

1. **French lavender (*Lavandula stoechas* subsp. *stoechas*)** – not quite so hardy as the English lavenders, but fairly upright plants to 45cm (18in) with small tufts of leaves along the stems and fat chunky flower heads, from the top of which sprout colourful 'ears'.
2. **'Hidcote'** – the classic, compact, deep-purple lavender frequently used for dwarf hedging; 45cm (18in) high and a neat, upright grower. Easily available and often sold in herb areas of garden centres.
3. **'Old English'** – it's debatable whether this is one variety or several going under the same name, but they are large, wide-spreading plants 90cm (3ft) in each direction, with light lavender-blue flowers and masses of scent. They are good for cottage gardens where you don't mind them leaning over paths and lawns.

1

2

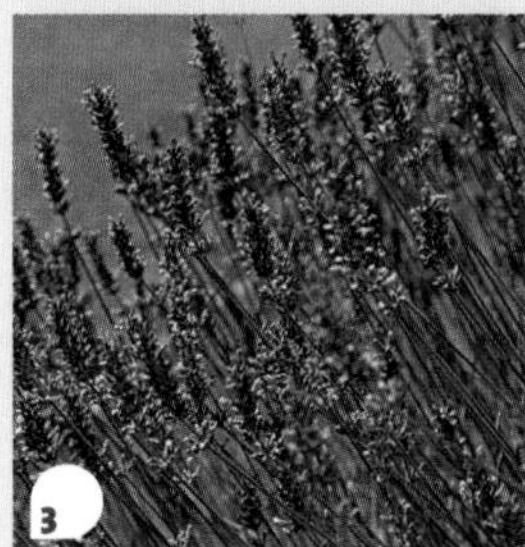
3

HARVEST

Pick some of the flowers when they are at their peak for drying; you can also use some of them fresh at this time. Lavender leaves can also be used, but pick only young, perfect leaves and do so reasonably sparingly – they aren't something you'll need a lot of.

STORAGE

Hang bunches of lavender flowers upside down to dry in a cool, airy place out of direct sunlight; use these as dried flowers for winter arrangements, or cut off the whole heads to use as potpourri. Alternatively, when the tiny individual flowers come away from the stalks easily, rub them off, lay them out thinly to complete drying, then use them as an ingredient in a mixture of dried leafy herbs. They team well with sunny southern European herbs such as thyme and oregano.

PROBLEMS

Rosemary beetle is one of the latest invasion of new bugs from the continent; it is mainly concentrated round the London area as yet, but it is expected to spread in the south and possibly elsewhere shortly. Look for small, oval, metallic green beetles with purple stripes eating rosemary, but also lavender, thyme and sage. Best picked off by hand.

Old plants often become craggy and woody and dry out and die off when they have been pruned back hard in an effort to tidy them up and encourage compact new growth. If you have old lavender plants that have not been properly trimmed after flowering for several years, and which have splayed out and look tatty, root some cuttings in June/July as insurance, then cut back carefully to just above the woodiest parts in September. New growth will sometimes grow out from there the following spring. Shorten this new growth back to produce bushier growth, which soon covers the bald, woody bases. If it doesn't work, you still have some rooted cuttings to replace the old plants. As a rule, lavender is best replanted every four or five years.

Sometimes plants become weak and floppy and splay open in the centre, and this often happens when they have been given over-lush growing conditions. Don't use high-nitrogen fertilisers for lavender; if you do want to feed them, use a light sprinkling of blood, fish and bonemeal when you plant them, and dust a half handful of sulphate of potash round the foot of each plant in late spring. Do this in rainy weather so it gets washed in. Splaying plants can also be a sign that they haven't been trimmed back in the autumn: trim correctly the following autumn, and repeat every year to keep in shape.

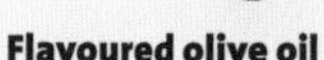

Flavoured olive oil

Take an attractive, clear glass bottle with a fairly wide neck and a screw top. Wash and dry some suitable herbs – such as a fresh sprig of rosemary, a whole bay leaf, some heads of lavender flowers and some stems of thyme and Greek oregano or marjoram. Push them into the bottle and fill it up with good-quality olive oil and after a few months the oil becomes infused with the volatile oils. (This particular mixture will give you a flavour of southern France.) The herbs can be left in place for as long as it takes you to use the oil.

If the ingredients are attractively arranged this makes a handy gift; just tie a ribbon round the neck and add a decorative hand-written label.

How to use lavender

Use dry flower heads in potpourri, or separate the individual florets from dried heads to fill lavender bags. Lavender is a good natural moth repellent that's making a comeback now, because so many natural fibres are being used in clothing and this has produced a population explosion among clothes moths. It's their caterpillars that do the damage, eating holes in clothes – they especially love cashmere. Use lavender bags to deter the adults daintily, instead of filling your bedroom cupboards with smelly, pesticide-based moth remedies.

Lavender pillows are said to help you sleep, so fill a large cotton bag with dried lavender flowers that have been detached from their stems, sew up the top and slide it inside a lacy pillowcase. Small ones also make relaxing scatter-cushions for bedrooms.

Now that lavender is fashionable amongst foodies, there are all sorts of recipes around for lavender ice cream, lavender jelly for eating with lamb (you can also add lavender flowers to a mixture of herbs to make a herb and apple jelly), or you can sprinkle a few fresh lavender flowers, removed from the stalk, in salads along with other edible flowers. Search the Internet and you'll find yet more novel ideas.

Lemon grass

Cymbopogon citratus

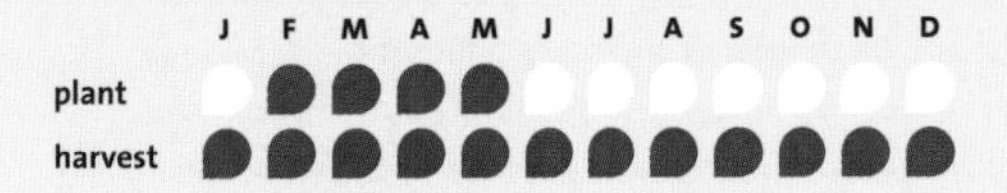

Lemon grass is one of several tropical plants that have joined the nation's larders since Thai and other exotic cuisines became popular among home cooks. When you buy it in shops what you get is a small bundle of neatly trimmed pieces that are the bases of grassy stems – these are the only bits that are any good for cooking. When it's grown at home, lemon grass makes quite a large, coarse, grassy (and it has to be said, not terribly attractive) plant with rather sharp-edged leaves. It is best grown indoors at room temperature in the winter, but it can be put in the conservatory or out on the patio in summer.

HOW TO GROW

Degree of difficulty: Easy and needs little attention.
Sow: Indoors at room temperature in spring or early summer (February to May). Pot the seedlings when they are big enough to handle; two or three will be plenty, even for serious users.
Plant: If you don't sow seeds, root 'bought' pieces of lemon grass stem in a glass of water (they send out roots quickly at any time of year except midwinter) and plant them in pots when the roots are 1cm (½in) long; they transplant best at this stage. If you leave them in water too long they don't settle well in soil.
Routine care: Keep the plants in a sunny situation; in winter they need a temperature of at least 13°C (55°F) or, more ideally, room temperature. In summer, move them to a conservatory, where they are quite heat-tolerant if they are kept well watered, or put them out on a warm, sunny patio. Lemon grass grows quickly; re-pot into larger containers as they fill the old ones with root. Water generously, particularly in summer.

VARIETIES

The species ***Cymbopogon citratus*** is the only form available. A native of India, it makes a dense clump of grassy leaves that can grow 75cm (2ft 6in) high and 30cm (12in) across within a couple of years. It needs careful positioning when it's brought indoors for the winter, since the sharp leaves can be a bit hazardous, especially if you have small children. A small pot keeps it slightly 'corsetted', but if it's re-potted into a large container it's capable of growing to about 1.2m (4ft) tall and 60–90cm (2–3ft) wide.

HARVEST

You can start using lemon grass as soon as the plants are sufficiently well established to have thick, off-white, slightly swollen stem bases – similar to the pieces you buy in the shops. (In practice it usually takes about a year for a young plant to reach this stage.) It is quite difficult to cut one complete, good-sized stem from a well-established clump, the easiest way is to tip the whole plant out of its pot and divide it up, re-potting one piece to continue growing. Divide the rest into individual stems, selecting those of a suitable thickness, and trim off the tops to leave spring-onion-like shoots similar to the pieces you buy in the shops. These stay fresh for some time in the fridge, or stood in a jar with their very ends in water.

PROBLEMS

None. How refreshing.

What to do with lemon grass

Peel the outer layers from the base of the stems and slice the most tender part very thinly and add to Thai or Malaysian curries, especially the sort cooked in coconut milk. Alternatively, fry thin slices then boil them with rice to add flavour and fragrance.

Since all parts of the plant have the distinctive lemon aroma which is only released when the leaves are cut or crushed, cut up the long leaves that are trimmed off when preparing the stems for cooking and put them in a bowl to act as a 'live potpourri' room freshener.

Lemon verbena

Aloysia triphylla

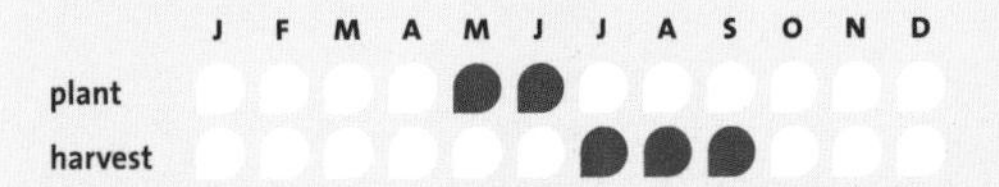

The most intensely lemon-scented of any herbs, lemon verbena leaves are mostly used as a fragrant 'decoration' in fruit salads and drinks such as punch or cordials, though if you are entertaining friends to something messy that they'll have to eat with their fingers – such as giant prawns – scatter a few leaves in warm water in fingerbowls. They are also good scattered on the table along with flower petals, or you can use sprigs in flower arrangements for indoor fragrance. Whole leaves can be dried for potpourri, as they keep their scent for a long time.

HOW TO GROW

Degree of difficulty: Easy and little effort needed, provided the site suits it.

Plant: Seeds are very rarely available, so start with a young plant from a herb farm, nursery or garden centre.

Routine care: Re-pot your plant into a larger pot or tub when it fills the original container with roots; use a free-draining compost made by mixing equal parts of potting grit, John Innes No.2 potting compost and multi-purpose compost.

Lemon verbena is a very tender perennial, so don't put it outside until some time after the last frost – from early June onwards. It can be planted in the garden for the summer, and in milder areas it can be planted there permanently. Cold weather may kill off the top growth, but it can usually be relied on to sprout from below ground if given a thick mulch of organic matter. In colder areas it does best in a pot. Stand it in a warm, sunny spot where there is fairly high humidity. Water sparingly at first and gradually increase the supply in hot weather, but be warned that it never wants to be waterlogged. Use well-diluted general-purpose liquid feed regularly from May until September.

If you cut shoots sporadically for the kitchen, nip out the tips of strong stems regularly to make the plant branch, otherwise it'll grow long and leggy. In cooler areas, come the autumn trim the plant back to reduce the size, cut down watering and move it back under cover. Alternatively, root some cuttings in late summer and keep a couple of young plants on a windowsill indoors through the winter, since they'll take up far less room. It's a good idea to take cuttings every year anyway, as old plants will sometimes die off suddenly for no good reason.

PROPAGATING LEMON VERBENA CUTTINGS

Treat lemon verbena cuttings in exactly the same way that you'd treat fuchsia cuttings, but make them a bit longer. Snip off the tips of shoots 7.5–10cm (3–4in) long, carefully strip off the lower leaves and make a clean cut with a sharp craft knife immediately under a leaf joint. Dip the cut end into hormone rooting powder as belt and braces (experienced propagators

will be able to get away without doing this). Push five cuttings in round the edge of a 10cm (4in) pot filled with multi-purpose compost and, after watering in, put a large, loose plastic bag over the top. Cuttings will take about eight weeks to root in summer; check them regularly during this time and remove any that turn black, dry out or are clearly dying off. At the end of eight weeks, make a few holes in the plastic bag to acclimatise them gently to fresh air, then after a few more weeks pot them up.

VARIETIES

Originally from Argentina and Chile, the botanical species, ***Aloysia triphylla* (syn. *Lippia citriodora*)**, is the only one grown: there are no cultivars. It makes a fairly attractive, bushy plant with long, narrow leaves topped with branching spikes of tiny, pale, lilac-white flowers throughout the summer. The plant usually reaches about 90cm (3ft) tall and 60cm (2ft) wide by the autumn.

HARVEST

Start picking a few leaves or the tips of shoots as soon as they are large enough; picking the tips regularly acts like 'finger-pruning' and keeps the plant in a neat, bushy shape.

PROBLEMS

Failure to grow well usually indicates that the growing conditions are not to the plant's liking. In a cold, wet summer it is better kept in a pot in the conservatory or a sunny porch. In very dry conditions it helps to spray the leaves or 'damp down' the ground round the plant; this creates humidity without falling into the trap of overwatering it.

Whitefly and/or red spider mite may be a problem with plants grown under glass. The best way to deter red spider mite is to maintain high humidity; the easiest way to get rid of both pests is to move the plants outside when the weather permits, if they are normally kept inside in summer.

Lovage

Levisticum officinale

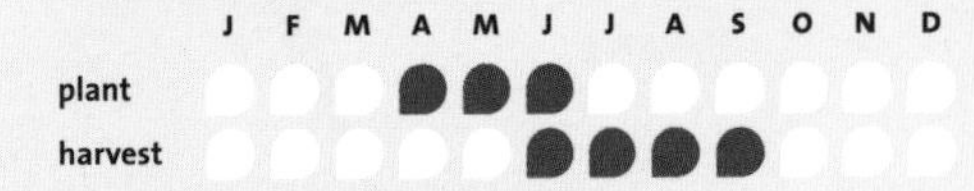

Lovage is a little-known herb with some peculiar uses and is worth growing if only for its novelty value. 'Love parsley', as it was once known, was revered as an aphrodisiac in medieval Europe (is it beginning to sound more interesting?). The seeds were chewed as a remedy for flatulence and the leaves had a great reputation for preventing smelly feet – since two or three placed into the bottoms of shoes acted as a herbal deodorizer. (It seems rather more prosaic now...) But more recently the thick stems were used rather like vegetable stock cubes – cut into pieces and boiled up in soups and stews – as both the leaf-stalks and leaves have a celery-like taste that enhances the flavour of meat. But lovage is a lot easier to grow than real celery and it's a hardy perennial, so you'll only need to plant it once.

HOW TO GROW

Degree of difficulty: Easy and no effort needed.
Plant: Buy a young plant from a herb farm or nursery in spring or summer and plant in rich, moisture-retentive soil containing plenty of organic matter. It's good for the back of a border or a herb garden. One plant is enough, as it grows quite big.
Routine care: Feed and mulch in spring to encourage new growth. In autumn, as old leaves turn brown and the stems die back, cut the plant close to ground level.

VARIETIES

Levisticum officinale is a botanical species and there are no named varieties. The plant looks rather like a tall, spreading celery plant, reaching up to 1.5m (5ft) high and topped by flat heads of yellow, cow-parsley-like flowers in mid- to late summer.

HARVEST

Cut individual leaves or a whole leaf complete with its stout leaf-stalk as soon as the plant is big enough. Cut little and often to avoid over-taxing the plant, at least while it is young. Split the leaf stalk to release more of the flavour when using it in soups and stews, and remove the herb before serving.

PROBLEMS

Like its relative, celery, lovage is rather prone to attacks of leaf miner, which tunnel into the leaves making wiggly transparent patterns. When this happens there's nothing you can do except pick off affected foliage, since the culprit is safe inside the leaf. Lovage is a big, tough plant and a few leaf miners won't do it any harm; just be sure to pick perfect leaves for cooking.

Lovage cordial

You will occasionally still see bottles of 'lovage' set alongside country wines in a rural pub, as there was once a very popular drink known as lovage cordial – although it was highly alcoholic. It's easily made if you grow your own. Harvest 15g (½oz) of fresh, ripe lovage seeds and crush them lightly with a pestle and mortar. Into a large screw-top bottle, put the crushed seeds, 100g (4oz) of sugar and 20 fl oz (1 pint) of brandy, shake well and store out of sunlight for two months, shaking occasionally. After this time, pour it through a coffee filter paper to remove the 'bits' and return it to a clean screw-top bottle. The resulting 'cordial' is said to have many of the same medicinal properties as the herb itself; a small glass-full taken as a nightcap to settle the stomach and sweeten the breath may prove surprisingly popular.

Marjoram and Oregano

Origanum species

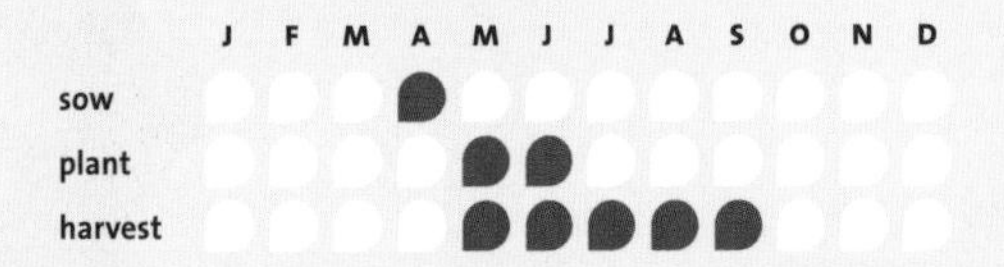

Some of our most popular traditional herbs, marjoram and its more pungent brother oregano, are available in several different types that vary in strength and flavour. It's worth growing a small collection if you use a lot of herbs in cooking – you'll certainly be able to fit them in. The plants are small, spreading, yet bushy and very aromatic, with tiny pink flowers clustered in dome-shaped heads. They are appealing to us and also very attractive to butterflies and bees.

HOW TO GROW

Degree of difficulty: Not difficult or time consuming.
Sow: Thinly in pots on a windowsill indoors in April. Rather than pricking out the seedlings individually, it's better to sow several potfuls and allow all the seedlings to grow as a clump, as this quickly gives you a much bushier plant. That said, most people buy young plants from a herb farm or nursery in early summer.
Plant: After the last frost is over, from mid-May onwards. Although most kinds (except sweet, or knotted, marjoram) are hardy perennials, they have usually been raised under cover and do best if not planted out until the weather is fairly kind. Marjoram and oregano need a warm, sunny, sheltered spot with well-drained soil; they also do well in raised beds and containers.
Spacing: Plant 23cm (9in) apart.
Routine care: Water new plants in and water sparingly in dry spells. After flowering, snip off the dead heads. Sweet marjoram does not survive the winter so pull out the old plants in late autumn, but other marjorams and oreganos are perennials that can stay put in winter. After the worst of the cold weather is over, around April, clip or prune plants to tidy them up and remove any dead stems.

PROPAGATING MARJORAMS

Stems often root where they touch the ground, so you can find young plants round the edge of older ones, which only need digging up and potting. Otherwise, take stem cuttings in July or August; snip off 5cm (2in) lengths, strip off the lower leaves and insert six cuttings round the edge of a clay flowerpot filled with gritty compost (made by mixing half-and-half multi-purpose compost and potting grit). Don't cover with a plastic bag, because marjorams dislike humidity, but stand the pot on a windowsill indoors out of direct sun. Young plants should be ready for potting on next spring.

VARIETIES

1 **Sweet or knotted marjoram (*Origanum majorana*)** – a small, rounded, strongly aromatic, slightly floppy, bushy plant usually reaching no more than 30cm (12in) high by the end of summer, with clusters of tiny white flowers growing tightly together in knot-like clusters. This species is grown as an annual.

2 **Pot marjoram (*Origanum onites*)** – a low-spreading, evergreen perennial, roughly 30cm (12in) wide and as much high, with deep mauve flowers in midsummer, usually grown from cuttings. It is probably the best for general use.

3 **Wild marjoram or oregano (*Origanum vulgare*)** – a small, evergreen, spreading plant which grows to 45cm (18in) in each direction and has slightly hairy leaves and very small, rather tubular, mauve flowers. It grows wild in this country but does not develop a particularly strong flavour. Needs a hot dry situation.

4 **Greek oregano** – thought to be a subspecies of ***Origanum vulgare,*** this variety has slightly larger, broader and bristlier leaves and a strong pungent aroma and flavour. It is good for Mediterranean cookery, barbecues and robust lamb dishes. Needs a hot, dry situation and is not reliably winter hardy, so bring pots under cover for the winter.

HARVEST

Start snipping short lengths from the tips of the young stems to use for cooking as soon as the plants are large enough to stand it.

PROBLEMS

A lack of aroma or flavour is usually caused by 'soft' growing conditions. Marjorams and oreganos are plants that originate from sunny, stony hillsides, so for best results you need to recreate those conditions by growing them in pots or tubs with plenty of drainage material in the bottom and, ideally, by siting them on a hot patio.

Recipe – Marinated roast or barbecued lamb

Pick half a cup of marjoram and/or oregano leaves, bruise them gently with a pestle and mortar and crush two cloves of garlic. Mix the herbs with 10 fl oz (½ pint) of olive oil and leave to infuse for a few hours. Add salt and pepper to taste. Take a shoulder or leg of lamb and remove the bone (or ask the butcher to bone it for you), then leave the meat to marinate in the mixture overnight. Shortly before cooking, lay the meat skin-side down and spoon the marinade mixture over the upperside before rolling up the meat with the herbs sandwiched in the centre. Secure with skewers and cook over a barbecue or in a hot oven. As it cooks the flavour really permeates the meat.

Mint

Mentha species

	J	F	M	A	M	J	J	A	S	O	N	D
plant				●	●	●						
harvest					●	●	●	●	●			

Mint is one of the classic English herbs that few gardeners are ever without, but when it's not well grown it spreads everywhere and is covered with unsightly mildew in summer. However, that's easily remedied by giving it a little TLC. While you're at it, why not try some of the more unusual mint varieties that can be found at herb farms? That way you can have at your disposal a selection of mints with slightly different flavours to suit different purposes. Bunches of mint were traditionally put in jam jars of water in butchers' shops to deter flies, and at home one of the more aromatic mints can be used in the kitchen in summer as a living fly-spray-cum-air-freshener. Alternatively, use flowering or foliage sprigs of mint in jugs of mixed country flowers, especially hardy annuals.

HOW TO GROW

Degree of difficulty: Easy, but benefits from some attention to detail.

Plant: Buy young plants of named varieties from herb farms, nurseries or garden centres in late spring or early summer and plant them out in fertile, moisture-retentive ground that's been enriched with plenty of well-rotted organic matter. One plant of each of your favourite varieties is enough to get a good crop of leaves. If you want to avoid mint spreading, sink a bottomless bucket or large plastic flowerpot with the bottom cut out into the ground, leaving 5cm (2in) of rim standing proud. Fill this with rich compost and plant the mint within it to keep the roots and runners contained.

Spacing: Space 45cm (18in) apart if you're growing a collection of mint; allow 60–90cm (2–3ft) between a patch of mint and other herbs.

Routine care: Water new plants in well, and water regularly in dry spells. Mulch and feed with a general-purpose fertiliser each spring, around April, when new growth first appears. A lot of people cut mint back hard when it starts trying to flower in midsummer, to encourage leafy shoots, but flowering doesn't harm the plant and mint flowers are attractive to bees and some butterflies, so personally I'm all in favour of leaving them be. When the flowers have faded and the insects have lost interest, cut the flowered stems down to 5–7.5cm (2–3in) above ground and the plants will soon throw up strong new shoots.

Mint quickly exhausts the soil, so for best results dig up the old plant in spring every second year, just as it starts growing again, then divide it up and replant one strong young section in a new site with freshly prepared ground. Throw the rest of the old plant away, and don't put it on the compost heap – unless you want wall-to-wall mint growing everywhere.

HOW TO HAVE FRESH MINT IN WINTER

For fresh mint sauce right through the winter, long after mint has died down outside in the garden, grow a plant in a pot indoors.

The traditional way of doing this was to dig up some mint roots in late autumn, lay them out flat in a pot or seed tray half-filled with compost, then *just* cover them with a little more compost before keeping them on a windowsill indoors while they started back into growth. You can still do that if you like, but if you think of it in time, root a few cuttings instead.

Sometime during late summer, pick three strong, leafy, non-flowered 10cm (4in) long sprigs of mint and sit in them in a glass of water on the windowsill until they take root. It won't take long. Then pot them all in the same pot as one clump (which gives you a decent-sized plant straight away) with some multi-purpose compost.

Kept at room temperature and in good light on a sunny windowsill, the plant will keep growing right through the winter and the more you cut it, the shorter and bushier it will be.

VARIETIES

1 Spearmint (*Mentha spicata*) – also known as garden mint, this is the commonest kind, with smooth, almost shiny, pointed, deep green, serrated-edged leaves and small, roundish heads of mauve flowers. It's the best mint for mint sauce or to cook with new potatoes or peas. The curly form, *M. spicata* 'Crispa', can be used similarly and is particularly attractive.

2 Apple mint (*Mentha suaveolens*) – another familiar species which bears large oval leaves with a slightly furry, crinkly texture and flowers that are branching lilac spikes at the tops of stems. The delicate flavour reminds you of a combination of apples and mint. It is good for making a mild, aromatic mint sauce or jelly, and can be used when cooking potatoes or peas.

3 Eau de cologne mint (*Mentha* x *piperita* f. *citrata*) – dark green, smooth, oval leaves and a very strong and distinctive eau de cologne scent. It's good cooked with potatoes and peas, but the very best for cutting to put in jars in the kitchen or in jugs with cut flowers.

4 Buddleja mint (*Mentha longifolia*) – an unusual species with long, narrow, grey-green leaves slightly felted in appearance. Its long, lilac flower spikes are beloved of butterflies, bees and beneficial insects, so it's worth growing for that reason, despite having little culinary value. The variegated form is particularly attractive, with pale yellow blotches over grey felty leaves.

5 Pineapple mint (*Mentha suaveolens* 'Variegata') – a rather short, compact mint which is mainly grown for its small, slightly hairy, green and white variegated foliage. It's an attractive species that's good in containers or herb gardens as a contrast with other herbs, despite having no great culinary use. Bruise the leaves *very* gently to detect the aroma of pineapple.

6 Ginger mint (*Mentha* x *gracilis* 'Variegata') – another short, slow-spreading, compact mint, but with smooth, oval, green leaves splashed with gold. It makes a good contrast in tubs and hanging baskets of mixed herbs. The 'ginger' refers to the variegation, not the flavour or scent – this is purely a decorative variety.

HARVEST

Start snipping 2.5cm (1in) long sprigs as soon as the plants are large enough; this can continue throughout the summer. Hard cutting prevents mint plants from flowering and keeps them leafy instead.

PROBLEMS

Powdery mildew is a common problem in mid- to late summer, just after mint has flowered. Cut affected stems or the whole plant down to 5–7.5cm (2–3in) above the ground, feed and water the plant well and apply a 2.5cm (1in) thick mulch of well-rotted garden compost and strong new shoots should soon appear. (You can also use 'second hand' compost for this, meaning the stuff that's left over when re-potting plants or after pricking out seedlings from a tray.)

Mint rust appears as small orange-red spots on the foliage. Leaves should be picked off straight away and destroyed or disposed of, well away from other mint. There's no cure for this disease, so if the same patch is regularly affected, remove and destroy the whole plant and start again with new stock some distance away, to be on the safe side.

Parsley

Petroselinum crispum

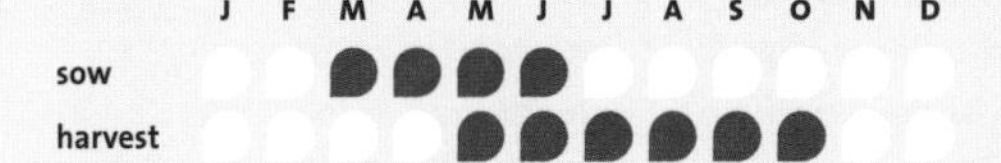

This is one of the classic, must-have herbs that somehow we never seem to use all that much these days, now that steamed fish with parsley sauce has fallen off most menus; but it's still worth using in stocks and sauces, or raw as a garnish. The non-frilly, flat-leaved type (also called French parsley) is reckoned to be the best flavoured by serious cooks, who use the frilly sort more for decoration. Usually grown as an annual, parsley plants do come up again in their second year but they run to seed early, so the crop is best sown afresh each spring. Parsley makes a good decorative edging round veg beds, or you can grow one or two plants in pots on the kitchen windowsill – which is a great way to keep them going all winter, since they die down outdoors.

HOW TO GROW

Degree of difficulty: Easy and virtually no work, provided the soil is reasonably rich and not too cold at sowing time.

Sow: March to June where you want it to crop, or else in small pots which can then be planted out with the entire 'plug' of roots. It doesn't transplant well.

Plant: Plant or thin out as soon as seedlings are large enough.

Spacing: Space 15cm (6in) apart with 15cm (6in) between rows.

Routine care: Water regularly, feed every few weeks with general-purpose liquid or soluble feed. Remove yellowing leaves. Replace plants when they seem worn out or if they start running to seed.

GROWING IN CONTAINERS

For indoor windowsill supplies, sow a pot or half-tray of seeds every three months all year round – plants keep growing indoors even through the winter. Sprinkle seeds thinly over the entire surface of a pot of multi-purpose compost, then lightly cover with more compost. Thin out the resulting seedlings, leaving them 2.5cm (1in) or so apart. For a fast potful that lets you cut frequently, allow all of them to grow. They are ideal for windowsill supplies. For a patio tub, replant several whole pots of parsley raised early on a windowsill indoors, without breaking them up or disturbing the roots, into a large tub – or simply sow outside in a tub from March onwards.

Parsley alternative

If you like the idea of parsley but don't quite like the taste, consider growing leaf celery instead. Plants look very much like parsley but have a distinctive, not-too-strong celery flavour instead. Grow in exactly the same way, in pots or in the ground, and use as for parsley.

Look for leaf celery or Par-cel in the herbs section of catalogues and seed racks.

VARIETIES

1 Plain-leaved – the non-frilly, flat-leaved type (sometimes called French parsley) believed to be the best tasting by serious cooks.

2 'Moss Curled'– an old-favourite, frilly type, used for parsley sauce and as decoration.

HARVEST

Individual leaves can be removed, complete with stalks, when large enough to use throughout the growing season, but avoid picking the very oldest or the very youngest, besides obviously damaged or discoloured leaves.

Recipe – Real parsley sauce

Instead of just opening a packet of white sauce mix and shoving in a handful of chopped parsley, try making the real thing using home-grown herbs. Parsley sauce is wonderful over freshly cooked garden vegetables, especially broad beans or calabrese, but it's also delicious with a mixture of baby veg such as leeks, carrots and Florence fennel.

Melt a generous knob of butter and 'sweat' a small chopped onion in it until soft. Stir in 2 tablespoons of cornflour (or use sauce flour, which is even less likely to go lumpy) and, as it thickens, slowly add a few tablespoons of milk every now and then, stirring constantly. Save the vegetable stock from the pan in which you cooked your vegetables, and add it gradually until you have a thick pouring sauce. Bring this up to the boil then simmer for a few minutes more so that the flour is cooked. Stir in a handful of freshly chopped parsley and finish with a dollop of cream and/or another knob of butter.

STORAGE

Parsley does not dry very well; it's far better frozen. Freeze a large bunch inside a big plastic bag – the leaves are easy to rub off the stalks when frozen – or make parsley 'ice-cubes' by chopping it up finely and freezing in ice-cube trays with two teaspoonfuls of water per 'cube'. When you need parsley for casseroles and sauces, you can just drop a cube straight into the cooking pot, with no need to defrost.

PROBLEMS

Germination can be very slow, and as a result people often abandon potfuls that don't seem to do anything. Low temperature is almost always the problem; sown in pots indoors above 16°C (60°F) parsley germinates quickly, but since seedlings can't be transplanted, thin them out slightly if they are overcrowded and pot-on the whole clump when it fills the original container with roots, without breaking up the rootball.

Old wives' tales

Parsley is said to grow well if the woman of the household wears the trousers. To my embarrassment, I seem to be able to grow it quite well. Its slow germination is often said to be as a result of the fact that the seeds go 'nine times to the devil' before they germinate. In reality, cold, wet soil is likely to blame. Many gardeners water the base of the seed drill with boiling water before sowing to raise the soil temperature, but it is doubtful if this has much effect on germination.

Rosemary

Rosmarinus officinalis

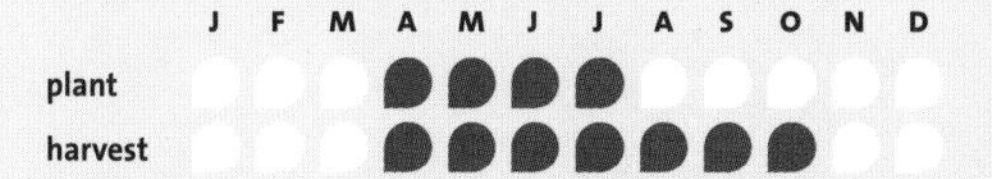

Rosemary is a dual-purpose plant that most people know as a culinary herb but tend to use more as an aromatic, ornamental garden shrub – so pick a bit and start using it. It does make an excellent garden plant, since the blue flowers contrast so well with the deep green, glossy 'needles'. Nowadays most rosemaries flower twice a year, in late spring and early summer, then again in late summer or autumn until bad weather shuts them down for the year. Whether edible or ornamental, they really earn their keep in any garden.

HOW TO GROW

Degree of difficulty: Not difficult or time consuming.

Plant: Start with a young plant from a herb farm, nursery or garden centre. Plant it in late spring or early to midsummer in a warm, sunny, sheltered spot with well-drained soil, since this plant needs Mediterranean conditions to thrive. Ideally, allow it time to establish before it has to face its first winter outside. Plant a single specimen in a border (it teams very well with lavenders and ornamental sages) and allow it to grow bushy, or grow a row and clip them as a decorative dwarf edging to a veg patch or herb garden.

Spacing: Allow 60–90cm (2–3ft) between a potentially bushy rosemary and its neighbours; when growing it as a dwarf edging, plants can be positioned 30cm (12in) apart.

Routine care: Water new plants in and keep them watered in dry spells until they are well established. In May or June, when the first flush of flowers is over, clip dwarf edgings or prune free-standing plants to improve their shape.

TRAINED ROSEMARY

Rosemary is often sold in garden centres during the summer trained as mini-standards or small topiary shapes for growing in pots on the patio; but it's very easy to style your own plant.

Choose a strong, straight shoot to root as a cutting and cut several 7.5cm/3in lengths from the tips of rosemary shoots. Strip off the leaves from the lower half and push them in round the edge of a pot of multi-purpose compost to about halfway. Stand the pot on a windowsill indoors, out of direct sun, to root, watering sparingly. When they've rooted, choose the strongest cuttings to pot up individually.

To grow a mini-standard; train the single stem straight up a cane until it reaches 30cm (12in) high then nip out the growing tip to make it branch. Remove any side shoots that grow lower down and only allow three or four that emerge from the top of the shoot to develop. When they are a few centimetres long, nip out their growing tips to make them branch again and repeat this process until a nice solid round 'head' has built up, about 15cm (6in) in diameter.

To grow a topiary shape; choose a suitable two-dimensional topiary frame from the garden centre, push this into a pot and plant one rosemary plant alongside it. Nip out the growing tip to make it branch, and as shoots develop tie them up to the topiary frame so that the stems outline the shape. Continue doing this until it's all covered, then nip out any unwanted shoots.

VARIETIES

1 ***Rosmarinus officinalis*** – the basic rosemary you'll find in garden centres; it grows into a 90cm x 90cm (3ft x 3ft) evergreen shrub with the familiar slender, needle-like leaves and mid-blue flowers. It's a good all rounder for garden use and cooking.

2 **'Miss Jessopp's Upright'** – a tall, rather vertical form which can reach up to 1.5m (5ft) high and maybe 90cm (3ft) wide with pale violet flowers speckled with deeper blue. It is good for training as a hedge of medium height, or as an 'accent' in a herb garden or border. When stripped of their needles the long, straight stems make superb herbal kebab skewers, ideal for using on the barbecue.

3 **'Severn Sea'**– a very aromatic, mound-shaped plant that reaches roughly 75cm (2ft 6in) in each direction, with mid-blue flowers.

4 **'Fota Blue'** – a really attractive, spreading variety that reaches 45cm (18in) high and 1.2–1.5m (3–4ft) wide with very deep blue flowers speckled indigo with a faint white throat.

5 ***Rosmarinus officinalis* Prostratus Group** – (also known as *R. lavandulaceus*), is a compact, prostrate rosemary which was grown by the Victorians as a hanging basket plant in the conservatory, since it is slightly less hardy than other kinds. It has light blue flowers and highly aromatic leaves which have a rather resinous scent, making it less suitable than most varieties for cooking. Great by a front door or on the patio in tubs, hanging baskets or window boxes. Can be clipped formally or allowed to grow naturally.

HARVEST

Start picking sprigs when the plants are large enough. Being evergreen, you can cut for the kitchen at any time of year, though the flavour is most intense during the growing season.

STORAGE

When you prune or clip the plants, save the pieces to dry. Cut off the best bits – the 15cm (6in) long tips of the shoots – wash to remove dust or dirt then dip briefly in boiling water and then cold water before hanging up small bunches to dry in an airy place out of sunlight. Alternatively, if you only have a few pieces to dry, lay them flat in an airing cupboard or shelf above a radiator and turn them over regularly. When completely dry, rub the 'needles' off the stems and store them in dry jars. That said, I prefer fresh any day.

PROBLEMS

Rosemary beetle is one of our newest pests which has recently come over from the continent. As yet this pest is mainly found round the London area. Look out for small, oval, metallic-green beetles with purple stripes eating the foliage; the same bugs also feed on other Mediterranean herbs, especially lavender, thyme and sage. There's no chemical deterrent, and it's too soon to tell which of our native insects may take a fancy to eating the newcomer, so pick them off by hand.

Sage

Salvia officinalis

	J	F	M	A	M	J	J	A	S	O	N	D
plant				●	●	●	●					
harvest				●	●	●	●	●	●	●		

Sage is best known for its starring role in sage and onion stuffing, but it has a far larger repertoire. In the kitchen the leaves can be used in any of the ways you'd use a bay leaf, and in the garden ornamental sages with coloured leaves make a great contribution to beds and borders – especially now that the Mediterranean look is so firmly 'in'.

HOW TO GROW

Degree of difficulty: Not difficult or time consuming.

Plant: Plants can be raised from seed, but since one plant one of any variety is usually all you need, most people prefer to save time by buying a plant from a nursery or garden centre. Plant in a sunny and reasonably sheltered spot with well-drained soil, or in a tub with plenty of drainage material in the base, in late spring or early to midsummer. Although sage is hardy it *is* a Mediterranean plant, so it's a good idea to get plants well established before they have to face a winter outside.

Spacing: Allow 60cm (2ft) between culinary sage and it's neighbours, a smaller distance of 45cm (18in) in the case of ornamental sages, which aren't so large or strong growing.

Routine care: Water in new plants and from then on water in dry spells; once established, sage is fairly drought tolerant. Prune plants in spring to tidy their shape.

TAKING CUTTINGS

It's a good idea to take cuttings of the tricolor sage, since this variety is least likely to survive a wet winter, but it's also worth taking cuttings of any sages because old plants tend to die off once they develop thick woody bases. Replace plants roughly every 5–6 years.

Take cuttings in late spring or early summer; cut off several 10cm (4in) lengths from the tips of strong, healthy stems. Peel away the lower leaves, then dip the bare end in hormone rooting powder (optional) and push five cuttings in round the edge of a 10cm (4in) pot filled with gritty compost (mix equal parts potting grit and multi-purpose compost). Stand the pots on a windowsill indoors out of direct sun – don't put them inside a plastic bag or the leaves will rot. They may take time to root, so remove any cuttings that die and pot the rest at the end of the summer. Keep young plants under cover for the winter.

VARIETIES

1 **Common sage (*Salvia officinalis*)** – the best-known kind for culinary use, is a low, bushy plant which reaches 60cm (2ft) high and 90cm (3ft) wide. It has strongly aromatic, grey-green oval leaves with a deeply textured surface and purplish-blue flowers in summer which attract bees.

2 **Purple sage (*Salvia officinalis* 'Purpurascens')** – a smaller plant (45cm/18in high and 60cm/2ft wide) with very attractive, purple-grey leaves that are just as good for cooking as 'real sage', though they have less aroma. They are also good as foliage in small jugs of country flowers. Mauvish flowers in summer.

3 **Tricolor sage (*Salvia officinalis* 'Tricolor')** – the very pretty mauve, cream and green variation with blueish flowers, which is quite a small, slow and weak grower since only part of each leaf contains chlorophyll. They reach roughly 30cm (12in) high and 45cm (18in) wide. It can be used for cooking, but frankly it doesn't have a lot of taste and it's much better used in flower arrangements. Needs rather better growing conditions than the other ornamental sages. Plants are sometimes lost in a cold wet winter, so take cuttings as security.

4 **Golden sage (*Salvia officinalis* 'Icterina')** – a stronger growing variety with gold and light green variegations, also good for cooking and as foliage to go with cut flowers. Height 45cm (18in), spread 75cm (2ft 6in).

HARVEST

Start picking leaves or short shoots as soon as the plants are large enough; choose young unblemished leaves during the growing season – April to October. During the winter, although a lot of foliage remains on the plant, sage dries out and is in no fit state to use.

STORAGE

Dry some sage leaves during the summer for winter use. Pick perfect leaves, dip them briefly into boiling water then cold (this helps them to keep their colour when dried), and lay them out thinly on baking sheets to dry in the airing cupboard or similar warm, dry place out of sunlight, turning them over occasionally. When completely dry, store them in screw-top jars.

PROBLEMS

Rosemary beetle is common – see rosemary, page 296.

After a long spell of wet weather, particularly in winter, it's not uncommon to find that some stems have died off – in summer it's easier to spot because the leaves turn grey and dry looking, but most of them look like that in winter anyway. Prune out affected stems in spring when you're tidying up plants.

Old sage bushes tend to behave like lavender and splay open, revealing thick woody stems which don't produce new growth when pruned. If this happens, cut them back in spring and hope for the best – old plants often die at this stage. This is easily prevented by giving plants an annual spring trim while they are young to keep them bushier and in good shape.

How to use sage

Put half a lemon and three sage leaves in the cavity before roasting a chicken to give it more flavour, or put 3–4 whole sage leaves in the water when boiling broad beans or cooking haricot or butter beans.

Or try sprinkling 2–3 teaspoonfuls of chopped fresh sage over sliced, par-boiled potatoes, layered with grated cheese in a shallow baking dish. Pour over single cream and top with more grated cheese, then bake in a moderate oven for an hour until browned on top.

Salad burnet

Sanguisorba minor

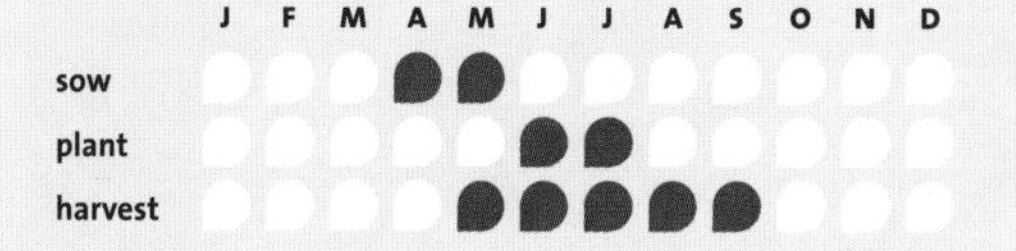

Given the current penchant for unusual herbs, I'm surprised we haven't heard more of salad burnet. It's a close cousin of the garden perennial with the spectacularly shaggy, pink, bottlebrush-type flowers, *Sanguisorba obtusa*. The salad version has dull red-brown blobs for flowers, but it's grown for its cucumber-flavoured leaves, which are normally used as a herb but they also make great salad leaves. It's evergreen, so it can be harvested all year, and it's very drought-tolerant and doesn't run to seed.

HOW TO GROW

Degree of difficulty: Not difficult or time consuming.
Sow: Plants are not easy to find, but seeds are available from organic suppliers. Sow in spring or autumn in pots, under cover. Thin out seedlings, but don't bother pricking out.
Plant: In spring or summer in a sunny spot with well-drained soil with some organic matter. Plant a bought plant or a potful of seedlings as one clump without dividing them – they'll make a leafier plant so you get more salad leaves. If you grow several potfuls, salad burnet makes an attractive edging plant for a path or potager, and it also grows well in troughs or pots.
Spacing: Space 30cm (12in) apart in a salad bed; 20cm (8in) apart as an edging.
Routine care: Water in new plants and in dry spells until they are established, after which they should be self-sufficient. Don't feed plants grown in the ground. Water container-grown plants sparingly and feed them only very occasionally, ideally with an organic liquid feed such as seaweed extract, since very 'soft' lush growth loses its delicate flavour.

Strong clumps can be divided in autumn. Cut back to 5–7.5cm (2–3in) above ground level then dig it up and divide it into several pieces. This is the quickest way to get more plants and reduce overgrown clumps.

If you want top-quality leaves in winter, grow a plant in a tub and move it under cover for protection. This will keep the plants growing for longer in spring and autumn and give you more tender leaves. You could also cut from outdoor-grown plants if you wish.

VARIETIES

Salad burnet (*Sanguisorba minor*) is a botanical species with no named varieties. A young plant begins as a rosette shape, but new growth from the centre soon turns it into a low 'tuft' of ferny foliage roughly 30cm (12in) high and as wide, with flower stems up to about 45cm (18in) high in summer.

HARVEST

Start cutting whole, young, unblemished leaves, complete with their stalks, as soon as the plants are large enough. This is a plant that wants cutting back hard (to about 10cm/4in above ground level) several times throughout the summer if you don't crop it much, as this encourages strong new growth of tender young leaves and discourages it from flowering.

PROBLEMS

None.

Savory

Satureja species

	J	F	M	A	M	J	J	A	S	O	N	D
sow			●	●								
plant					●	●						
harvest					●	●	●	●	●	●		

It's mostly only serious herb fanciers who have heard of savory, but it's a herb that healthy eaters should discover as it's the one that goes with beans – by which I mean the dried sort, such as haricots. It's also brilliant with lentils and most kinds of veg. Real fans might like to grow both summer and winter savory and experience the benefits all year round.

HOW TO GROW

Degree of difficulty: Not difficult or time consuming.
Sow: In pots on a windowsill indoors in March/April. There's no need to prick out the seedlings individually, it's much better just to thin them a bit if they are very overcrowded and leave the whole lot to grow on as a clump; this gives you a large bushy plant much faster.
Plant: Plant in well-drained and rather impoverished soil or in troughs, tubs or windowboxes filled with John Innes seed compost in a sunny, sheltered spot, after the risk of frost has passed. Summer savory is best planted any time from mid- to late May to late June. Winter savory can be planted from June until late summer.
Spacing: Space 20cm (8in) apart.
Routine care: Water in new plants and water again in dry spells until they are established; water plants grown in containers sparingly. Don't feed savory, since lush growth makes them too soft to survive. If you grow winter savory long-term in the same container, then liquid feed sparingly with an organic feed, such as seaweed extract, during the growing season – May to mid-August.

VARIETIES

1 **Summer savory (*Satureja hortensis*)** is a small, bushy, half-hardy annual, growing roughly 20cm (8in) high and 15cm (6in) wide, with aromatic, narrow, linear leaves and tiny mauve-pink or off-white flowers in summer. It's fit to pick from late June until mid-September. If you want summer savory out of season, sow it in pots in late summer and keep it indoors – at room temperature – throughout the winter.

2 **Winter savory (*Satureja montana*)** looks very similar but it's a semi-evergreen perennial that lives for several years, growing 30cm (12in) high and about 20cm (8in) wide with pinkish flowers in summer. If you want to use it through the winter, bring a pot-grown plant under cover from September to April so it keeps its leaves and grows for longer in spring and autumn.

HARVEST

Start snipping young shoots when the plants are large enough. Keeping the plants cut back makes them grow bushier so they produce yet more snippable material. Happily, the more you cut, the more you get.

PROBLEMS

None.

Using savory

Chop fresh summer or winter savory into cooked bean salads, lentil dishes or onto cooked vegetables such as French beans. The flavour is rather hot and spicy; it acts almost as a seasoning, so it's handy for anyone wanting to cut down on salt or who can't eat spices. Use sparingly.

Sweet cicely

Myrrhis odorata

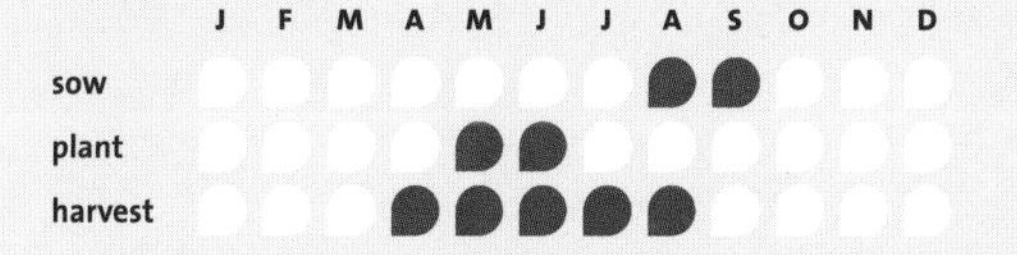

A large member of the cow parsley family, sweet cicely is an unlikely looking 'herb', but its mild, sweet, aniseed flavour makes it invaluable for anyone trying to lose weight or cut down on sugar or artificial sweeteners. It is also prized by adventurous gardener–cooks in search of new flavours.

HOW TO GROW

Degree of difficulty: Not difficult or time consuming.
Sow: Sweet cicely seeds have a very short period of viability, so it's best to start with a plant from a herb farm. If you want more, collect your own ripe seeds straight from the plant at the end of summer when they turn deep brown and sow them into pots straight away. (Keep them outdoors, though, as they need a cold spell before they'll germinate the following spring.) Easier still, watch out for the self-sown seedlings underneath your plant once it's flowered.
Plant: Plant out after the last frosts have passed, in late May or June, in reasonably well-drained soil containing plenty of well-rotted organic matter. Sweet cicely prefers light or dappled shade.
Spacing: Space plants 60cm (2ft) apart.
Routine care: Water in new plants and water again in dry spells until well established. If you want to keep the plants leafy and avoid lots of unwanted seedlings – and in ideal conditions it can self-seed rather over-enthusiastically – cut the flower stems down as soon as you see them. Once plants start to die down naturally in autumn, cut the stems and foliage down to a few centimetres above ground level and clear away the remains.

VARIETIES

Myrrhis odorata is a botanical species and there are no named varieties. It has delicate-looking, feathery foliage and large, flat-topped, white flowers that are very similar to cow parsley; when they are over, the skeletal shape of the seed heads is outlined with long, thin, dark seeds that stand upright and are heavily ridged.

HOW TO USE SWEET CICELY

Add a few whole leaves to the pan when stewing apples or pears and taste before adding sugar or sweeteners when it's cooked – you should find you only need half the usual quantity, and the mild aniseed taste of the herb adds a new dimension to the flavour.

You can use it in the same way when cooking plums, rhubarb or soft fruit, or you can add some finely chopped sweet cicely leaf to cream, crème fraîche or cream cheese to give it a delicate flavour that goes well with both sweet and savoury dishes. It is also superb with parsnips or cabbage: put a couple of leaves in the pan when cooking and remove before serving, or sprinkle a little freshly chopped leaf on to coleslaw.

HARVEST

Start picking whole, unblemished leaves as soon as they are large enough to use; the plant starts to grow quite early in the spring, so you could be picking from April until the end of August.

PROBLEMS

None.

Tarragon

Artemisia dracunculus

	J	F	M	A	M	J	J	A	S	O	N	D
plant					●	●						
harvest						●	●	●	●	●		

Tarragon enjoys a certain reputation; it's one of the most highly rated herbs for flavour, but it's not all that often grown because it's a perennial and a little on the large side for most small herb gardens. Added to this, it is not something you'll ever need a lot of. Many gardeners who think they grow the superior French tarragon actually have the far larger and coarser Russian tarragon, which is much easier to propagate, resulting in there being more of it circulating 'in the trade'. But if you are going to give up space for tarragon, it's worth going to some trouble to obtain and grow the one with the better culinary qualities. One plant is all you'll need.

HOW TO GROW

Degree of difficulty: Russian tarragon grows like a weed, but French tarragon is a lot fussier and needs attention to detail – neither variety demands a lot of work, though.

Plant: Buy a plant from a reliable herb farm or nursery. (French tarragon isn't available as seeds, though you can sometimes obtain seeds of the Russian sort.)

Plant French tarragon in a warm, sheltered, sunny situation and well-drained soil. It also does well in a large pot or tub and, being portable, you can easily move the pot into a cold greenhouse for winter protection and to extend the growing season slightly at either end.

If you are going to grow the Russian sort (though I must have put you off by now), it is quite happy with normal garden conditions, and perfectly hardy.

Spacing: Allow 60cm (2ft) between French tarragon and its neighbours and 90cm–1.2m (3–4ft) in the case of Russian tarragon as it is tall, spreads more rampantly, and the stems will splay out a bit.

Routine care: After planting, water in and keep new plants watered in dry spells until they are well established. During the summer, pinch out the flowering ends of the stems as soon as you see them, to keep plants leafy.

As the plants start to die down naturally in autumn, cut all of the stems down to within a few centimetres of ground level and remove the debris. Protect French tarragon crowns with a deep insulating layer of bracken, bark chippings or similar materials, or return tub-grown plants to a greenhouse.

Replace your old plant every three years, since young ones are far more vigorous. To do this, dig up a dormant plant in late March and gently tease the roots apart with your hands to divide the clump and replant the best portion, ideally in a new site.

VARIETIES

1 **French tarragon (*Artemisia dracunculus* French)** – (sometimes listed as *Artemisia dracunculus* subsp. *sativa*), which originates from southern Europe, this is very much the species to go for; the plant grows 45–60cm (18in–2ft) high and makes a modest, semi-bushy clump that barely spreads. The long, narrow, bright green leaves have a strong, sharp-peppery flavour (a good means of identifying it). Although it produces sprays of barely noticeable tiny yellow flowers in summer, it doesn't set seed.

2 **Russian tarragon (*Artemisia dracunculus* Russian)** is a native of Siberia, growing 1.2m (4ft) high and wide with strong stems; the leaves are a tad coarser and have a rather bitter flavour, though this plant also has tiny yellow flowers in summer.

1

2

Using tarragon

Tarragon is traditionally used with chicken, in mayonnaise, rice dishes and in sauce béarnaise. It is also a vital ingredient of *omelette fines herbes*, which is made with a mixture of tarragon, parsley, chives and chervil.

To make it, take 1 tablespoonful of each of the chopped fresh herbs and add half to beaten eggs before the mixture is cooked and sprinkle over the rest afterwards. (This is especially brilliant when made with home-grown herbs and newly laid eggs from your own hens.)

Tarragon is also delicious when finely chopped and mixed with cream cheese in sandwiches, as a filling for jacket potatoes, or mashed into softened butter and melted over cooked vegetables such as baby broad beans or asparagus tips.

HARVEST

Start cutting 5cm (2in) long pieces from the tips of young shoots as soon as the plants are big enough. You can continue cutting 'little and often' through the summer from roughly early June to early October, when new growth stops and stems start toughening up in preparation for dying off.

PROBLEMS

Rust can occur on tarragon; look out for tiny orange dots on the foliage when buying a plant and reject it if you see any. If it occurs on a plant growing in the garden, cut down affected stems at once and remove and dispose of them in the hope that it won't spread. Remove a badly affected plant entirely and start again with a new one some distance away.

Recipe – Tarragon vinegar

Find an attractive, clear glass bottle with a screw top, and feed into it two or three washed and dried perfect stems of French tarragon, long enough to almost reach from the bottom of the bottle to the top. Fill it up with good-quality white wine vinegar. Leave it to steep for a couple of months before using it; the flavour of the herb will infuse the vinegar. There's no need to remove the tarragon stalks once you start using the bottle's contents. Use the flavoured vinegar for making French dressing to go with salads. A bottle also makes a good gift, especially if paired with a matching bottle of home-made herb oil.

Thyme

Thymus species

	J	F	M	A	M	J	J	A	S	O	N	D
plant				●	●	●	●					
harvest	●	●	●	●	●	●	●	●	●	●	●	●

Thymes are a large group of attractive, drought-tolerant, evergreen plants that include some classic must-have herbs and several more that are 'dual-purpose' (ornamental as well as edible), while a few – often treated as rock plants and grown for their flowers – are not really culinary plants at all. So be sure you have the right varieties for the job. As an added bonus for wildlife-friendly gardeners, bees love thyme flowers.

HOW TO GROW

Degree of difficulty: Needs the right conditions to thrive, but when given these it is not difficult or time consuming.

Plant: Plant in a sunny spot with poor, very well-drained soil, and if it's not like that naturally, work in lots of grit, or grow thymes in a raised bed or containers. The best time to plant is in late spring and early summer, after the last frosts are over. Even though the plants are hardy perennials, they'll get off to a far better start if they can establish themselves when conditions are comfortable; they like a bit of time to find their feet before they have to face their first winter outside. Trim plants to improve their shape in spring, and remove any dead stems.

Spacing: Space bushy, upright varieties 23cm (9in) from their neighbours and spreading varieties 30cm (12in) apart. They are relatively slow growing, but if you are aiming for a carpeting effect at these spacings you'll usually be able to cover the ground within the first season.

Routine care: Water in and water again in dry spells while young plants are becoming established, but thereafter don't water unless absolutely essential. Don't feed them either, since thymes need to work hard. Take cuttings from July to late September, in the same way as for rosemary cuttings (though those of thyme will be much smaller), and replace old plants after three years.

GROWING THYME IN CONTAINERS

To get the best from thyme varieties in containers, grow them in terracotta tubs, troughs or wide shallow pans, in the correct soil conditions. Put plenty of coarse drainage material such as small stones in the bottom, then fill with John Innes No.1 potting compost mixed with up to 50 per cent potting grit to make a well-drained mix that's not too rich. Plant one thyme per 13cm (5in) pot, for a bushy, upright variety, and one to a 25cm (10in) diameter terracotta 'pan' for a spreading type. What looks even better is to group a small collection of different culinary and dual-purpose varieties together in a larger tub to make an attractive, aromatic, potted herb garden, or else grow them as a path edging.

VARIETIES

1 **Common or garden thyme (*Thymus vulgaris*)** – the best known culinary thyme, with very aromatic and strongly flavoured short, thin, needle-like leaves and deep mauve-pink flowers in June/July. It is an upright and bushy variety that reaches up to 30cm (12in) high and 20cm (8in) across.

2 **Lemon thyme (*Thymus* x *citriodorus*)** – a bushy species with lemon-scented leaves and a slightly milder flavour than common thyme; pink flowers June/July. It grows 15cm (6in) high and about the same across; good for pots.

3 **Caraway thyme (*Thymus herba-barona*)** – a low, creeping species 5cm (2in) high and spreading up to about 30cm (12in), with unusual, small, shiny leaves that are caraway scented. It develops rosy-purple flowers in June/July. Plants are good for trailing down hollow-topped, dwarf walls and growing in tubs. The herb is traditionally tucked under beef while it is roasting, which is unusual since thymes usually go with lamb.

4 **Orange thyme (*Thymus* x *citriodorus* 'Fragrantissimus')** – a bushy variety with small, grey, needle-like, orange-scented leaves and pale pink or white flowers in June/July; good for putting under duck while it's roasting.

5 ***Thymus* x *citriodorus* 'Bertram Anderson'** – a spreading, dual-purpose variety 10cm (4in) high, eventually making a low mound 60cm (2ft) wide, with mild-flavoured, gold-variegated leaves and mauve-pink flowers in June/July.

HARVEST

Start snipping young shoots a few centimetres long from the tips of the stems as soon as the plants are large enough to take this, but don't over-cut young plants. That said, though, regular cutting does help to keep older plants of upright varieties neat and bushy, and prevents them succumbing to their natural inclination to turn woody at the base and splay open at the top, becoming shapeless.

Being evergreen, thymes can be picked all year round, though many cooks consider that the flavour is better in summer when plants are growing actively and bathing in bright sunlight.

STORAGE

Dry surplus thyme in summer, either before or well after the flowering season. Dip whole stems briefly in boiling water then cold, to wash off dust and preserve the colour, then hang them upside down to dry in an airy place out of direct sunlight. When perfectly dry, rub the small leaves off the stems with your fingers and store them in screw-top jars in a cool, dark place.

PROBLEMS

Failure to grow well is usually a sign of unsuitable growing conditions – most often too much water or insufficiently sharp drainage, too much organic matter or excess nutrients. Aim for poor, dry conditions and you'll do far better. Or grow in pots, or in a raised bed with 2.5cm (1in) of gravel over the soil.

WHERE TO SEE GREAT FRUIT AND VEG GARDENS

West Dean Gardens
West Dean, Chichester, West Sussex PO18 0QZ
Set in the grounds of this internationally renowned college is a carefully restored, walled kitchen garden, complete with spectacular Victorian glasshouses and packed with a huge, mouthwatering array of plants; from peaches and peppers to aubergines and orchids. The garden is open year round, as is the beautiful parkland and arboretum with views over the Sussex Downs. Check website or call for opening hours and admission prices: 01243 818210; www.westdean.org.uk.

Chatsworth
Bakewell, Derbyshire DE45 1PP
The home of the Duke and Duchess of Devonshire, the kitchen garden is just one stunning element of this breathtaking garden, and is as well executed as the rest of the estate. The house and gardens at Chatsworth are both well worth a visit; the 105 acres contain rare trees, temples, sculptures, the famous 300-year-old Cascade, maze, rock garden and Pinetum. Check website or call for opening hours and admission prices: 01246 565300; www.chatsworth.org

Audley End
Saffron Walden, Essex CB11 4JF
English Heritage work with Garden Organic (HDRA) to run the kitchen garden in the grounds of this originally Tudor mansion as a working organic plot, which looks very much as it would have done in late Victorian times. The Vine house is one of the earliest in the country, which contains some vines over 200 years old. The rest of the landscaped gardens have also been restored, including the lake in 18th-century parkland, the temple and 19th-century parterre. Check website or call for opening hours and admission prices: 01799 522399; www.english-heritage.org.uk

Tatton Park
Knutsford, Cheshire WA16 6QN
This National Trust-owned property is home to the RHS Flower Show every summer. The house is set amongst extensive gardens that include an acclaimed fernery and formal Italian garden designed by Joseph Paxton, a Japanese garden and follies. The walled kitchen gardens have been recently extensively restored and original varieties of fruit and veg are cultivated using traditional techniques. Check website or call for opening hours and admission prices: 01625 374435; www.tattonpark.org.uk

Heligan
Pentewan, St Austell, Cornwall PL26 6EN
The famous 'Lost Gardens of Heligan' were rescued in 1990 and have been spectacularly restored to their former glory. The Jungle Valley, Alpine Ravine, Italian and New Zealand Gardens are impressive, but the five walled gardens are an incredible insight into the way Victorians produced fruit and veg. Heligan grows over 300 cultivars of fruit and veg, including many rare varieties, and the pineapple pits, potting sheds, melon houses and gardeners' bothies show what went on behind the scenes. Check website or call for opening hours and admission prices: 01726 845100; www.heligan.com

Calke Abbey
Ticknall, Derbyshire DE73 7LE
This National Trust property is held up as an example of a property in decline, but the gardens are carefully tended and have undergone some restoration work. Calke's grounds contain a pretty flower garden (including the only surviving original Auricula Theatre), a bothy, an ice house, restored peach house and a working, walled kitchen garden. Check website or call for opening hours and admission prices: 01332 863822; www.nationaltrust.org.uk

Barrington Court
Barrington, nr Ilminster, Somerset TA19 0NQ
Tended by the National Trust, Barrington Court is home to a Jekyll-inspired English–Italian garden, orchards, an arboretum and tree-lined avenues as well as a working kitchen garden growing a good range of fruit and veg varieties. Check website or call for opening hours and admission prices: 01460 241938; www.nationaltrust.org.uk

Garden Organic Yalding
Benover Road, Yalding, nr Maidstone, Kent ME18 6EX
A series of 14 historically themed formal and informal gardens takes the visitor on an inspiring tour of gardening through the ages. All are organically tended by HDRA. The Apothecary's Garden, 1950s Allotment and Organic Vegetable Display Garden are of particular interest to veg growers. Check website or call for opening hours and admission prices: 01622 814650; www.gardenorganic.org.uk

Garden Organic Ryton
Coventry, Warwickshire CV8 3LG
Ten acres of beautifully tended grounds demonstrate the successful practice of organic gardening at Ryton. There are flowering gardens as well as displays of fruit, veg and herbs; a new development is the first biodynamic garden open to the public in the UK. Check website or call for opening hours and admission prices: 024 7630 3517; www.gardenorganic.org.uk

Titsey Place
Titsey Hill, Oxted, Surrey RH8 0SD
This historic garden has been carefully restored since 1993. Formal gardens sit side by side with an old walled kitchen garden that has been restored as an illustration of Victorian techniques. It includes a good selection of fruit and veg varieties grown in beds and inside glasshouses. Check website or call for opening hours and admission prices: 01273 715359; www.titsey.org

SUPPLIERS AND SPECIALISTS

Garden centres and nurseries stock a good range of popular veg seeds, young plants and fruit trees, canes and bushes.

The big seed firms have seeds on sale on the racks in garden centres, but they also issue several catalogues at different times of year which list a much larger range of veg seeds, onion sets, shallots and garlic, seed potatoes, young veg plants and fruit and nut trees and bushes. These catalogues are sent free to anyone on their mailing list. You can also see what they sell and buy what you need via their websites.
Suttons, Woodview Road, Paignton, Devon TQ4 7NG; 24-hour orderline: 0844 922 0606; www.suttons.co.uk
Thompson and Morgan UK Ltd, Poplar Lane, Ipswich IP8 3BU; 01473 688 821; www.seeds.thompson-morgan.com
S.E. Marshall & Co, Alconbury Hill, Huntingdon, Cambs PE28 4HY; 01480 443390; www.marshalls-seeds.co.uk
Samuel Dobie & Son, Long Road, Paignton, Devon TQ4 7SX; 0844 701 7625; www.dobies.co.uk
Mr Fothergill's Seeds, Kentford, Suffolk CB8 7QB; 01638 751161; www.mr-fothergills.co.uk

Specialist seed firms have slightly different ranges, sometimes angled to organic growers or particular types of enthusiasts. These companies sell the more unusual gourmet and exotic varieties which aren't generally available in garden centres, and some also supply ranges of tools, equipment and gadgets for organic growers or other enthusiasts.
Simpson's Seeds, The Walled Garden Nursery, Horningsham, Warminster, Wiltshire BA12 7NQ; 01985 845004; www.simpsonsseeds.co.uk (tomatoes, peppers, chillies and some other unusual veg).
W. Robinson & Son, Sunny Bank, Forton, Nr Preston, Lancs PR3 0BN; 01524 791210; www.mammothonion.co.uk (giant and exhibition veg).
Suffolk Herbs, Monks Farm, Kelvedon, Essex CO5 9PG; 01376 572456; www.suffolkherbs.com (unusual veg varieties, herbs, wildflowers, and organic gardening gear).
Medwyn's of Anglesey, Llanor, Old School Lane, Llanfair PG, Anglesey, LL61 5RZ; 01248 714851; www.medwynsofanglesey.co.uk (vegetable seeds, especially those for showing, from Medwyn Williams, winner of ten consecutive Gold Medals at the RHS Chelsea Flower Show).
Chiltern Seeds, Bortree Stile, Ulverston, Cumbria LA12 7PB; 01229 581137; www.chilternseeds.co.uk (unusual veg varieties and large list of unusual plants).
The Organic Gardening Catalogue, Riverdene, Molesey Road, Hersham, Surrey KT12 4RG; 0845 130 1304; www.OrganicCatalogue.com (seeds, some plants, organic gardening gear).
Plants of Distinction, Abacus House, Station Yard, Needham Market, Suffolk IP6 8AS; 01449 721720; www.plantsofdistinction.co.uk (range of unusual and heritage veg in their 'Simply Vegetables' catalogue).
Seeds of Italy, C3 Phoenix Industrial Estate, Rosslyn Crescent, London Borough of Harrow HA1 2SP; 0208 427 5020; www.seedsofitaly.com (continental veg, salad and sprouting seeds also some continental kitchen equipment).

Specialist nurseries will have the biggest range of varieties and often unusual varieties that aren't widely available.
Ken Muir Ltd, Rectory Road, Weeley Heath, Clacton-on-Sea, Essex CO16 9BJ; 01255 830181; www.kenmuir.co.uk (fruit and nuts, also some specialist sundries).
Keepers Nursery, Gallants Court, East Farleigh, Maidstone, Kent ME15 0LE; 01622 726456; www.fruittree.co.uk (a wide range of usual and unusual fruit).
Reads Nursery, Hales Hall, Loddon, Norfolk NR14 6QW; 01508 548395; www.readsnursery.co.uk (unusual fruit including citrus and greenhouse figs and grapes).
Deacon's Nursery, Moor View, Godshill, Isle of Wight PO38 3HW; 01983 522243; www.deaconsnurseryfruits.co.uk (fruit).

Mail-order equipment and biological control suppliers can often be handy for finding specialist gear that's very useful for fruit and veg growing, especially for organic gardeners.
Defenders Ltd., Occupation Road, Wye, Ashford, Kent TN25 5EN; 01233 813121; www.defenders.co.uk (natural, biological pest control).
Agralan Ltd, The Old Brickyard, Ashton Keynes, Swindon, Wiltshire SN6 6QR; 01285 860015 www.agralan.co.uk (organic gardening gear, natural pest control, biological control).
Wiggly Wigglers, Lower Blakemere Farm, Blakemere, Herefordshire HR2 9PX; 01981 500391; www.wigglywigglers.co.uk (composters, wormeries, green gardening sundries).
Chempak Products, Hillgrove Business Park, Nazeing Road, Nazeing, Essex EN9 2BB; 01992 890770; www.chempak.co.uk. To buy products online: www.gardendirect.co.uk (specialist feeds including ericaceous).

Organizations
The National Vegetable Society A charity set up to advance the culture, study and improvement of vegetables, and which offers help and advice to growers at all levels – from newcomers to experienced show growers. Their website lists shows, gives advice on growing, showing and equipment – including sales.www.nvsuk.org.uk
National Society of Allotment and Leisure Gardeners Limited Offers advice on all aspects of allotments and allotment gardening. NSALG Ltd, O'Dell House, Hunters Road, Corby, Northants NN17 5JE; 01536 266576; www.nsalg.org.uk
Common Ground A charity devoted to local communities; organizers of Apple Day, the annual celebration of orchards and apples, and their website lists Apple Day events all over the country. Common Ground, Gold Hill House, 21 High Street, Shaftsbury, Dorset SP7 8JE; 01747 850820; www.commonground.org.uk
The Royal Horticultural Society Free advice including individual answers to cultural and pest and disease problems, (there is a charge for soil analysis and fruit identification) to members only via the garden at Wisley (details of how to access the service are in the members' handbook). The society also offers useful advice of all sorts on its website, and many other benefits including free magazine and entry to RHS gardens, discounted tickets for RHS shows and access to the members' library. The Royal Horticultural Society, 80 Vincent Square, London SW1P 2PE; 0845 260 5000; www.rhs.org.uk

INDEX

Note: Page numbers in *italics* refer to photographs/illustrations.

Actinidia
A. arguta 'Issai', *see* Mongolian gooseberry
A. deliciosa, see kiwi fruit
Agaricus bisporus 133
alfalfa 167, 168, *168*
Allium
A. cepa, see onion; shallots
A. porrum, see leek
A. sativum, see garlic
A. schoenoprasum 272–3, *272–3*
A. s. 'White Form' 272, *272*
A. tuberosum, see chives
allotments 7, 23, 27, 75, 79, 80, 84, 86, 89,107,140
almond 11, 46, 234–7, *236*
'Mandaline' 236, *236*
Aloysia triphylla, see lemon verbena
Anethum graveolens, see dill
Anthriscus cerefolium, see chervil
aphids 51–2, *51*, 76, 115, 153, 194, 255
see also blackfly; greenfly
Apium graveolens
A. g. var. *dulce, see* celery
A. g. var. *rapaceum, see* celeriac
apple 16, 54, 188–94
'Ashmead's Kernel' 192, *193*
'Blenheim Orange' 192, *193*
'Braeburn' 192, *193*
'Bramley's Seedling' 192, *193*
container-grown 188
cooking 23
cordons 188, 190, *190*
'Cox's Orange Pippin' 192, *193*
crab apple 194
'Discovery' 192, *193*
dwarf pyramids 190, *190*
dwarfing rootstocks 188, 191
'Egremont Russett' 192, *193*
'Ellison's Orange' 192, *193*
espaliers 188, 190, *190*
'family trees' 188, 191
'Fiesta' 192, *193*
'Greensleeves' 192, *193*
growing tips 188
harvesting 191
'Howgate Wonder' 192, *193*
'James Grieve' 188, 192, *193*
'Jonagold' 192, *193*
'Lane's Prince Albert' 192, *193*
pollination 191
problems 194
pruning 190–1
'Queen Cox' 192, *193*
step-over trees 188, 190, *190*
storage 191
'Sunset' 192, *193*
'Tydeman's Early Worcester' 192, *193*
'Worcester Pearmain' 192, *193*
apricot 54, 195–7
container-grown 196
dwarf 195, 196
Flavorcot 197, *197*
growing tips 195
harvesting 196
'Moorpark' 197, *197*
patio apricots 196
pollination 197
preserving 195
problems 197
pruning 196
'Tomcot' 197, *197*
'Torinel' rootstock 197
Armoracia rusticana, see horseradish
Artemisia dracunculus 302–3, *302–3*
A. d. French 302, 303, *303*
A. d. Russian 302, 303, *303*
artichoke 41
globe 20, 45, 68–9, *68*
'Green Globe' 69, *69*
'Gros Camus de Bretagne' 69, *69*
'Gros Vert de Lâon' 69, *69*
'Violetto di Chioggia' 69, *69*
Jerusalem 70, *70*
'Fuseau' 70
asparagus (*Asparagus officinalis*) 20, 71–2, *71*
'Connover's Colossal' 72, *72*
'Jersey Giant' 72, *72*
'Jersey Knight' 72, *72*
asparagus beetle 72
asparagus pea 73, *73*
aubergine 49, 74, *74*
'Black Enorma' 74, *74*
'Moneymaker' 74, *74*

'baby veg' 27, 89, 93–4, 96, 107, 125–6, 138, 140, 182
bacteria, soil 33
Barbarea vulgaris, see land cress
basil 41, 159, *159*, 264–5, *264*
bush basil 265, *265*
lemon basil ('Kemangie') 265, *265*
'Neapolitana' (lettuce-leafed) 265, *265*
'Purple Ruffles' 265, *265*
Spice basil 265, *265*
'Sweet Genovese' 265, *265*
bay 266, *267*
beanpole frameworks 80, *80*
beans 46
aduki 168, *168*
broad 24, 49, 75–6, *75*
'Aquadulce' 76, *76*
'Green Windsor' 76, *76*
'The Sutton' 76, *76*
'Witkiem Manita' 76, *76*
dried 78, 79
French 12, 24, 27, 36, 77–9, *77*
'Blue Lake' 78, *78*
'Borlotto Lingua di Fuoco' 78, *78*
'Cobra' 78, *78*
dwarf 27
'Golden Teepee' 78, *78*
growing tips 77, 79
harvesting 79
'Hunter' 78, *78*
'Opera' 78, *78*
'Pea bean' 78, *78*
problems 79
'Purple Teepee' 78, *78*
'Soissons' 78, *78*
'Sonesta' 78, *78*
mung 167, 168, *168*
runner 12, 16, 27, 36, 54, 79, 80–1, *80*
'Enorma' 81, *81*
'Lady Di' 81, *81*
'Sunset' 81, *81*
'Wisley Magic' 81, *81*

beds 14–15
cutting 20
raised 14–15
bees 230, 268, 272, 288, 304
beetroot 36, 82–3
'Bull's Blood' 83, 159
'Burpee's Golden' 83, *83*
'Cylindra' 83, *83*
'Detroit 6-Rubidus' 83, *83*
'Egyptian Turnip Rooted' 83, *83*
beetroot leaves 83
Beta vulgaris
B. v. subsp. *cicla* var. *cicla* 166, *166*
B. v. subsp. *vulgaris* 82–3, *82*
B. v. var. *flavescens* 175–6, *175*, *176*
big bud 203
biological controls 50–1, 112, 153, 181, 239
bird netting 46, *47*
birds 46, 52, 81, 161, 217, 243, 246, 251, 254, 258, 259
black bean aphid 76
see also blackfly
black beetles 52
blackberry 11, 16, 20, 27, *198*, 199–201, 253
'Black Butte' 200, *220*
growing tips 199
harvesting 201
'Helen' 200, *200*
hybrids 200–1, *200–1*
'Loch Tay' 200, *200*
pruning 201
'Veronique' 200, *200*
blackcurrant 202–3, *202*, 253, 255
'Ben Connan' 203, *203*
'Ben Hope' 203, *203*
'Ben Lomond' 203, *203*
'Ben Sarek' 203, *203*
blackfly 51–2, *51*, 69, 209
see also black bean aphid
blanching 113, 117
blueberry 16, 19, 34, 41, 204–5
'Bluecrop' 205, *205*
'Chandler' 205, *205*
container-grown 204, 205
'Earliblue' 205, *205*
'Herbert' 205, *205*
pollination 204
'Sunshine Blue' 205, *205*
'Top Hat' 205, *205*
bolting 122, 125, 127, 131, 137, 156, 159, 164
Borago officinalis (borage) 115, *115*, 268, *268*
B. o. 'Alba' 268
Bordeaux mixture 53, 242
botrytis (grey mould) 53, *53*, 181, 261
boysenberry 200, *200*
Brassica 14, 26–7, 34, 37, 41, 44, 46, 52
B. Juncea, see Chinese mustard
B. napus Napobrassica Group, *see* swede
B. oleracea
B. o. Acephala Group, *see* kale
B. o. Botrytis Group, *see* cauliflower
B. o. Capitata Group, *see* cabbage
B. o. Gemmifera Group, *see* Brussels sprouts
B. o. Gongylodes Group, *see* kohl rabi
B. o. Italica Group, *see* broccoli, sprouting
see also calabrese
B. rapa
B. r. Pekinensis Group, *see* Chinese cabbage
B. r. Rapifera Group, *see* turnip
B. r. ssp. *chinensis, see* pak choi, 'Mei Quing'
see also oriental leaves
broccoli
sprouting 24, 27, 84–5
'Bordeaux' 84
'Claret' 85, *85*
'Late Purple Sprouting' 85, *85*
'Rudolph' 85, *85*
'Summer Purple' 84
'White Sprouting' 85, *85*
Brussels sprouts 24, 27, 86–7, *86*, *87*
'blown' 86, 87
'Falstaff' 87, *87*
'Trafalgar' 87, *87*
bullace 248
butterflies 85, 91, 288

cabbage *88*, 89–91, *91*
'Autoro' 90
'Cuor di Bue' (Bullsheart) 90, *90*
'Golden Acre' 90, *90*
growing tips 89
harvesting 91
'Hispi' 90, *90*
'January King' 90, *90*
'Maestro' 90
'Offenham Flower of Spring' 90, *90*
problems 91
red 89, 90, *90*, 91
'Red Drumhead' 90, *90*
spring-hearting 89, 90, 91
summer/autumn-hearting 89, 90, 91
'Tundra' 90, *90*
'Wheeler's Imperial' 90, *90*
'Winter King' 90, *90*
winter-hearting/Savoy 89, 90, 91
cabbage root fly 52, 91
cabbage white butterfly 85, 91
calabrese 92
'Chevalier' 92, *92*
'Crown and Sceptre' 92, *92*
'Trixie' 92, *92*
Calendula officinalis 115, *115*
cane blight 253
cane spot 253
canker 141, 194
bacterial 209
Cape gooseberry 206–7, *206–7*
Capsicum annuum, see chilli; pepper
caraway 269, *269*
caraway thyme 305, *305*
carrot 11, 36, 54, 93–5, *93*, *95*, 141
'Autumn King' 94, *94*
'Flyway' 94, *94*
growing tips 93
harvesting 94
'Healthmaster' 94, *94*
juice 93
'Nantes' 94, *94*
problems 95
'Purple Haze' 94, *94*
'Resistafly' 94
storage 95
'Sytan' 94
'Yellowstone' 94, *94*
carrot fly 93, 94, 95, 101
Carum carvi, see caraway
Castanea sativa, see sweet chestnut
caterpillars 52, *52*
cauliflower 24, 96–7
'All Year Round' 97, *97*
'Autumn Giant' 97, *97*
'Avalanche' 96
'Igloo' 96
'Purple Cape' 97, *97*
'Romanesco' 97, *97*
'Trevi' 97, *97*
'Violet Queen' 97, *97*
celeriac 54, *98*, 99, 141
'Monarch' 99
celery 42, 100–1, *101*
self-blanching 100, 101
trenching 100
'Victoria' 101
celery fly 99, 101
Chaenomeles japonica, see ornamental quince 208, 209
chalcid wasp 50, *50*
Chamaemelum nobile 270, *270*
C. n. 'Flore Pleno' 270, *270*
chamomile 270, *270*
double Roman chamomile 270, *270*
Roman chamomile 270, *270*
'Treneague' 270
cherry 208–9
cordons 208, 209
dwarfing rootstocks 208
fan-trained 208, 209
'Kanzan' 208
'Morello' 209, *209*
'Stella' 209, *209*
'Sunburst' 209, *209*
'Sweetheart' 209, *209*
chervil 36, 159, 271, *271*
chicken manure pellets 34, 41
chicons 103

- chicory 102–3, *102*
  - 'Rossa di Treviso' 103, *103*
  - 'Rossa di Verona' 103, *103*
  - 'Sugar Loaf' 103, *103*
  - 'Witloof Zoom' 103, *103*
- chilli 16, 27, 49, 104–5, *104*
  - 'Apache' 105, *105*
  - 'Jalapeño' 105, *105*
  - 'Joe's Long Cayenne' 105, *105*
  - 'Thai Dragon' 105, *105*
- Chinese cabbage 106, 159
  - 'Green Rocket' 106, *106*
  - 'Wong Bok' 106, *106*
- Chinese gooseberry, *see* kiwi fruit
- Chinese lantern flower 207
- Chinese mustard 138, *138*, 139, 159
  - 'Red Giant' 138, *138*
- chitting 149
- chives 41, 115, *115*, 272–3, *272–3*
  - garlic/Chinese 272
  - white 272
- chocolate spot 76
- *Cichorium*
  - *C. endivia, see* endive
  - *C. intybus, see* chicory
- citron 227, *227*
- *Citrus*
  - *C. aurantiifolia, see* lime
  - *C. hystrix, see* kaffir lime
  - *C. limon, see* lemon
  - *C. medica, see* citron
  - *C. sinensis, see* orange
- 'clamps' 95
- *Claytonia perfoliata* 154, *154*, 159
- clubroot 85, 91, 124, 125, 172, 183
- Cobbett, William 172
- cobnuts 234, 235, 236–7, *236*
- codling moth 194
- cold frames 37
- common/garden thyme 305, *305*
- compost 16, 33, 35
- containers 16–19, 27
  - growing beans in 79, 81
  - growing berries in 204, 205, 258, 259
  - growing brassicas in 89
  - growing chillies in 105
  - growing fruit trees in 188, 196, 212, 213, 226
- copper tape 51, 261
- cordons 14, 20, 39, 188, 190, *190*, 208–9, 215, 243
- coriander (*Coriandrum sativum*) 274–5, *274–5*
  - 'Cilantro' 275, *275*
  - 'Leisure' 275, *275*
  - 'Moroccan' 275, *275*
- corn salad 159
- *Corylus*
  - *C. avellana, see* cobnuts; hazel
  - *C. maxima*
    - *C. m., see* filberts
    - *C. m.* 'Purpurea', *see* filberts, red
- courgette 16, 24, 26, 27, 36, 37, 41, 53, 107–9, *107*
  - 'Clarita' 108, *108*
  - 'Defender' 108, *108*
  - 'Gold Rush' 108, *108*
  - growing tips 107
  - harvesting 108
  - 'Parthenon' 108, *108*
  - problems 109
- Cox, Richard 192
- crab apple
  - 'Evereste' 194, *194*
  - 'John Downie' 194, *194*
- cranberry 16, 19, 34, 41, 210, *211*
  - 'Early Black' 210
- crop rotation 26, *26*
- cucumber 53, 110–12, *110*, *111*, 230
  - 'Crystal Lemon' 112, *112*
  - 'Cum laude' 112, *112*
  - 'Flamingo' 112, *112*
  - growing tips 110–11
  - harvesting 112
  - indoor 111
  - 'Long White' 112, *112*
  - outdoor 111
  - problems 112
- cucumber mosaic virus 109, 112, 171
- *Cucumis*
  - *C. melo, see* melon
  - *C. sativus, see* cucumber
- *Cucurbita*
  - *C. maxima* 169–71
  - *C. moschata* 169–71
  - *C. pepo* 107–9, *107–8*, 169–71
- cuttings
  - herbs 281, 284–5, 288, 290, 295, 297
  - root 29, 290
  - stem 288
- cutworms 52, *52*
- *Cydonia oblonga, see* quince
- *Cymbopogon citratus, see* lemon grass
- *Cynara scolymus, see* artichoke, globe
- damson 246, 248
- dandelion 113, *113*, 159
  - 'Pissenlit a Coeur' 113
- *Daucus carota sativus, see* carrot
- dehydrators 54, 195
- desserts 205, 214, 215, 230, 253
- dewberry 200
- *Dianthus, see* pinks
- dill 276–7, *276–7*
  - 'Mammoth' 276
  - 'Vierling' 277
- *Dipsotaxis tenuifolia* 157, *157*
- disease 44, 50, 52, 53
  - *see also specific conditions*
- disease resistance 27
- 'double guyot' system 222, *222*
- dried beans 78, 79
- dried chillies 104, 105
- dried flowers 69, 281
- dried herbs 54, 265, 296, 298, 305
- drinks
  - alcoholic 201, 248, 250, 268
  - soft 93, 111, 202, 228, 270, 286
- early cropping 27
- 'earthing up' 100–1
- edible flowers 114–15
- eelworms 153
- Ellison, Rev 192
- *Encarsia formosa* 181, 239
- endive *116*, 117
  - 'Moss Curled' 117
- Epsom salts 34
- ericaceous fertiliser 34, 41
- ericaceous (lime-hating) plants 16–17, 204, 210, 226
- *Eruca sativa, see* rocket
- espaliers 20, 39, 188, 190, *190*, 243
- exotics 11
- F1 hybrids 27
- 'family trees' 188, 191, 243
- fans 20, 39, 208, 209, 212, 246
- feeding 41
  - *see also* fertiliser
- fennel 178
  - bronze 278, *278*
  - Florence 118, *119*
- fenugreek 168, *168*
- fertiliser 29, 30, 33–5
  - ericaceous 34, 41
  - general purpose 33, 34, 41
  - inorganic 33
  - liquid/soluble 34
  - solid 33
- fig (*Ficus caria*) 16, 212–14, *212*
  - 'Brown Turkey' 214, *214*
  - 'Brunswick' 214, *214*
  - container-grown 212, 213
  - cuttings 213, *213*
  - fan-trained 212
  - growing tips 212
  - harvesting 214
  - pruning 212, 213
  - root restriction 212
  - 'White Marseilles' 214, *214*
- filberts 234, 235, 236, *236*, 237
  - red *236*, 237
- fireblight 194, 245, 250
- flea beetle 139, 156, 183
- Florence fennel 118, *119*
  - 'Victorio' 118
- flowers
  - cut 277
  - dried 69, 281
  - edible 114–15
- *Foeniculum vulgare* 178
  - *F. v. purpureum* 278, *278*
  - *F. v.* var. *azoricum* 118, *119*
- forcing 102, 103, 113, 257
- *Fortunella japonica, see* kumquat
- *Fragaria*
  - *F. vesca* Semperflorens Group, *see* strawberry, alpine
  - *F.* x *ananassa, see* strawberry
- freezing crops 54, 142, 217, 294
- freezing weather/frost 19, 46
- fruit-fly 174
- fruit cages 20, 46
- fuzzy-leaved marigold 181
- gall mite 203
- garden calendar 56–61
- garden design 12
- garden planning 24–7
- 'garden-within-a-garden' concept 20
- garlic 27, 120–2, *120*
  - Elephant 121, *121*
  - growing tips 120
  - harvesting 121
  - 'Lautrec Wight' 121, *121*
  - plaits 122, *122*
  - problems 122
  - 'Purple Modovan' 121, *121*
  - 'Purple Wight' 121, *121*
  - 'Solent Wight' 121, *121*
  - storing 121
- ghost spot 53, 181
- glyphosate 279
- gooseberry 20, 39, 42, 215–17, *216*
  - cordons 215
  - cuttings 215
  - growing tips 215
  - harvesting 217
  - 'Hinomaki Yellow' 217, *217*
  - 'Invicta' 217, *217*
  - 'Pax' 217, *217*
  - problems 217
  - pruning 217
  - storage 217
  - 'Whinham's Industry' 217, *217*
  - *see also* jostaberry
- gooseberry sawfly 217
- grape 218–23, *218*
  - 'Black Hamburgh' 220, *220*
  - 'Brandt' 223, *223*
  - 'Buckland Sweetwater' 220, *220*
  - 'Dornfelder' 223, *223*
  - 'Flame' 220, *220*
  - growing tips 218, 221
  - harvesting 220, 223
  - indoor 218–20
  - 'Muller Thurgau' 223, *223*
  - 'Muscat of Alexandria' 220, *220*
  - outdoor 221–2
  - 'Perlette' 220, *220*, 223, *223*
  - problems 220, 223
  - pruning 219, 221–2
  - 'Strawberry' 223, *223*
  - thinning fruit 220
  - training 219, *219*, 220, 221–2, *222*
- greenfly 51–3, *51*, 74, 105, 128, 131, 147, 194, 239, 253, 261, 265, 271
- greenhouses 49
  - growing fruit in 218–20, 229–30
  - growing tomatoes in 177, 181
  - growing vegetables in 79, 110–11, 118, 125, 143
- grey mould, *see* botrytis
- Hamilton, Geoff 15
- 'hardening off' 37
- hazel 234–7, *236*
- heartsease 115, *115*
- heeling-in 39
- *Helianthus tuberosus, see* artichoke, Jerusalem
- herbaceous borders 20
- heritage varieties 27
- hoeing 44–5
- hoof and horn 34
- horseradish 279, *279*
- horticultural fleece 46, 49, 197
- hosepipe bans 42
- hoverflies 52
- insect-proof mesh 49
- intercropping 24
- James I 232
- jostaberry 203
- *Juglans regia, see* walnut
- juicers 54
- juices 93, 111, 202
- kaffir lime 227, *227*
- kale 123–4, *123*, *124*
  - 'Black Tuscany' 123, *123*
  - 'Pentland Brig' 123, *123*
  - red 159
  - 'Redbor' 123, *123*
- kiwi fruit 224–5, *225*
  - 'Atlas' 224
  - 'Hayward' 224
  - 'Jenny' 224, *224*
  - Siberian 224, *224*
- kohl rabi 125
  - 'Blusta' 125, *125*
  - 'Logo' 125, *125*
  - 'Supershmelze' 125, *125*
- kumquat 227, *227*
- lacewing 52
- *Lactuca sativa, see* lettuce
- ladybird 52
- lamb recipes 289
- lamb's lettuce (corn salad/mache) 128, 159
  - 'Cavallo' 128
  - 'Verte de Cambrai' 128, *128*
  - 'Vit' 128, *128*
- land cress 159, 184, *184*, *185*
- *Laurus nobilis, see* bay
- *Lavandula* (lavender) 280–1, *280–1*
  - 'Hidcote' 280, *280*
  - *L. stoechas* subsp. *stoechas* 280, *280*
  - 'Old English' 280, *280*
- Lawrence, D. H. 214
- Lawson, Nigella 239
- leaf celery 293
- leaf chicory 159
- leaf miner 286
- leek 23, 27, 37, 126–7, *126*
  - 'Apollo' 127, *127*
  - 'King Richard' 126, 127, *127*
  - 'Musselburgh' 127, *127*
  - 'Neptune' 127, *127*
- lemon 226–8
  - container-grown 226
  - growing tips 226
  - harvesting 228
  - 'Lemonade' 227, *227*
  - 'Meyer's Lemon' 227, *227*
  - problems 228
  - pruning 228
  - 'Variegated' 227, *227*
- lemon grass 282, *283*
- lemon verbena 284–5, *284–5*
- lettuce 24, 27, 37, 42, 52, 129–31, *129*, *130*, 159
  - 'Cocarde' 130, *131*
  - cut-and-come-again 131
  - growing tips 129
  - harvesting 131
  - 'Little Gem' 130, *131*
  - 'Lobjoits Green Cos' 130, *131*

'Lollo Rossa' 130, *131*, 159
'Marvel of Four Seasons' 130, *131*
'May King' 130
mini cos 16
out-of-season 129–30
problems 131
'Rouge D'Hiver' 130
'Tom Thumb' 130, *131*
'Valdor' 130
'Webb's Wonderful' 130, *131*
'Winter Density' 130
Levisticum officinale, see lovage
lime 227, *227*
lingonberry 210
loganberry 27, 199, 200, *200*, 201
lovage 286, *287*
LY654 200, *200*
Lycopersicon esculentum, see tomato

maggots 142, 145, 194
magnesium 34
magpie moth 217
Malus, see apple
mangetout 142, 143, 144, 145
'Oregon Sugar Pod' 143
manure 33, 34
marjoram 281, 288–9, *288–9*
pot 289, *289*
sweet/knotted 289, *289*
wild 289, *289*
marrow 107–9
'Tiger Cross' 108, *108*
Medicago sativa, see alfalfa
medlar 250, *250*
'Nottingham' 250
melon 49, 229–31, *229*, *230*
'Amber Nectar', syn. `Castella' 231, *231*
'Galia' 231, *231*
growing tips 229–30
harvesting 231
musk melon 231
pollination 230
problems 231
'Sweetheart' 231, *231*
watermelon 231
Mentha 290–2, *291–2*
M. longifolia 292, *292*
M. spicata 292, *292*
M. suaveolens 292, *292*
M. s. 'Variegata' 292, *292*
M. x *gracilis* 'Variegata' 292, *292*
M. x *piperita* f. *citrata* 292, *292*
'Mesclun' 159
Mespilus germanica, see medlar
mice 76, 145, 171
mildew 137, 194
American gooseberry 217
downy 164
powdery 109, 112, 145, 172, 220, 223, 292
mint 290–2, *291*
apple mint 292, *292*
buddleja mint 292, *292*
eau de cologne mint 292, *292*
ginger mint 292, *292*
pineapple mint 292, *292*
spearmint 292, *292*
mizuma 138, 139
mizuna 159, *159*
Mongolian gooseberry 224, *224*
Morus
M. alba, see mulberry, white
M. nigra, see mulberry
mulberry 232, *233*
'Chelsea' 232, *232*
white 232
mulches 45
mushroom flies 133
mushroom logs 133
mushrooms *132*, 133
mycorrhizal fungi 204
Myrrhis odorata, see sweet cicely

nasturtium 115, *115*
Nasturtium officinale, see watercress
nectarine 11, 16, 27, 46, 240–2
'Lord Napier' 241, *241*
'Pineapple' 241, *241*
nematodes 51
nitrogen 33, 34
nuts 234–7, *235–6*
harvesting 237
problems 237
pruning 235
root restriction 235
storage 237

Ocimum basilicum, see basil
oils
flavoured olive 281
herb-infused 54
onion 27, 45, 95, 134–7, *135*
'Electric' 136, *136*
giant 137
growing tips 134
harvesting 137
'Radar' 136, *136*
'Red Baron' 136, *136*
'Robinson's Mammoth Improved' 137
'Senshyu' 136, *136*
'Showmaster' 137
spring onion 16, 27, 134, 136, *136*, 137
'Sturon' 136, *136*
'Stuttgarter Giant' 137
'The Kelsae' 137
orange 227, *227*
oregano, Greek 281, 289, *289*
organic gardening 7, 14, 29–30, 33–4
organic matter 11, 14, 29, 30, 33
oriental leaves 138–9, *138*, *139*
Origanum 281, 288–9, *288–9*
O. majorana 289, *289*
O. onites 289, *289*
O. vulgare 289, *289*
ornamental quince 248

pak choi 138, 139, 159
'Mei Quing' 138, *138*
parsley 36, 54, 121, 141, 293–4, *293–4*
Hamburg 141
'Moss Curled' 294, *294*
plain-leaved 294, *294*
parsnip 140–1, *140*
'Arrow' 140
'Avonresister' 141, *141*
'Dagger' 140
'Tender and True' 141, *141*
parthenocarpy 107, 108
passion fruit (*Passiflora edulis*) 238–9, *238*
'Crackerjack' 239, *239*
purple granadilla 239, *239*
Pastinaca sativa, see parsnip
pea weevil 76, 145
peach 11, 16, 27, 46, 240–2
'Jalousia' 240, *241*
'Peregrine 240, *241*
'Rochester' 240, *241*
peach leaf curl 242
pear 11, 16, 194, 243–5
Asian 245
'Beth' 244, *244*
'Beurre Superfin' 244, *244*
'Concorde' 244, *244*, 245
'Conference' 243, 244, *244*, 245
cordons 243
'Doyenne du Comice' 243, 244, *244*
espaliers 243
'family trees' 243
growing tips 243
harvesting 245
pollination 243
problems 245
pruning 243
'Red Williams' 244, *244*
step-over trees 243
storage 245
'Williams Bon Chretien' 244, *244*
peas 37, 46, 142–5, *143*
'Alderman' 144, *145*
'Carouby de Mausanne' 144, *145*
continuous cropping 142–3
eat-all 142, 143, 144, 145
'Feltham First' 144, *145*
growing tips 142
harvest 144
'Hurst Green Shaft' 144, *145*
'Kelvedon Wonder' 144, *145*
marrowfat 144
'Oregon Sugar Pod' 144, *145*
petit pois 144
problems 145
shelling type 142, 143, 144, 145
snap 142, 143, 144, 145
'Sugar Ann' 144, *145*
'Sugar Snap' 144, *145*
'Waverex' 144, *145*
pepper 41, 49, 54, 146–7, *146*
'Bell Boy' 147, *147*
'Big Banana' 147, *147*
'Gypsy' 147, *147*
'Redskin' 147, *147*
perennial vegetables 20
pests 14, 29, 37, 44, 49, 50–2
see also specific pests
Petroselinum crispum 293–4, *293–4*
P. c. var. *tuberosum* 141
Phaseolus
P. coccineus, see beans, runner
P. vulgaris, see beans, French
phosphates 33
Physalis
P. alkekengi 207
P. ixocarpa 206, 207
P. peruviana syn. *Physalis edulis* 206–7, *206–7*
Phytosieulus persimilis 112, 239
pigeons 76, 85, 86, 91, 143
Pimm's 268
pinks 115, *115*
Pisum sativum, see peas
planning gardens 24–7
planting fruit 38–9, *38*
plum 246–8
biennial bearing 246, 248
cherry plum/myrobalan plum 247, *247*
'Czar' 247, *247*
fan-trained 246
growing tips 246
harvesting 248
hedgerow 248
'Marjorie's Seedling' 247, *247*
'Merryweather' 247, *247*
Mirabelle 247, *247*
'Old Green Gage' 247, *247*
'Oullin's Golden Gage' 247, *247*
problems 248
pruning 246
thinning fruit 246
'Victoria' 246, 247–8, *247*
pollination, hand 46–9
polyanthus 115, *115*
polytunnels 49, 92, 107, 111, 118, 125–6, 143, 181
Portulaca oleracea var. *sativa, see* purslane
pot marigold 115, *115*
potagers 12, 15
potash 33
sulphate of 34, 41
potassium 33
potato 23–4, 27, 30, 41, *148*, 149–53, *152*, 172
baby new for Christmas 153
'Belle de Fontenay' 150, *151*
'Charlotte' 150, *151*
'Duke of York' 150, *151*
'Edzell Blue' 150, *151*
'Estima' 150, *151*
'Foremost' 150, *151*
'Golden Wonder' 150, *151*
growing tips 149, 152
harvesting 152
'International Kidney' 150, *151*
'Kestrel' 150, *151*
'King Edward' 150, *151*
'Mayan Gold' 150, *151*
'Mimi' 150, *151*
'Pentland Javelin' 150, *151*
'Pink Fir Apple' 150, *151*
problems 153
'Rocket' 150, *151*
seed potatoes 149, 153
'Smile' 150, *151*
storage 152
'Yukon Gold' 150, *151*
potato blight 53, *53*, 153, 181
preserves 54, 114, 194, 254
Primula hybrids 115, *115*
protecting crops 46–9
Prunus
P. armeniaca, see apricot
P. avium, see cherry
P. cerasifera, see plum, cherry plum/myrobalan plum
P. cevasus, see cherry
P. domestica, see plum
P. dulcis var. *dulcis, see* almond
P. persica, see peach
P. persica var. nectarina, *see* nectarine
P. spinosa, see sloe
pumpkin 23, 27, 36, 169–71, *171*
'Becky' 170, *170*
giant 170–1
purslane 154, 159
Golden Purslane 154, *154*
Green Purslane 154, *154*
Winter purslane (miner's lettuce) 154, *154*, 159
Pyrus communis' see pear

quince 194, 249–50
'Meech's Prolific' 249, *249*
'Vranja' 249, *249*

radicchio 159
radish 27, 155–6, *155*, 168, *168*
'Beauty Heart' 156, *156*
'Black Spanish' 156, *156*
edible-podded 155
'French Breakfast' 156, *156*
'Mantanghong' 156, *156*
'Mooli' 156, *156*
'Munchen Bier' 155
'Scarlet Globe' 156, *156*
'Sparkler' 156, *156*
Raphanus sativus, see radish
raspberry 16, 42, 201, 251–3
'All Gold' 252, *252*
fruit care 251
'Glen Ample' 252, *252*
growing tips 251
harvesting 253
'Joan J' 252, *252*
'Malling Admiral' 252, *252*
'Malling Jewel' 252, *252*
problems 253
support and pruning 252
recipes
alfalfa salad 167
asparagus pea 73
baked beetroot 82
baked squash 169
blackcurrant juice 202
carrot juice 93
carrot-slaw 95
celeriac 99
chamomile tea 270
chard 176
chervil soup 271
chicken stuffed with garlic 122
chicory 103
courgettes 109
crab apple jelly 194
cucumber juice 111
cucumber raita 111
easy cranberry (or cowberry) sauce 210
easy dill cucumbers 276
easy redcurrant jelly 254
endives 117
flavoured olive oil 281

Florence fennel 118
garlic as veg 122
gingered peaches 242
glazed fig tart 214
globe artichoke 69
gooseberry fool 215
Greek tomato salad 181
home-grown green salad 159
kale 124
lovage cordial 286
marinated roast/barbecued lamb 289
melon baskets 230
Mexican-style salsa 206
mild and spicy Thai-style vegetables 275
mixed berries 205
neeps and tatties 172
pear salad 245
Pimm's 268
quick lemon syllabub 205
quince vodka 250
real lemonade 228
real parsley sauce 294
roast garlic 122
roast squash 169
roast winter root veg 141
rose petal jelly 114
rosemary potatoes 150
Rumtopf 201
sloe/damson gin 248
sorrel parcels 162
spinach in filo parcels 163
squash soup 169
stuffed cabbage leaves 91
stuffed peppers 147
stuffed vine leaves 223
succotash 173
summer pudding 253
tarragon omelette 303
tarragon vinegar 303
turnips in cream 183
red clover 168, *168*
red mustard 138, *138*, 159
red spider mite 50, *50*, 74, 112, 239, 285
redcurrant 20, 254–5
'Jonkheer van Tets' 255, *255*
'Red Lake' 255, *255*
'Rovada' 255, *255*
redcurrant blister aphid 255
rhubarb (*Rheum* x *hybridum*) 256–7, *257*
'Hawke's Champagne' 256
'Stockbridge Arrow' 256
'Timperley Early' 256
Ribes
R. nigrum, see blackcurrant
R. rubrum, see redcurrant
R. uva-crispa, see gooseberry
rocket 157, *157*, 159
wild 157, *157*
root restriction 212, 235
rosemary 41, 295–6, *295–6*
'Fota Blue' 296, *296*
'Miss Jessopp's Upright' 296, *296*
'Severn Sea' 296, *296*
rosemary beetle 281, 296, 298
Rosmarinus
R. lavandulaceus 296, *296*
R. officinalis 41, 295–6, *295–6*
rot 79
blossom end 181
brown 250
crown 257
neck 231
root 112, 145
white 137
Rubus
R. caesius, see dewberry
R. idaeas, see raspberry
see also blackberry
Rumex
R. acetosa 162, *162*
R. scutatus 162
Russell, Ken 214
rust 76, 122, 127
mint 292
sage 41, 297–8, *297–8*
golden 298, *298*
purple 298, *298*
tricolor 298, *298*
salad burnet, *see Sanguisorba minor*
salad leaves, mixed 14, 16, 24, 27, 30, 41, 44, 158–9, *158*, *159*
salad recipes 159, 167, 181, 245
Salvia officinalis 41, 297–8, *297–8*
S. o. 'Icterina' 298, *298*
S. o. 'Purpurascens' 298, *298*
S. o. 'Tricolor' 298, *298*
Sanguisorba minor 299, *299*
Satureja 300, *300*
S. hortensis 300, *300*
S. montana 300, *300*
scab 153, 194, 245
scale insect 228, 266
seaweed extract 34, 204
seed sowing 36–7
self-seeding 278
shallots 160–1, *161*
'Golden Gourmet' 160, *160*
'Griselle' 160, *160*
'Hative de Niort' 160, *160*
'Jermor' 160, *160*
'Red Sun' 160, *160*
silver leaf disease 197, 209
site and situation 11, 12
sloe 248
slugs 14, 44, 50–1, *50*, 72, 91, 101, 106, 117, 131, 139, 152–3, 171, 261
snails 14, 44, 51, *51*, 106, 117, 131, 139, 176, 261
soil 11, 14
pests 52
preparing a new patch 29
routine preparation 30
Solanum
S. melongena, see aubergine
S. tuberosum, see potato
sooty mould 228, 266
sorrel 162, *162*
buckler-leaved sorrel 162
spiders 52
spinach 16, 24, 27, 41, 159, 163–6, *163*
'Bordeaux' 164, *164*
'Galaxy' 164, *164*
growing tips 163
harvesting 164
New Zealand 165, *165*
perpetual (spinach beet) 166, *166*
problems 164
'Tornado' 164, *164*
Spinacia oleracea, see spinach
spring onion 16, 27, 134, 136, 137
'North Holland Blood Red' 136, *136*
'Overwintering White Lisbon' 136, *136*
'White Lisbon' 136, *136*
'sprout tops' 87
sprouters 167
sprouting seeds 167–8, *167*
squash 27, 169–71, *171*
'Butternut' 170, *170*
'Crown Prince' 170, *170*
summer 107–9, *108*
'Sweet Dumpling' 170, *170*
'Turk's Turban' 170, *170*
'Vegetable Spaghetti' 170, *170*
squirrels 237
standards 266, 295
step-over trees 20, 188, 190, *190*, 243
storing crops 54
see also freezing crops
strawberry 14, 16, 20, 42, 44, 54, 201, 258–61
alpine 261, *261*
'Aromel' 260, *260*
bare-root runners 259, *259*
'Cambridge Favourite' 260, *260*
'Cambridge Late Pine' 260, *260*
container-grown 258, 259
'Elsanta' 260, *260*
'Flamenco' 260, *260*
'Florence' 260, *260*
growing tips 258
harvesting 261
'Honeoye' 260, *260*
'Mae' 260, *260*
'Mara des Bois' 260, *260*
perpetual-fruiting 261
problems 261
'Royal Sovereign' 260, *260*
sulphate of potash 34, 41
summer savory 300, *300*
summer squash 107–9
'Patty Pan' 108, *108*
'Sunburst' 108, *108*
supports 38–9
swede 141, 172
'Invitation' 172, *172*
'Marian' 172, *172*
sweet chestnut 234, 235, *236*, 237
'Regal' *236*, 237
sweet cicely 301, *301*
sweetcorn 23, 24, 26, 27, 36, 37, 173–4
'Applause' 174, *174*
'Golden Bantam' 174, *174*
growing tips 173
harvesting 174
'Incredible' 174, *174*
mini 173
problems 174
'Sundance' 174, *174*
sweet varieties 174
Swiss chard 175–6, *175*, *176*
rainbow chard 159, 175, *175*
ruby chard 175, *175*

Tagetes minuta 181
Taraxacum officinale see dandelion
tarragon 41, 302, *302*
French 302, 303, *303*
Russian 302, 303, *303*
tayberry 27, 199, 200, *200*, 201
'Buckingham' *200*, 201
Tummelberry *200*, 201
teas, herbal 270
tender plants 36–7
Tetragonia expansa 165, *165*
Tetragonolobus purpureus, see asparagus pea
Thymus (thyme) 41, 281, 304–5, *304–5*
T. herba-barona (caraway thyme) 305, *305*
T. vulgaris (common/garden thyme) 305, *305*
T. x *citriodorus* (lemon thyme) 305, *305*
T. x *citriodorus* 'Bertram Anderson' 305, *305*
T. x *citriodorus* 'Fragrantissimus' (orange thyme) 305, *305*
tomatillo 206, 207
tomato 16, 27, 37, 41, 49, 54, 177–81, *180*
'Ailsa Craig' 178, *179*
'Brandywine' 178, *179*
buying plants 179
'Gardener's Delight' 178, *179*
'Green Grape' 178, *179*
growing tips 177–8
harvesting 179
'Marmalade' 178, *179*
problems *181*
recipes 206
ripening late fruit 178
'Roma' 178, *179*
sun-dried 54
'Sungold' 178, *179*
'Tigerella' 178, *179*
'Tornado' 178, *179*
'Tumbler' 178, *179*
tomato blight 53, *53*
tomato feed 34, 41
traditional plots 20
Trifolium pratense 168, *168*
Trigonella foenum-graecum 168, *168*
Tropaeolum majus 115, *115*
tunnel cloches 258
turnip 182–3, *183*
'Atlantic' 182, *182*
'Golden Ball' 182, *182*
'Snowball' 182, *182*
'Tokyo Cross' 182, *182*
turnip gall weevil 183

Vaccinium
V. corymbosum, see blueberry
V. macrocarpon, see cranberry
V. vitis-idaea var. *minus, see* lingonberry
Valerianella locusta, see lamb's lettuce
Vicia faba, see beans, broad
Vigna
V. angularis 168, *168*
V. radiata 167, 168, *168*
vine weevil 50, *50*
Viola 115, *115*
V. tricolor 115, *115*
viruses 53, *53*, 253, 261
Vitis vinifera, see grape

walnut 234, 235, *236*, 237
'Broadview' 237
'Franquette' 237
wasps 50, *50*, 242, 248
parasitic 52, 181
traps 242, 248
water butts 42
watercress 159, 184, *184*, *185*
watermelon 231
weather conditions 19, 46–9
weedkiller 30, 279
weeds 23, 29, 30, 44–5, 137
perennial 29, 30
weevils 50, *50*, 76, 145, 183, 237
whitecurrant 255
'Blanka' 255
'White Versailles' 255, *255*
whitefly 50, *50*, 74, 181, 239, 285
winter purslane (miner's lettuce) 154, *154*, 159
winter savory 300, *300*
wireworms 52
woolly aphid 194

Zea mays, see sweetcorn

ACKNOWLEDGEMENTS

My thanks to all those who have helped in the creation of *The Kitchen Gardener*, and who have made the journey from a germ of an idea to the production of what I hope will be an indispensable guide for anyone who wants to grow their own food. To my good friend Sue Phillips for her help and advice, to Nicky Ross, Helena Caldon and Lorna Russell for their editorial skills, and to Jonathan Buckley who prods me into shape on photoshoots and manages to make other vegetables – and fruits – look even more appetizing than I do; I am deeply grateful. They are also very good company when I am let out into the wide world, having been shut away on my own with a laptop for too long! Thanks also go to Smith & Gilmour for all their work on the design and to Lizzie Harper for her lovely illustrations.

PICTURE CREDITS

Illustrations by Lizzie Harper

BBC Books and Ebury Publishing would like to thank the following for their help in sourcing and providing photographs and for permission to reproduce copyright material. While every effort has been made to trace and acknowledge all copyright holder, we would like to apologize should there be any errors or omissions.

Jonathan Buckley: 6, 8–9, 11, 14 (Design: Sue and Wol Staines/Glen Chantry), 17, 18, 20 (Design: Marilyn Abbott/West Green House), 21, 25, 28, 31, 32, 34, 35, 36, 39, 40, 43, 44, 48, 51 (2), 62–3, 66–7, 69 (1), 69 (3), 69 (4), 71, 72 (tr), 76 (1) (Design: Sarah Raven), 78 (3), 78 (6), 81 (2), 82, 83 (1) (Design: Sarah Raven), 88, 90 (3), 94 (5) (Design: Sarah Raven), 98, 102, 103 (1), 103 (4), 108 (2), 110, 112 (1), 114 (Design: Jean Goldberry/Catalyst Television), 115 (2), 115 (3), 115 (4), 115 (5), 115 (6), 117, 123 (2), 123 (3), 124 (Design: Sarah Raven), 129, 131 (2), 131 (3), 131 (5), 135, 136 (2) (Design: Sarah Raven), 136 (3), 136 (4), 138 (tl) (Design: Sarah Raven), 138 (1), 138 (2), 141 (1), 143, 145 (4), 145 (8), 146, 148, 151 (3) (Design: Sarah Raven), 151 (9), 151 (12), 151 (16), 152, 154 (1), 156 (2), 156 (6), 157 (1), 157 (2), 160 (tl), 163, 167, 170 (4), 170 (5), 171, 174 (2), 175 (1), 175 (3), 176, 179 (2), 179 (7), 179 (8), 180, 183, 184 (1), 185, 186–7, 193 (16), 198, 200 (6), 202, 207, 212, 214 (1), 218, 223 (1), 225, 229, 236 (5), 254, 257, 260 (6), 260 (8), 262–3, 265 (2), 265 (6), 267, 268 (2), 271, 273, 274, 276 (1), 277, 278 (1), 279, 283, 284, 288, 297, 298 (1), 300 (1), 301, 302, 304, 305 (5); Justina Burnett: 78 (2), 205 (2), 239 (2); Nick Dunn (Trees for Life): 209 (4); Nigel Cattlin/FLPA: 50 (4), 53 (4); GAP Photos/Pernilla Bergdahl 193(4), 203 (3), 272 (1); GAP Photos/Richard Bloom 194 (1), 220 (2), 296 (2); GAP Photos/Mark Bolton 164 (1), 179 (3), 220 (5), 280 (1), 287, 289 (1), 289 (4), 292 (6), 296 (3); GAP Photos/Elke Borkowski 244 (4); GAP Photos/Lynne Brotchie 68, 80, 291; GAP Photos/Nicola Browne 70, 93, 227 (5); GAP Photos/Leigh Clapp 227 (5); GAP Photos/Sarah Cuttle 168 (6); GAP Photos/Amanda Darcy 292 (3), 299; GAP Photos/Claire Davies 108 (6); GAP Photos/Paul Debois 45, 50 (1), 72 (3), 105 (3), 125 (1), 189, 193 (11), 200 (8), 203 (1), 203 (4), 216, 217 (3), 252 (2), 255 (1); GAP Photos/FhF Greenmedia 47, 52 (5), 73, 78 (5), 81 (4), 113, 151 (4), 155, 158, 236 (3), 247 (3), 247 (4), 247 (6), 256 (3), 264, 298 (3), 305 (1); GAP Photos/Tim Gainey 51 (1); GAP Photos/John Glover 13, 55, 86, 97 (4), 104, 105 (1), 136 (8), 138 (3), 139, 166, 179 (6), 193 (5), 193 (7), 200 (9), 207 (2), 209 (1), 209 (2), 211, 228, 230, 233, 249 (2), 289 (3; GAP Photos/Jerry Harpur 22, 175 (2), 240, 298 (2); GAP Photos/Marcus Harpur 159 (3), 194 (2), 247 (8),250, 272 (3),292 (1); GAP Photos/Neil Holmes 193 (13), 220 (4), 289 (2), 292 (4), 292 (5), 296 (5); GAP Photos/Michael Howes 53 (1), 200 (2),255 (2), 255 (4), 294 (1); GAP Photos/Dianna Jaswinski 206, 280 (2); GAP Photos/Andrea Jones 75, 85 (1), 87 (br), 90 (5), 126, 200 (3), 276 (2); GAP Photos/Lynn Keddie 120, 168 (5); GAP Photos/Geoff Kidd 115 (7), 130, 168 (2), 168 (3); GAP Photos/Zara Napier 51 (3), 94 (1), 261, 305 (4); GAP Photos/Clive Nichols 106 (1), 131 (1), 154 (2), 179 (10), 293, 294 (2); GAP Photos/Sharon Pearson 174 (1); GAP Photos/Julie Pigula 236 (7); GAP Photos/Howard Rice 168 (1), 247 (5), 260 (3), 260 (4), 305 (2), 305 (3); GAP Photos/Rice/Buckland 30; GAP Photos S & O 87 (1), 90 (6), 90 (9), 121 (1), 121 (2), 121 (3), 121 (4), 179 (5), 193 (18), 200 (10), 214 (2), 220 (1), 241 (2), 244 (7), 260 (2), 260 (10), 265 (4); GAP Photos/J S Sira 115 (1), 217 (2), 268 (1), 292 (2); GAP Photos/Jason Smalley 91; GAP Photos/Friedrich Strauss 95, 125 (3), 175 (tl), 207 (1), 223 (3), 224 (2), 227 (2), 236 (9), 236 (10), 247 (2), 300 (2), 303 (2); GAP Photos/Graham Strong 50 (5), 50 (6), 74 (2), 81 (3), 105 (tr), 106 (2), 275 (1); GAP Photos/Maddie Thornhill 119, 170 (3),200 (7), 265 (3); GAP Photos/Visions 84, 132, 145 (2), 161, 193 (8), 193 (9), 193 (10), 227 (4), 227 (6), 227 (8), 231 (2), 236 (1), 238, 244 (3), 244 (5), 244 (6), 247 (7); GAP Photos/Juliette Wade 123 (tl), 123 (1), 140, 170 (2), 249 (1), 278 (2); GAP Photos/Jo Whitworth 53 (2), 78 (1), 78 (7), 105 (4), 107, 147 (1), 159 (2), 165, 193 (2), 193 (3), 236 (4), 285; GAP Photos/Rob Whitworth 78 (8), 131 (4), 160 (4), 193 (6), 235, 236 (11); Torie Chugg/The Garden Collection: 76 (3), 145 (3), 162; The Organic Gardening Catalogue: 78 (9), 112 (2); Garden Picture Library (Photolibrary Group): 50 (2), 52 (4), 78 (10), 83(4), 90 (7), 97 (2), 103 (2), 128 (1), 145 (1), 151 (11), 159 (1), 160 (2), 193 (15), 193 (17), 196, 205 (4), 205 (5), 220 (3), 227 (1), 241 (5), 244 (1), 247 (1), 252 (5), 265 (5), 303 (1); Garden World Images: 244 (2); Gardeners' World Magazine/Tim Sandall :50 (3); ***Gardeners' World Magazine***/Torie Chugg: 53 (3); Gardeners' World Magazine: 76 (2); Gardeners' World Magazine/Sarah Cuttle: 127 (1); ***Gardeners' World Magazine***/Tim Sandall: 145 (6); ***Gardeners' World Magazine***: 223 (5); Andrew Lawson: 193 (12), 200 (5), 221, 269, 270 (1); 147 (2), Torie Chugg: 193 (1), 241 (3), 252 (1), 260 (5), 260 (7), 272 (2), 296 (1), 296 (4); Greenvale AP: 151 (15); Marshalls Seeds: 121 (5), 136 (7), 136 (8), 172 (2), 232 (1), 236 (8), 241 (1); Ken Muir 200 (4), 203 (2), 205 (6), 209 (3); Derek St Romaine: 69 (2), 72 (1), 72 (2), 74 (1), 77, 85 (2), 92 (2), 97 (1), 147 (3), 151 (10), 154 (3), 184 (2), 270 (2), 295, 298 (4), 214 (3), 217 (4), 252 (4), 260 (9); Reads Nursery 197 (2), 227 (3), 227 (7); Sea Spring: 83 (tr), 90 (8), 92 (1), 92 (3), 94 (4), 94 (6), 97 (3), 97 (6), 101, 108 (1), 112 (3), 125 (2), 127 (2), 128 (2), 145 (9), 151 (5), 155 (br), 156 (1), 156 (3), 156 (4), 164 (3), 174 (3), 182 (1), 182 (4), 275 (3), 280 (3), 236 (6), 239 (1); Suttons Seeds: 76 (4), 81 (1), 83 (2), 85 (3), 85 (4), 90 (1), 90 (2), 90 (4), 94 (4), 97 (5), 103 (3), 105 (2), 108 (5), 108 (7), 127 (4), 131 (6), 131 (7), 136 (1), 136 (5), 141 (2), 145 (5), 147 (4), 151 (2), 151 (6), 151 (7), 151 (8), 151 (14), 156 (5), 160 (1), 160 (3), 160 (5), 164 (2), 168 (4), 170 (1), 170 (6), 172 (1), 174 (4), 179 (1), 179 (4), 179 (9), 182 (2), 182 (3), 197 (1), 197 (3), 200 (1), 205 (1), 205 (3), 217 (1), 223 (2), 223 (4), 224 (1), 231 (3), 236 (2), 241 (4), 255 (3), 256 (2), 260 (1), 265 (1); Thompson and Morgan: 78 (4), 83 (3), 87 (2), 94 (2), 112 (3), 127 (3), 145 (7), 151 (1), 151 (13), 151 (17), 231 (1), 256 (1); Wonderful PR 151 (15).